# Detectors and Sources
# for THz and IR

Fedir F. Sizov

Published by **Materials Research Forum LLC**
Millersville, PA 17551, USA

Published as part of the book series
**Materials Research Foundations**
Volume 72 (2020)
ISSN 2471-8890 (Print)
ISSN 2471-8904 (Online)

Print ISBN 978-1-64490-074-1
ePDF ISBN 978-1-64490-075-8

Distributed worldwide by

**Materials Research Forum LLC**
105 Springdale Lane
Millersville, PA 17551
USA
http://www.mrforum.com

Printed in the United States of America
10 9 8 7 6 5 4 3 2 1

# Table of Contents

# Preface

The spectral ranges from infrared $\lambda \approx 1...30$ μm (radiation frequency $\nu \approx 300...10$ THz) and to terahertz one from $\lambda \approx 30$ μm to $\lambda = 3$ mm ($\nu \approx 10...0.1$ THz) are the areas for researches that encompass the different spheres of applications. Out of these applications, a few are in security screening and surveillance, astronomy, spectroscopy, biomedicine (*e.g.*, oncology, dermatology, dentistry), food and package inspection, detection of concealed weapons, vision through camouflage, *etc.* The increasing demand for the fast transmission of large amounts of data will lead to the extension of operation frequencies in communications, toward the THz frequency range. Been nonionizing in nature, THz radiation penetrates many dielectric materials like plastics, ceramics, cardboard or dry gypsum plasterboard allowing a non-destructive and contact-free testing. The IR and THz medical imaging technologies can provide guidance for surgeons in delimiting the margins of tumors, help clinicians to visualize diseased area, *etc.*

Nowadays, a shift from the scientific to more application-oriented research and development is observed. However, despite the potential wide area of applications, especially for the development of THz technologies, the latter ones still meet limitations that are connected, *e.g.*, with such factors as the cost and extensionality of systems, and availability of imaging and spectroscopic systems able to probe large areas in a limited time. In addition, the relatively low output power of compact THz sources and the low sensitivity of compact (uncooled) detector should be improved. For example, in the radiation frequency range $\nu \sim 1...4$ THz, these demands cause, as a rule, the uncomfortably slow getting of images and spectroscopic information, which is strengthen by the strong atmosphere absorption in this spectral range.

Scientific and particularly application activity in the THz and IR technologies have increased significantly in the last two decades, and it is expected that the trends, especially in the THz science and technology, will be continued and extended. However, though promising for a wide range of applications (*e.g.*, imaging, communication, spectroscopy, sensing), the THz instrumentation has not yet gained its potential due to a lack of cost-effective, portable, and efficient technologies.

The purposes of topics in this book are connected with the attempt to describe and provide a representative, though by no means exhaustive, overview of the current state of the technologies of IR and THz detectors and sources, and to show the difference in some aspects of their use. Therewith, it is impossible to do a full assessment to any of them, and it is inevitable that some areas important for different categories of readers have been neglected. For the domain of THz technologies, in which the spectral range lies in between electronics and optics, there exists some disconnect between those who is engaged in the field from the high-frequency (optical) end of the spectra and those who came from the low-frequency end (radio waves). It is supposed that this small book can be an introductory guide for new-coming researchers working in these spectral ranges.

# Abbreviations

**AC** – alternating current;
**ATT** – Avalanche transit time;
**BARITT** – barrier injection transit time;
**BBR** – black body radiation;
**BIB** – block impurity band;
**BLG** – bilayer grapheme;
**BLIP** – background limited performance;
**BST** – barium strontium titanate;
**BTC** – bound to continuum;
**BWO** – backward wave oscillator;
**CEB** – cold electron bolometer;
**CMB** – cosmic microwave background;
**CP** – Cooper pairs;
**CSL** – chirped superlattice;
**CW** – continuous wave;
**DC** – direct current;
**DDR** – double drift region;
**DSB** – double sideband;
**DQW** – double quantum well;
**DWELL** – dot-in-a-well;
**EM** – electromagnetic;
**EO** – electro-optical;
**FEL** – free electron laser;
**FET** – field effect transistor;
**FIR** – far-infrared;
**FOV** – field of view;
**FSS** – frequency selective surface;
**FPA** – focal plane array;
**FTS** – Fourier transform spectroscopy;
**HBV** – heterostructure barrier varactor;
**HEB** – hot electron bolometer;
**HEMT** – high electron mobility transistor;
**HIS** – high impedance surface;
**HOPG** – highly ordered pyrolic graphite;
**HPCVD** – hybrid physical-chemical deposition;
**HTS**[1] – high-temperature superconductivity;
**HTS**[2] – high impedance surface;
**IC** – integrated circuit;
**IF** – intermediate frequency;
**IMPATT** – impact avalanche and transit time;

**IR** – infrared;
**KID** – kinetic inductance detector;
**LNA** – low-noise amplifier;
**LO** – local oscillator;
**LOP** – local oscillator power;
**LPE** – liquid phase epitaxy;
**LSB** – lower side band;
**LTG** – low-temperature grown;
**LTSC** – low-temperature superconductivity;
**LWIR** – long wavelength infrared;
**MCT** – mercury cadmium telluride;
**MEMS** – microelectromechanical systems;
**MITATT** – mixed avalanche tunnelling transit time;
**MM** – metamaterial;
**MMIC** – monolithic microwave IC;
**MOSFET** – metal-oxide-semiconductor field effect transistor;
**MQWs** – multiple quantum wells;
**MTF** – modulation transfer function;
**MWIR** – medium wavelength infrared;
**NDT** – non-destructive;
**NEDT** – noise equivalent difference temperature;
**NEP** – noise equivalent power;
**NIR** – near infrared;
**OR** – optical rectification;
**OTF** – optical transfer function;
**PC** – photoconducting;
**PCA** – photoconducting antenna;
**PTF** – phase transfer function;
**Radar** – radio-detection and ranging;
**RF** – radiofrequency;
**ROIC** – read-out integrated circuit;
**RP** – resonant photon;
**RTD** – resonant tunneling diode;
**SBD** – Schottky barrier diode;
**SDR** – single drift region;
**SEM** – scanning electron microscope;
**SET** – single electron transistor;
**SHEB** – semiconductor hot electron bolometer;
**SIN** – superconductor-insulator-normal;
**SISP** – semi-insulating surface plasmon;
**S/N** – signal-to-noise ratio,
**SIS** – superconductor-insulator-superconductor;
**SNS** – superconducting-normal-superconducting

**SL** – superlattice;
**SSB** – single sideband;
**SSD** – self-switching diode;
**STJ** – superconductor transition junction;
**SQUID** – superconducting quantum interference device;
**SWIR** – short-wavelength infrared;
**TCR** – temperature coefficient of resistance,
**TDS** – time-domain spectroscopy;
**TES** – transition edge sensor;
**THz** – terahertz;
**TI** – topological insulator;
**TRAPATT** – trapped plasma avalanche triggered transit;
**TUNNETT** – tunnel injection transit time negative resistance;
**QCL** – quantum cascade laser;
**QDIP** – quantum dot infrared photodetector;
**QE** – quantum efficiency;
**QR** – quantum ring;
**QD** – quantum dot;
**QP** – quasiparticle;
**QW** – quantum well;
**QWIP** – quantum well infrared photodetector;
**USB** – upper side band;
**UV** – ultraviolet;
**VHTS** – very high-temperature *superconductivity*;
**VPE** – vapor phase epitaxy;
**WHO** – World Health Organization.

# Chapter 1. Introduction (background)

## 1.1 Introduction

The objects in the Universe and on the Earth surface emit varying amounts of the electromagnetic energy in a broad frequency range. Both the terahertz/sub-terahertz (THz/sub-THz) and infrared (IR) detectors, comprising relatively narrow bands of the electromagnetic spectrum, are now used in a large number of applications not limiting themselves to what is convenient for human retina operating in a narrow wavelength band $\lambda \approx 0.4...0.75$ μm. Therefore, without the instruments with different kinds of detectors tuned across different parts of the spectrum, we would be ignorant of many properties of objects and events in the environment. The vast majority of research works are mostly dependent on the availability of detectors and on the sources as well in the case of THz range under Earth's environmental conditions because of the weakness of THz radiation.

The THz and IR technologies have a wide range of applications in security, food control, biomedical and medical imagings, telecommunication, non-destructive testing, *etc*. As for the THz and IR systems, they are mainly of two types: spectroscopic systems and systems for imaging. Among others in the THz/sub-THz range are radars, communication means, and some other emerging systems.

The frequency gap between the microwave ($\nu < 100$ GHz) and IR ($\nu \approx 10...400$ THz) radiation frequencies is frequently called the THz gap (100 GHz $\leq \nu \leq$ 10 THz, 3 mm $\geq \lambda \geq 30$ μm). Till recent time, the THz range of the spectrum was complicated to a great degree for its potential applications because of the difficulty in providing suitable sources and detectors. The cause was in that if the THz range is approached from the low-energy (mm waves) side, then there is a difficulty of realizing the appropriate circuits to handle signals at this high frequency range. If the THz range is approaching from the IR side, then one of the problems is the design of optical systems with suitable materials and dimensions noticeably exceeding the wavelength.

Most of the IR and THz/sub-THz technologies are linked with the astronomy, imaging, and spectroscopy. Till the middle of the twentieth century, the imaging, *e.g.*, with telescopes was limited to visible light. In astronomy applications, the driving forces of research are connected with that the substantial fraction of the luminosity of the Big Bang lies in the THz region. It is worth noting that Earth's atmosphere is rich in the presence of a large number of atomic and molecular excitation spectral lines (rotations and vibrations) in the THz range of the electromagnetic spectrum, thus providing chemical "signatures". Therefore, it is precisely this wealth of spectral lines making these ranges valuable for the astrophysics, planetary, and Earth-observing studies. Since then, THz technologies have been continuously developed and applied, *e.g.*, to biomedicine, non-destructive testing in industry, defense, security, spectroscopy, *etc.*

The indicator of the importance of any technology is the number of scientific and engineering publications. The number of scientific publications on THz science demonstrates a growth since the 1970s and a considerable (exponential) growth since the 1990s (see Fig. 1.1.a). The ratios of publications in scientific and engineering journals to the total number are plotted in Fig. 1.1.b). Till 1980, the research was focused on fundamental aspects, as is evidenced by the large number of articles in scientific journals. After that, more applications-oriented engineering publications remain. The five countries with the most publications are the USA, China, Germany, Japan, and United Kingdom [1].

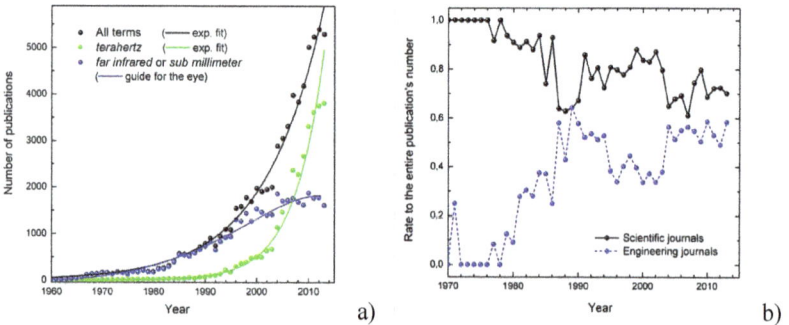

*Figure 1.1. (a) Number of publications between 1960 and 2013 with the topics on terahertz or far infrared and submillimeter waves. (b) Number of publications in scientific (fields: chemistry, physics, materials science) and engineering journals (engineering, chemical engineering) to the total number [1]. (By permission from Th. Hochrein).*

THz radiation penetrates well many visually opaque materials and suffers less the Rayleigh scattering as compared to the IR radiation. It interacts strongly with water, but penetrates through up to a few mm in biological tissue, and passes a large distance in mist (in the latter case, in the low-frequency THz spectral range). Because of the strong water vapor absorption (Fig. 1.2), THz radiation is greatly attenuated, and the remote sensing in

Materials Research Forum LLC
https://doi.org/10.21741/9781644900758

the THz spectral range is restricted for this cause to several or tenths meters (in dependence on the spectral range).

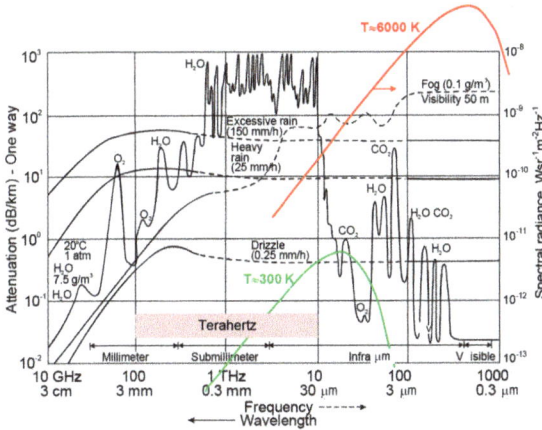

*Figure 1.2. Transparency of the Earth atmosphere from the visible to radiofrequency band region (modified after [2]). The spectral radiances for a black body at the temperatures T ≈ 6000 K (Sun) and T ≈ 300 (Earth surface) are also shown. It is seen that the strong water and oxygen absorptions in the atmosphere make high-altitude platforms essential for the good seeing of Universe's objects. (By permission from SPIE).*

Due to low photon energies, THz radiation is considered safe for a person (operator), because it is non-ionizing. However, the exposure to relatively large powers can be harmful due to thermal effects. For example, the US Federal Communications Commission established (for active THz systems) maximum permissible exposure limits of 1 mW/cm$^2$ for 6 min at 30...300 GHz [3]. At THz frequencies, due to specific rotational and vibrational "optical" transitions in molecules, the spectroscopic fingerprinting by THz technologies can be successfully applied.

Because of the peak radiation emission of bodies and objects on the Earth surface (average T ≈ 293 K) is at about the wavelength λ ≈ 10 μm, the main driving forces of the IR technology development were conditioned by military applications, reconnaissance, surveillance, and then by applications in the industry, crop yield, and medicine diagnostics. In the spectral ranges of 3...5 and 8...14 μm, the atmosphere is well transparent (Fig.1.2).

Materials Research Forum LLC
https://doi.org/10.21741/9781644900758

In astronomy and astrophysics, the earlier research and applications were carried out preferably under the older terms "far infrared" and "sub-millimeter waves" (see, *e.g.*, [4, 5]). This has been subsumed under the more modern term "terahertz" and "sub-terahertz" (see, *e.g.*, [1, 6, 7]). The radiation from astronomical objects during the process of formation of stars and planets is obscured by gas and dust, and, therefore, they are only "visible" in the THz range [8].

For the THz spectral range, over the past three decades, some new THz technologies have emerged and developed substantially. These technologies give the possibility to get information in regions of the electromagnetic spectra, which are inaccessible to the human visual perception. They have shifted from the astronomy and astrophysics applications, where they were mainly initially developed. At $v \sim 0.3...10$ THz, about a half of the energy in the Universe from the stellar radiation and accretion processes is concentrated [9]. The radiation in the visible range from astronomical objects during the process of forming and evolving is obscured by gas and dust, but the THz radiation can be detected from them quite well.

Nowadays, the THz technologies are shifted, to a great extent, to the high-speed broadband communication, by promising a higher transmission rate for information transfer, the high-resolution molecular spectroscopy, identification of hazardous chemicals, surveillance and security imaging, medical applications, drug and gas detections, security screening, food control, THz 3D tomography, *etc*.

The vast majority of research works in such applied fields are mostly dependent on the availability of THz detectors and sources. That is why the interest in the THz research activities up to now has been focused on the development of the technique of generation and/or detection of THz radiation. Much of the recent efforts have been related to exploiting techniques, which are compatible with solid-state III–V and silicon-based semiconductor technologies that can operate under room-temperature conditions and can be manufactured as low-cost efficient electronic or hybrid electronic-photonic systems [10].

The available THz detectors required for practicable civil applications have a number of drawbacks. The cold ones are very sensitive but bulky because of non-practical temperatures to be used. If uncooled, they are not very sensitive to operate in day-to-day civil applications, or they are slow. Combined with the complexity to realize THz emitters, this explains the difficulties for THz radiation technologies to ensure the market penetration with the THz systems for cost-effective civil applications.

Every bit of information comes on a beam of light in any spectral range to be registered by a detector. Therefore, it is important to gather every photon by a detector in any spectral range. The ideal photon detectors can count each photon in the radiation flow principally providing the information from the radiative object (after the Big Bang, the energy of photons becomes lower and lower, but they never stopped, being photons).

The conventional detectors with superconductive NbN nanowires for the single-photon detection were demonstrated in [11] for the visible and IR ranges. They used nano-bridge structures patterned via the electron beam lithography and etching in a 5-nm-thick NbN superconducting film. Later on, such detectors were realized in the wider spectral ranges (see, *e.g.*, [12, 13] – NbN superconductor; [14] – transition edge sensor). Single-photon detection can also be realized using the superconductor tunnel junction (STJ) detectors. The facilities for a single photon detection are important for the quantum information processing, astronomy, space communication, remote sensing, *etc*.

The IR (the wavelengths $\lambda \approx 0.75...30$ μm, $\nu \approx 400...10$ THz) and THz (the radiation frequency range $\nu \approx 0.1...10$ THz, $\lambda \approx 3$ mm...30 μm) [15–17] technologies have become one of the major fields of applied researches that are driven by potential applications. These technologies can provide information not available from any other techniques.

The THz spectral range separates the electronic ($\nu < 100$ GHz) and photonic domains of the spectrum (see Fig. 1.3). The technological gap between the microwave ($\nu < 100$ GHz) and IR ($\nu \approx 10...400$ THz) radiation frequencies is frequently called the THz gap ($\nu = 0.1...10$ THz).

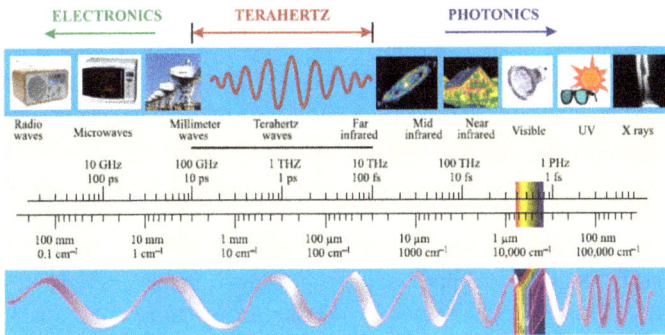

*Figure 1.3. The electromagnetic spectrum [18]. The typical terahertz frequency range (0.1–5 THz) is frequently used for the THz imaging and spectroscopy.*

In the THz spectral range, the photon energies $h\nu$ (at $\nu \approx 1$ THz, $h\nu \approx 4$ meV) are lower or comparable to the thermal energy even for biological systems at low temperatures (T $\approx$

78 K). Moreover, this radiation is nonionizing as, *e.g.*, X-ray radiation having the energy of photons exceeding the ionization energy of atoms (several eV).

Though the THz waves are non-ionizing, this does not mean that they are safe for people. The IEEE standard for safety levels with respect to human exposure (for the general public) to radiofrequency (RF) electromagnetic fields established a maximum permissible exposure limit of 1 mW/cm$^2$ for all regions that are not under the control of the company RF safety program [3]. The exposure to higher powers can be harmful due to thermal effects [19].

High RF power densities can result in the heating of biological tissue and an increase in the temperature that can cause damage due to the body's inability to quickly dissipate the generated excessive heat. Two areas of the body, eyes and testes, are known to be particularly vulnerable to the heating by the RF energy because of the relative lack of a needed blood flow to dissipate the excessive heat load. Therewith, the interaction mechanism with biological systems implies that intense THz pulses can induce significant non-thermal biological effects. Such effects were observed at the molecular, cellular and tissue levels [20–22].

However, even the low-intensity radiation can produce a variety of bio-effects [23]. Other health effects can be caused by thermal effects (temperature changes during the irradiation by a powerful THz radiation). This radiation is able to penetrate a large number of organic or inorganic materials opaque in visible or IR ranges without causing any damage but it is strongly absorbed by conductive materials and polar liquids such as water. Thereby, such characteristic as the absorption is suitable for THz spectroscopy and the applications of imaging methods to such substances. Currently, there are yet no health and safety standards for the exposure to THz radiation.

Today, the optoelectronic instrumentation in IR and THz (still and thermovision cameras, surveillance and observation systems, ground and rocket-borne telescopes, *etc.*) significantly expands the body of coming information about the Earth and space environment. The basic components of such instrumentation are both single detectors (receivers) and detectors composed of matrices of up to $10^6...10^8$ (in the IR range) sensitive elements. This number of sensitive elements in arrays are comparable with the number of sensitive receptors in such perfect instrumentation as human eye ($\sim 2 \cdot 10^8$).

The THz devices can be mentioned now as tools in the high-speed broad-band communication, high resolution molecular spectroscopy, identification of hazardous chemicals, surveillance and security imaging, biomedical applications, drug and gas detections, security screening of the envelopes, dry food control, water-level control in plants, paper industry applications, THz 3D tomography, *etc.* (see, *e.g.,* [15, 16, 24–29]).

Materials Research Forum LLC
https://doi.org/10.21741/9781644900758

The biomedical field is one of the rapidly growing areas in terahertz applications [30, 31]. But, of many promising terahertz technologies, only a few may currently be suitable for biomedical applications which require effective but harmless terahertz sources and highly sensitive detectors operating at elevated temperatures, as compared to low-temperature ones.

Researches and developments in the THz and IR technologies are not very new, and the literature on these questions is voluminous. In the past, a lot of works had already been done in military and surveillance researches and applications (see, *e.g.*, [25, 32...43]). There exist reviews dealing with the effects of the RF radiation on biological systems (see, *e.g.*, [44]).

The THz radiation is heavily absorbed by water (see Fig. 1.2) in the Earth atmosphere, mainly because of water vapor vibration-rotational states (see, *e.g.*, [45]). This catalog includes the information about $\sim 10^3$ water molecule lines in the ground vibration state and about 56,000 in the excited states of water molecules. The fundamental frequencies, namely, rotational and vibration modes of molecules like proteins or DNA, are located in this region as well (see, *e.g.*, [46–48]).

Because of the high atmosphere absorption in the radiation frequency range at $v \sim 1...10$ THz (Fig. 1.2), the THz technology is suitable for short stand-off-range applications (*e.g.*, the passive vision, not more than several meters). At the radiation frequency range of $v < 0.5$ THz, the applications of the close to real-time imaging are feasible for several tens meters. As in the IR, the atmosphere is much more transparent, and the passive IR technologies can be applicable for much longer distances (up to several and tens kilometers, by depending on the atmosphere conditions).

The THz waves are short enough to provide a spatial resolution of less than 1 mm (far field approximation). The resolution at tens and hundred nanometers can be achieved for solid-state objects [49, 50] (near-field approximation). The spatial resolution can be also improved several times beyond the Abbe limit, using lenses made of a material with high refractive index (*e.g.*, that of Si) compared to soft biological tissues [51]. The reduction in the size of a beam spot by several times, as compared to that observed in the case of an ordinary focusing in a free space, was achieved. However, the THz waves are long enough to penetrate most nonmetallic substances such as the materials used to make clothing, rucksacks, and tarps [52].

Plenty of materials, which are opaque in the visible and IR spectral ranges, are transparent in the THz range (see, *e.g.*, [53–55]). Those are paper, foams, plastic, textiles, *etc.* that can be used for packaging. The attenuation values (in dB) for some clothing, fabric, and building materials in the radiation range $v = 94...1042$ GHz are presented in

Ref. [52]. The THz imaging was applied to the non-destructive check of a quality of hidden damages in foams after the accident in 2003 with the space shuttle Columbia [56].

The THz radiation is of great importance in the fundamental researches, as well as in the technology and life sciences, because the rotational and vibrational lines of many substances are situated in this region. All the applications require different THz detector sensitivities closely related to the temperature exploitation and different spectral band applicabilities. These circumstances should be taken into account in estimations of the cost of THz systems which may differ by several orders.

Although the THz radiation is typically considered as propagated free-space beams, unlike the traditional optics in the visible and IR, the THz beams may be only a few wavelengths in the beam diameter, and the diffraction effects can become important in a low THz frequency range of ~0.1…0.5 THz.

The typical applications of IR technologies can be divided in two major groups. On the one hand, there are the near IR (NIR) and short wavelength IR (SWIR) spectral regions, which are commonly employed for the assessment of artworks (*e.g.*, paintings and frescoes), since some painting pigments are semitransparent for the IR radiation in these spectral bands and some other ones are not (*e.g.*, carbon-based). In this region, the IR spectroscopy is applied to biomedical science. On the other hand, there is the IR thermography, which involves the detection of surface and subsurface layers of objects by the differences in the thermal signatures in the medium wavelengths IR (MWIR) and long wavelengths IR (LWIR) spectral bands. In biomedicine, the thermography and spectroscopy are mostly used in any IR band [31]. High-resolution thermal images and the surveillance over large areas can be applied to the quantitative evaluation of land surface temperatures [57].

In spite of the great efforts in the past two or three decades resulting in an exponential growth in the number of publications [1, 6, 58] reaching thousands of hits, the THz applications in general are still at an early stage of development. Many other potential applications are likely to be added in the near future.

The THz technologies are useful to security agents and military personnel alike for revealing the concealed weapons, chemical explosives, and biological agents. Besides the security applications such as airport screeners, higher-resolution THz sensors could provide the enhanced identification of battlefield targets, better missile guidance, and other combat advantages. Soldiers, marines, and fighter pilots are increasingly trained to use not only the visible wavelengths that their eyes can process, but the IR and sub-THz wavelengths, as well their widespread fields of applications in military and civil domains.

Materials Research Forum LLC
https://doi.org/10.21741/9781644900758

Contrary to the IR range, where the imaging is passive, as a rule, it is necessary in most cases to use the THz illumination (THz sources) for the imaging.

The increasing demand for unoccupied and unregulated bandwidths for wireless communication systems will lead to the extension of operation frequencies toward the lower THz frequency range. Higher carrier frequencies, compared to the currently used ones, will allow the fast transmission of huge amounts of data needed for new emerging applications. Despite the tremendous hurdles to be overcome with regard for the sources and detectors, circuits, antennas, and architecture of systems in order to realize the ultrafast data transmission without extensive losses, a new communication area of researches is beginning [16, 59…61].

The development of the IR and THz technologies is important in the early cancer diagnostics, since cancer is one of the leading worldwide causes for death. In 2012, there were 8.2 million deaths from various forms of cancer [62]. Cancer is the second leading cause for death and was responsible in 2015 for 8.8 million deaths – nearly 1 in 6 global deaths [63]. The total number of deaths, due to cardiovascular disease reached 17.3 millions per year according to the WHO summary tables in 2008 [64]. Thus, the death data vs. different forms of cancer are comparable to cardiovascular diseases and will continue to rise to over 13.1 million in 2030 [65]. So, the economic impact of cancer is significant and is increasing. The total annual economic cost of cancer in 2010 was estimated to be approximately US $ 1.16 trillion [63].

Among women, the breast cancer decease is one of the prime causes for their death worldwide [64, 66]. Breast cancer remains the most commonly diagnosed type of cancer among women in the USA [67]. When detected early, it is possible to perform a lumpectomy surgery with minimum removal of normal breast tissue. That is why accurate techniques are needed to assess resection margins during surgery to avoid any secondary operations. In the case of THz technology for cancer diagnostics, it is related to the strong water absorption. The water concentration reveals a lot about the health of human tissue, since the water content in cancerous cells is higher than in healthy cells.

The thermography has a potential in the breast cancer diagnostics, by detecting the growth of a malignant tumor due to an increase in the internal temperature captured by thermograms. The IR thermography has emerged in recent years as an attractive reliable technique to address complex non-destructive (NDT) problems [68].

Over 3000 articles exist in which the medical IR thermal imaging was applied [69]. In medical applications, one should take into account that the medical infrared imaging can only be applied by physicians who have been educated and trained [70, 71].

Among the countries engaged in developing the THz technologies for diverse applications are the USA, UK, Germany, Netherlands, China, South Korea, Japan, Poland, Russia, and, to a certain degree, some others. Among the countries developing their own cooled and uncooled IR FPA (focal plane array) technologies for commercial and military applications, one can mention the USA, UK, Japan, Germany, South Korea, Canada, China, Italy, Russia, and some others. The THz and IR technologies, in spite of some instrumentation difficulties and atmosphere absorption problems, will be among one of the key technologies today for security systems, communications, biomedical and food control applications, *etc.*

For insight to the THz and IR applications, there exist a certain number of contemporary books and reviews in which the above-mentioned questions are considered as well (see, *e.g.*, [6, 7, 25, 30, 31, 39, 42, 66, 72–78].

## 1.2 Summary

The THz and IR technologies offer the possibility to collect information in ranges of electromagnetic spectra, which are inaccessible to human visual perception. The recent progress in applying the THz technologies to different domains of the human activity was possible due to the development of high-performance THz detectors (*e.g.*, for the cosmic microwave background and interstellar matter formation in astronomy; high-resolution molecular spectroscopy; detectors and sources for the active imaging, high speed and broadband communications, biomedical applications, surveillance, identification of hazardous chemicals, *etc.*).

However, with the complexity of highly sensitive uncooled or slightly cooled detectors and complexity of handheld or, at least portable, THz sources, this validates yet the difficulties of THz radiation technologies to ensure the market penetration with THz systems for cost-effective civil applications. For the short-range THz communication technique, under the environmental Earth conditions, one can mention the development of active silicon photonics integrated technologies.

Over the last decades, the successful development of IR detectors, arrays, and cameras lead to a significant progress in monitoring environmental pollution, surveillance and reconnaissance, security imaging, IR astronomy, medical diagnostics, car driving, *etc.* In many applications, this is due, to a certain degree, to the rapid development of uncooled microbolometer arrays, which are produced now in larger volumes, as compared to all other IR devices.

# References to Ch. 1

[1] S. Kremling, Th. Hochrein, Terahertz-technology approaches to markets: Survey about current developments, 19th World Conference on Non-Destructive Testing, 13-17 June 2016, Munich, Germany, 2016, Session: Terahertz/Microwaves, p. 1–11.

[2] A.H. Lettington, I.M. Blankson, M. Attia, D. Dunn, Review of imaging architecture, Proc. SPIE, 4719 (2002) 327–340, Volume title: Infrared and Passive Millimeter-wave Imaging Systems: Design, Analysis, Modeling, and Testing. https://doi.org/10.1117/12.477457

[3] IEEE Standards Interpretations for IEEE Std C95.1™-2005. IEEE Standard for Safety Levels with Respect to Human Exposure to Radio Frequency Electromagnetic Fields, 3 kHz to 300 GHz, C95.1-2005, http://standards.ieee.org/findstds/interps/C95.1-2005_interp.pdf.

[4] P. Richards, Bolometers for infrared and millimeter waves, J. Appl. Phys. 76 (1994) 1–24. https://doi.org/10.1063/1.357128

[5] J.M. Lamarre, F. X. Desert, T. Kirchner, Background limited infrared and submillimeter instruments, Space Sci. Rev. 74 (1995) 27–36. https://doi.org/10.1007/978-94-011-0363-3_4

[6] Mittleman D., 2003. Sensing with Terahertz Radiation. Berlin–Heidelberg–New York, Springer. https://doi.org/10.1007/978-3-540-45601-8

[7] Y.-S. Lee, Principles of Terahertz Science and Technology. Springer: Science+Business Media, New York, 2009.

[8] S. Withington, Terahertz astronomical telescopes and instrumentation, Philosoph. Trans. Royal Soc. London A. Math., Phys. and Eng. Sci. 362 (2004) 395–402. https://doi.org/10.1098/rsta.2003.1322

[9] H. Dole, G. Lagache, J.L. Puget, K.I. Caputi, et al., The cosmic infrared background resolved by Spitzer contributions of mid-infrared galaxies to the far-infrared background, Astron. Astrophys. 451 (2006) 417–429. https://doi.org/10.1051/0004-6361:20054446

[10] K. Sengupta, T. Nagatsuma, D.M. Mittleman, Terahertz integrated electronic and hybrid electronic–photonic systems, Nature Electr. 1 (2018) 622–635. https://doi.org/10.1038/s41928-018-0173-2

[11] G.N. Gol'tsman, O. Okunev, G. Chulkova, A. Lipatov, et al., Picosecond superconducting single-photon optical detector, Appl. Phys. Lett. 79 (2001) 705–707. https://doi.org/10.1063/1.1388868

[12] W.H.P. Pernice, C. Schuck, O. Minaeva, M. Li, *et al.*, High-speed and high-efficiency travelling wave single-photon detectors embedded in nanophotonic circuits,

Nature Comm. 3 (2012) 1325. https://doi.org/10.1038/ncomms2307

[13] K. Smirnov, A. Divochiy, Yu. Vakhtomin, P. Morozov, et al., NbN single-photon detectors with saturated dependence of quantum efficiency, Supercond. Sci. Technol. 31 (2018) 035011. https://doi.org/10.1088/1361-6668/aaa7aa

[14] T. Gerrits, N. Thomas-Peter, J.C. Gates, A.E. Lita, *et al.*, On-chip, photon-number-resolving, telecommunication-band detectors for scalable photonic information processing, Phys. Rev. A 84 (2011) 060301. https://doi.org/10.1103/PhysRevA.84.060301

[15] S.S. Dhillon, M.S. Vitiello, E.H. Linfield, A.G. Davies, *et al.*, The THz science and technology roadmap, J. Phys. D: Appl. Phys. 50 (2017) 043001.

[16] D.M. Mittleman, Perspective: Terahertz science and technology, J. Appl. Phys. 122 (2017) 230901. https://doi.org/10.1063/1.5007683

[17] A.M. Walther, B.M. Fischer, A. Ortner, A. Bitzer, *et al.*, Chemical sensing and imaging with pulsed terahertz radiation, Anal. Bioanal. Chem. 397 (2010) 1009–1017. https://doi.org/10.1007/s00216-010-3672-1

[18] A. Rogalski, F. Sizov, THz detectors and focal plane arrays, Opto-Electr, Rev. 19 (2011) 346–404. (Open Access). https://doi.org/10.2478/s11772-011-0033-3

[19] Committee on Airport Passenger Screening: Millimeter Wave Machines National Materials and Manufacturing Board Division of Engineering and Physical Sciences, 2017. Airport Passenger Screening Using Millimeter Wave Machines: Compliance with Guidelines, Washington, DC: The National Academies Press, 185 p.

[20] L.V. Titova, A.K. Ayesheshim, A. Golubov, R. Rodriguez-Juarez, *et al.*, Intense THz pulses down-regulate genes associated with skin cancer and psoriasis: a new therapeutic avenue? Sci. Rep. 3 (2013) 2363. https://doi.org/10.1038/srep02363

[21] H. Huntzsche, H. Stopper, Effects of terahertz radiation on biological systems, Critical Rev. Environmental Sc. Technol. 42 (2012) 2408–2434. https://doi.org/10.1080/10643389.2011.574206

[22] C.M. Hough, D.N. Purschke, Ch. Huang, L.V. Titova, *et al.*, Global gene expression in human skin tissue induced by intense terahertz pulses, Terahertz Sci. Technol. 11 (2018) 28–33.

[23] A.G. Pakhomov, Y. Akyel, O.N. Pakhomova, B.E. Stuck, M.R. Murphy, Current state and implications of research on biological effects of millimeter waves: A review of literature, Bioelectromagnetics, 19 (1998) 393–413. https://doi.org/10.1002/(SICI)1521-186X(1998)19:7<393::AID-BEM1>3.0.CO;2-X

[24] P.U. Jepsen, D.G. Cooke, M. Koch, Terahertz spectroscopy and imaging – Modern techniques and applications, Laser Photon. Rev. 5 (2011) 124–166.

https://doi.org/10.1002/lpor.201000011

[25] Saeedkia D., 2013. Handbook of Terahertz Technology for Imaging, Sensing and Communications. Oxford: Woodhead Publishing. https://doi.org/10.1533/9780857096494

[26] Song H.-J., Nagarsuma T., 2015. Handbook of Terahertz Technologies: Devices and Applications. Boca Raton: CRC Press. https://doi.org/10.1201/b18381

[27] H. Kasban, M.A.M. El-Bendary, D.H. Salama, A comparative study of medical imaging techniques, Int. J. Inform. Sci. Intell. Syst. 4 (2015) 37–58.

[28] J.Y. Suen, Terabit-per-second satellite links: a path towards ubiquitous communication, J. Infrared, Millimeter, and Terahertz Waves, 37 (2016) 615–639. https://doi.org/10.1007/s10762-016-0257-x

[29] S. Brinkman, N. Vieweg, G. Gartner, P. Plew, A. Deninger, Towards quality control in pharmaceutical packaging: Screened folded boxes for package inserts, J. Infrared, Millimeter, and Terahertz Waves, 38 (2017) 339–346. https://doi.org/10.1007/s10762-016-0345-y

[30] Son J.-H., 2014. Terahertz Biomedical Science and Technology. Boca Raton: CRC Press. https://doi.org/10.1201/b17060

[31] Bronzino J.D., Peterson D.R., 2017. Biomedical Engineering Handbook – Biomedical Signals, Imaging, and Informatics, (4-th Edition), Boca Raton: CRC Press.

[32] R.D. Hudson, Infrared System Engineering, Wiley-Interscience, New York, 1969.

[33] N.S. Kopeika, A System Engineering: Approach to Imaging. Bellingham: SPIE Optical Eng. Press, 1998. https://doi.org/10.1117/3.2265069

[34] G.C. Holst, Electro-Optical Imaging Systems Performance. Bellingham: SPIE Optical Eng. Press, 2003.

[35] H. Tang, Z.-L. Li, Quantitative Remote Sensing in Thermal Infrared. Theory and Applications, Berlin: Springer, 2014. https://doi.org/10.1007/978-3-642-42027-6

[36] A. Richards, Alien Vision. Exploring the electromagnetic spectrum with imaging technology. SPIE Press, Bellingham, 2001. https://doi.org/10.1117/3.419855

[37] M. Henini, M. Razeghi, Handbook of Infrared Detection Technologies, Elsevier, Oxford, 2002.

[38] G.H. Rieke, Detection of Light: From the Ultraviolet to the Submillimeter, Cambridge University Press, Cambridge, 2003. https://doi.org/10.1017/CBO9780511606496

[39] Seeing Photons: Progress and Limits of Visible and Infrared Sensor Arrays, The National Academic Press, Washington, 2010.

[40] A. Rogalski, Infrared Detectors, Boca Raton: CRC Press, 2011.

[41] X.-C. Zhang, J. Xu, Introduction to THz Wave Photonics. Springer, New York, 2010. https://doi.org/10.1007/978-1-4419-0978-7

[42] K.-E. Peiponen, J.A. Zeitler, M. Kuwata-Gonokami, Terahertz Spectroscopy and Imaging, Springer, Heidelberg, 2013. https://doi.org/10.1007/978-3-642-29564-5

[43] Corsi C., Sizov F., 2014. THz and Security Applications. Detectors, Sources and Associated Electronics for THz Applications. Dordrecht: Springer. https://doi.org/10.1007/978-94-017-8828-1

[44] H. Hintzsce, H. Stopper, Effects of terahertz radiation on biological systems, Critical Rev. Environm. Sci. Technol. 42 (2012) 2408–2434. https://doi.org/10.1080/10643389.2011.574206

[45] H.M. Pickett, E.A. Cohen, B.J. Drouin, J.C. Pearson, Submillimeter, Millimeter, and Microwave Spectral Line Catalog, 2003, https://spec.jpl.nasa.gov/ftp/pub/catalog/doc/catdoc.pdf.

[46] P.H. Siegel, Terahertz technology in biology and medicine, IEEE Trans. Microw. Theory Tech. 52 (2004) 2438–2447. https://doi.org/10.1109/TMTT.2004.835916

[47] T. Loffler, K. Siebert, S. Czasch, T. Bauer, H. G. Roskos, Visualization and classification in biomedical terahertz pulsed imaging, Phys. Med. Biol. 47 (2002) 3847–3852. https://doi.org/10.1088/0031-9155/47/21/324

[48] M. Nagel, M. Forst, and H. Kurz, THz bio-sensing devices: fundamentals and technology, J. Phys. Condens. Matter. 18 (2006) S601–S618. https://doi.org/10.1088/0953-8984/18/18/S07

[49] H.-T. Chen, R. Kersting, G.Ch. Cho, Terahertz imaging with nanometer resolution, Appl. Phys. Lett. 83 (2003) 3009–3011. https://doi.org/10.1063/1.1616668

[50] M. Eisele, T.L. Cocker, M.A. Huber, M. Plankl, *et al.*, Ultrafast multi-terahertz nano-spectroscopy with sub-cycle temporal resolution, Nat. Photonics. 8 (2014) 841–845. https://doi.org/10.1038/nphoton.2014.225

[51] R.I. Stanchev, D.B. Philips, P. Hobson, S.M. Hornett, *et al.*, Compressed sensing with near-field THz radiation, Optica, 4 (2017) 989–992. https://doi.org/10.1364/OPTICA.4.000989

[52] A.J. Gatesman, A. Danylov, T.M. Goyette, J.C. Dickinson, *et al.*, Terahertz behavior of optical components and common materials, Proc. SPIE, 6212 (2006) 6212OE. https://doi.org/10.21236/ADA461642

[53] G. Zhao, M. Mors, T. Wenckebach, P. Planken, Terahertz dielectric properties of polystyrene foam, J. Opt. Soc. Am. B, 19 (2002) 1476–1479.

https://doi.org/10.1364/JOSAB.19.001476

[54] T.S. Hartwick, D.T. Hodges, D.H. Barker, F.B. Foote, Far infrared imagery, Appl. Opt. 15 (1976) 1919–1922. https://doi.org/10.1364/AO.15.001919

[55] S. Busch, M. Weidenbach, M. Fey, F. Schafer, *et al.*, Optical properties of 3D printable plastics in the THz regime and their application for 3D printed THz optics, J. Infrared, Millimeter, and Terahertz Waves, 35 (2014) 993–997. https://doi.org/10.1007/s10762-014-0113-9

[56] N. Karpowicz, H. Zhong, C. Zhang, K.-I. Lin, *et al.*, Compact continuous-wave sub-terahertz system for inspection applications, Appl. Phys. Lett. 86 (2005) 054105. https://doi.org/10.1063/1.1856701

[57] G. Bitelli, P. Conte, T. Csoknyai, F. Franci, *et al.*, Aerial thermography for energetic modeling of cities, Remote Sensing. 7 (2015) 2152–2170. https://doi.org/10.3390/rs70202152

[58] R.A. Lewis, A review of terahertz sources, J. Phys. D: Appl. Phys. 47 (2014) 374001. https://doi.org/10.1088/0022-3727/47/37/374001

[59] J. Federici and L. Moeller, Review of terahertz and subterahertz wireless communications, J. Appl. Phys. 107 (2010) 111101. https://doi.org/10.1063/1.3386413

[60] J.F. O'Hara, S. Ekin, W. Choi, I. Song, A Perspective on Terahertz Next-Generation Wireless Communications, Technologies. 7 (2019) 43. https://doi.org/10.3390/technologies7020043

[61] I. Akyildiz, J. Jornet, C. Han, Terahertz band: Next frontier for wireless communications, Phys. Commun. 12 (2014) 16–32. https://doi.org/10.1016/j.phycom.2014.01.006

[62] Information on http://www.cancerresearchuk.org/health-professional/cancer-statistics/worldwide-cancer.

[63] World Health Organization Fact Sheet, 2017, http://www.who.int/ features/factfiles/cancer/en/

[64] Information on http://www.world-heart-federation.org/cardiovascular-health/global-facts-map/.

[65] A.K. Panwar, A. Singh, A. Kumar, H. Kim, Terahertz imaging system for biomedical applications: Current status, Int. J. Eng. Technol. 13 (2013) 33–39.

[66] U. Raghavendra, U.R. Acharya, E.Y.K. Ng, J.-H. Tan, A. Gudiga, An integrated index for breast cancer identification using histogram of oriented gradient and kernel locality preserving projection features extracted from thermograms, Quantitative InfraRed Thermography, 13 (2016) 195–209.

https://doi.org/10.1080/17686733.2016.1176734

[67] American Cancer Society. Cancer facts & figures, https://www.cancer.org/ content/dam/cancer-org/research/cancer-facts-and-statistics/annual-cancer-facts-and-figures/ 2018/cancer-facts-and-figures-2018.pdf.

[68] F. Khodayar, S. Sojasi, X. Maldague, Infrared thermography and NDT: 2050 horizon, Quantitative InfraRed Thermagraphy. 13 (2016) 210–231. https://doi.org/10.1080/17686733.2016.1200265

[69] E. Sousa, R. Vardasca, S. Teixeira, A. Seixas, *et al.*, A review on the application of medical infrared thermal imaging in hands, Infrared Phys. Technol. 85 (2017) 315–323. https://doi.org/10.1016/j.infrared.2017.07.020

[70] R. Berz, H. Sauer, The medical use of infrared-thermography. History and recent applications, Thermografie-Kolloquium-2007, Vortrag 04, 1–12, 2007, www.ndt.net/search/ docs.php3?MainSource=61.

[71] I. Fernández-Cuevas, J.C.B. Marins, J.A. Lastras, P.M.G. Carmona, et al., Classification of factors influencing the use of infrared thermography in humans: A review, Infrared Phys. Technol. 71 (2015) 28–55. https://doi.org/10.1016/j.infrared.2015.02.007

[72] Committee on Developments in Detector Technologies; National Research Council, 2010.

[73] M. Golio, The RF and Microwave Handbook, Boca Raton: CRC Press, 2001. https://doi.org/10.1201/9781420036763

[74] E. Bründermann, H.-W. Hübers, M.F. Kimmitt, Terahertz Techniques. Springer, Heidelberg, 2012. https://doi.org/10.1007/978-3-642-02592-1

[75] D.T. Haynie, Biological Thermodynamics, Cambridge University Press, Cambridge, 2001. https://doi.org/10.1017/CBO9780511754784

[76] E.F. Ring, K. Ammer, Infrared thermal imaging in medicine, Physiol. Meas. 33 (2012) R33–R46. https://doi.org/10.1088/0967-3334/33/3/R33

[77] C. Corsi, New frontiers for infrared, Opto-Electr. Rev. 23 (2015) 3–25. https://doi.org/10.1515/oere-2015-0015

[78] M. Diakides, J.D. Bronzino, D.R. Peterson, Medical Infrared Imaging: Principles and Practices, Boca Raton: CRC Press, 2013. https://doi.org/10.1201/b12938

# Chapter 2. Brief history of THz and IR technologies

## 2.1 Introduction

A brief introduction to the history of THz and IR technologies for the learning by historical lessons is presented. These brief lessons learnt by historical highlights in the THz and IR science can be expected to be important for the future developments in these directions, since history frequently opens routes for new thinking. In this brief review, some important steps may have been missed. The Author apologizes for these possible shortcomings. The chapter is an attempt to present the Author's point-of-view to the history of developments in the THz and IR technologies.

## 2.2 Brief history of THz technologies

Some time ago, the THz region of the electromagnetic spectrum was often described as the final unexplored area of electromagnetic (EM) spectrum. Till about three to four decades ago, the potentiality of the THz spectral range for applications was unclear, largely because of the difficulty in providing suitable sources and detectors. However, during the last two decades, the THz science and technology are showing rapid growth. Now, these technologies are among the widely investigated research topics and applications (see, *e.g.*, [1, 2].

The research in this spectral range started from the discovery of radio waves by H. Hertz (1885–1989). Between 1885 and 1889, H. Hertz at Karlsruhe Polytechnic, generated electromagnetic waves and measured their wavelength $\lambda \approx 66$ cm [3] and velocity.

H. Hertz did not think that this transmission of electromagnetic waves can be used in the information communication [4]. It was Guglielmo M. Marconi, who thougt he could use Hertzian waves to send signals [5]. Marconi used a spark transmitter, as no amplitude modulation technique or active non-linear devices were known that time. G. Marconi's first achievement was the demonstration of sending successfully a wireless signal remotely on Salisbury Plain, U.K., that was in May 1897. Perhaps, it should be noted that J. Bose had given a public demonstration of wireless transmission over a mile in 1895 to remotely ring a bell and to explode gunpowder [3].

Despite to the thoughts that it is not feasible to transmit information through long distances as EM waves behave like light and thus couldn't bend to the Earth's curvature, in 1901, G. Marconi first demonstrated the trans-Atlantic radio waves transmission of about 2000 miles between Poldhu, UK and Newfoundland, St. Johns, Canada (likely it was at 820 KHz). To detect the first trans-Atlantic wireless signal (three dots representing the letter "S") G. Marconi used a 150 m long wire antenna put on a kite and "coherer" to convert an alternating signal into a direct current to measure by a telephone receiver to hear a signal. J.C. Bose at his lecture at Royal Society in 1899, proposed a similar coherer [6].

In Germany, Ch. Hülsmeyer was the first who uses radio waves to detect "the presence of distant metallic objects" (1904). He demonstrated the feasibility of detecting a ship in dense fog, but not the distance to it [7].

The history of THz technologies in different periods is presented, *e.g.*, in a number of papers and books [7…12].

In 1897, H. Rubens and E.F. Nichols first explicitly noted the existence of a gap in the electromagnetic spectrum, between the optical and electronic sources of radiation [12].

Although a lot of work in the THz spectral range has already been done under the older terms "far infrared" (see, *e.g.*, [13, 14]) or sub-mm waves (see, *e.g.*, [15, 16]), the term "terahertz" stands for a novel technique offering many potential applications of the THz technologies to industry, medicine, detection of drugs and explosives, telecommunications, *etc.* The term "terahertz" also represents a new generation of systems.

Historically, the THz technologies were used mainly within the astronomy community for studying the background of cosmic far-infrared radiation. They were utilized too by the laser-fusion community for the diagnostics of plasmas. The earliest investigations on measuring the energy content of the blackbody radiation in the far-infrared range (FIR) were presented by H. Rubens [17]. H. Rubens' experiments were mainly concentrated on the extension of the IR spectral range into the FIR and the absorption of water vapor in this spectral range (see, *e.g.*, [18]).

Now, as it is proven, the THz technologies are important for imaging and spectroscopy in spite of the highly absorbing environmental Earth conditions. The clothing, plastic packaging equipment, and microcircuit bodies are transparent in the THz spectral range ($v \le 1$ THz). This gives opportunity not only to reveal ceramic weapon and many types of explosives on the human body, as they are highly reflective substances due to the difference in dielectric permittivity or high water content, but also different types of diseases (as a rule in the reflection mode) – biomedical applications. This makes usage of THz technologies to be challenging in various types of applications.

For THz science and technology, important was the work of J.C. Bose at the Presidency College, Calcutta, with which he started the wireless mm-wave experiments in 1894 with new sensors, spark generators, polarizers and sources. The shortest wavelength used in Bose's experiments was about 5 mm. He invented contact detectors involving metals and semiconductors [19].

Important was the invention of a galena crystal detector [20]. In it J. Bose claimed "A coherer or detector of electrical disturbances, Hertzian waves, light-waves or other radiations, comprising contacting pieces of sensitive substance having a characteristic curve (giving the relation between an increasing impressed electromotive force and the resultant current passing through the sensitive substance), which is not straight but is either convex or concave". This was a point-contact semiconductor rectifier, involving a contact device with galena (lead sulfide), and the first patent for a semiconductor device in the world (see, *e.g.*, [7, 12]). Later, this device was used as a receiver for the demodulation of continuous-wave radio signals. This was the first semiconductor diode detector, although the terms "diode" and "semiconductor" were not known yet. In 1954, G. Pearson and W. Brattain [21] gave priority to Bose for the use of a semiconducting crystal as a detector of radio waves.

After World War I, E.F. Nichols performed a series of experiments with short EM waves developing improved radiometer receivers and Hertzian oscillators. E.F. Nichols and J.D. Tear (1923) succeeded in obtaining wavelengths down to 1.8 mm using the interferometric method [12]. In 1923, Glagolewa-Arkadiewa [22] showed a possibility to get a source (using Al sawdust in thick oil as Hertzian oscillators) operating within the wavelengths from 5 cm down to 82 μm ($v \approx 3.66$ THz). These works in producing such electromagnetic wavelengths filled the gap between spectra of the IR and radio wavelengths.

From the middle of the 1920s till the 1950s–1960s, there was a relatively steady flow of papers, when a rapid expansion of papers began growing up (see, *e.g.*, [1, 9]).

Radio-detection and ranging (radar) was invented in the 1930s, whereas a working radar system for the detection of ships was actually demonstrated in Germany in 1904 (Ch. Hülsmeyer) [7] not allowing to directly measure the distance to a target, but detecting only the presence of distant objects. This work did not materialize into applications, since there was no real need for radars at that time till the maturing of the airplane production in the 1930s. By 1939, England had established a chain of radar stations along its south and east coasts to detect aggressors in air or on the sea. Except for the UK, other states were developing independently systems of this type (Germany, the United States, the former USSR, Japan, the Netherlands, France, and Italy).

The designing and testing of a gun-aiming radar were performed in Ukraine (Kharkiv) in 1935–1940 [23]. As for the history of research and development at the sub-THz spectral range in the former USSR, which is relatively little known, this topic was considered, *e.g.*, in [24, 25]. These articles review the history and state-of-the-art of quasioptical systems based on various transmission-line technologies. The trace of the development of quasioptics back to the very early years of experimental electromagnetism touching the "Hertz waves" was considered. The role of the Ukrainian microwave, antenna, radar, and remote sensing in the worldwide development was indicated in 2015 by awarding the status of IEEE Milestone [26]. The development of radar systems led to new magnetron and klystron sources, and these were employed for microwave spectroscopic experiments.

Since then, it followed a continual progress and development in technology for both sources and detectors. Shortly after World War II, the Golay cell, which is still in use today, was proposed (M.J.E. Golay, 1946–1947) (see part "Brief history of IR technologies").

The genesis of microwave and ultimately THz spectroscopy was the war time development of microwave radars. By 1948, the field was mature enough [27]. The technology advanced rapidly, and the submillimeter threshold at 300 GHz had been passed by 1954 [28]. This drive toward even higher frequencies was aided by the rapidly increasing absorption strengths of the spectra of many of the most important small fundamental molecules (*e.g.*, carbon, chlorine, nitrogen, oxygen, *etc.*).

The prediction of the existence of "relict radiation" (the cosmic microwave background) remaining from the "Big Bang" was made in 1948 by R. Alpher and G. Gamow [29] developing G. Lemaître's Big Bang theory (late 1920s–1930s). R. Alpher and R. Herman estimated what temperature of the cosmic microwave background (CMB) ought to be. They obtained $T \approx 5$ K (the precise measured temperature of microwaves is $T = 2.725$ Kelvin degrees [30] that corresponds to $\lambda_{max} \approx 1.064$ mm ($\nu_{max} \approx 282$ GHz) of the blackbody radiation. Measurements of the CMB at various frequencies from different platforms showed it to have the spectrum of thermal blackbody spectrum, as predicted by the Big Bang model [31].

The first direct observations of the CMB were made in 1964 by A. Penzias and R. Wilson [32]. They were trying to measure the background microwave interference to enable a noise-free communication within this spectrum band for AT&T. They were conducting experiments with the sensitive Holmdel Horn Antenna, originally used to detect radio waves that were bounced off Echo balloon satellites, and later the Telstar, the first active communication satellite. They determined that the buzzing noise was coming from all parts of the sky at all times of day and night from outside of our galaxy.

During the 1950s, the foundations were laid from grating spectroscopy, in which narrow slits were required for a high resolution, to Fourier-transform spectroscopy (FTS) having large apertures, though it was progressing relatively slow mainly due to limitations of the computer technique. Of the papers published in the 1950s about 80% concerned applications connected with the study of semiconductor optical properties (*e.g.*, cyclotron resonance), absorption spectra of gases, diagnostics of high-temperature plasmas, *etc.* [9].

In the early 1950s, there appeared the first carcinotrons or backward wave oscillators (BWO) [33]. This high-power (mW range at 1 THz) source was demonstrated for the first time in 1952 offering a limited electronic tunability (~10 %). It was a French-made BWO tube [34] that was called carcinotrons – derived from the Greek word for Cancer (crab).

The late 1950s are characterized by starting the development of high-power gyrotrons (the USA, Australia and former USSR) [9]. The gyrotrons are members of a specific family of devices in the class of vacuum electronic sources of coherent microwave radiation.

Important for using in radar and alarm systems as GHz radiation sources are IMPATT (IMPact ionization Avalanche Transit-Time) diodes. In 1958, Read [35] proposed the working principles of this device. W. Read showed that, when the impact ionization is used to inject electrons, an avalanche diode with a significant transit time delay might exhibit a negative resistance characteristic of that is required for the beginning of oscillations. In 1959, there were devised the IMPATT diodes in Ge on the base of the generation of coherent oscillations in avalanche breakdown microwave frequency diffusion diodes (for Refs., see [36]). Later, the devices (oscillators) based on the phenomenon of microwave frequencies generation were produced with Si and GaAs diodes [37, 38]. The IMPATT diodes have received much attention in the last years as cost-effective and low-sized THz sources. Their device performance on the base of Si, GaAs, and InP is improved from year to year. The Si IMPATT diodes have more reliable and mature technology.

The experiments in the late 1950s and the early 1960s provided a considerable opportunity for THz science: the emission of THz radiation from a heated plasma delivered temperature and density information; the Golay cell detection and diffraction grating dispersion (G.N. Harding, *et al.*, 1961) were successfully used for the plasma diagnostics in the 100–1500 μm band. At that time, the operational frequency range of BWO sources was subsequently extended to 1 THz and higher [12].

The 1960s can be marked by the great progress in the detector and source development. Golay cells were widely replaced by pyroelectric detectors. New detectors, though deeply cooled (*n*-InSb hot electron bolometer, Ge bolometer, Ge:Ga extrinsic photoconductor, detector based on the Josephson effect), were developed. The important discovery in this decade was the water vapor laser (1964, followed by several other gas lasers, which provided the radiation of several continuous wave lines a little bit smaller than the radiation frequency of 1 THz [9].

At the early 1960s, the Gunn effect was invented [39]. Gunn diodes are used in Gunn oscillators. These sources are widely used today in the frequency range $v \leq 0.3$ THz for radio communications, military and commercial radar sources.

In the 1960s, there were developed the electrical-discharge-pumped THz sources [40]. They reveal the strong submillimeter wave emission from low-pressure water vapor, using a spectrometer. The optically pumped THz gas lasers were considered in [41]. At the same time, the high-resolution THz Fourier transform spectroscopy was developed [42].

Started in the late 1960s, there were developed the studies of the influence of THz and mm-wave radiation on biological systems. Researchers [43] have looked at effects on *E. coli*. They surveyed the growth of bacteria after the 0.136-THz irradiation with estimated power of 7 $\mu$W for 4 hours. The growth inhibition was observed after 2 hours, when cells were irradiated in the lag phase, and after 1.5 hours, when cells were irradiated in the log phase. Later, a different group tried to reproduce this experiment [44]. The same object (*E. coli*) was irradiated with the 0.136-THz radiation for 4 hours. In contrast to the former investigations, they did not find any evidence for the growth inhibition.

In the 1960s–1970s, the appreciable growth in the number of papers concerning the THz and microwave spectral regions was registered [1]. These years were important for astronomy at THz/sub-THz frequencies, due to the progress in receivers and high-altitude outboard platform observatories.

The semiconductors band structure conception and impurity properties of semiconductors were widely explored at that time using BWO, interferometer, and molecular-gas laser sources (J.M. Chamberlain *et al.*, 1969, 1972). Remarkable instrumental progress by D.H. Martin (1967) was augmented by the developments in computing techniques, which enabled the full advantages of Fourier transform spectroscopy to be realized [12].

At the end of the 1960s, the time domain spectroscopy (TDS) was realized [45]. In that paper, a new method for obtaining the complex permittivity and permeability of linear materials over a broad range of microwave frequencies was described by obtaining the reflected and transmitted transient responses of the component to an incident sub-nanosecond rise-time pulse and then performing the discrete Fourier transforms.

Materials Research Forum LLC

https://doi.org/10.21741/9781644900758

There, the progress in new detector technologies for the astronomy based on the use of uncooled and cooled fast Shottky barrier diodes (SBDs) for both direct detection and mixers applicable at sub-THz and mm-wave spectral regions[1] (for Refs., see, *e.g.*, [47]) can be noted.

Other detectors important, *e.g.*, in the astronomy for heterodyne systems were deeply cooled superconductor-insulator-superconductor (SIS) mixers [48], although the physics of these devices was developed almost two decades earlier.

Important for applications THz sources were realized by the optical-to-THz conversion using nonlinear materials, photoconductors, and photodiodes as interaction media. In the late 1960s–early 1970s, the tunable sub-THz radiation obtained by the mixing procedure in nonlinear crystals was used [49, 50].

In 1971, the operation principle of quantum cascade lasers (QCLs) was proposed [51], and the first demonstration of QCL was realized at Bell Labs in 1994 [52]. QCLs now are among the most used solid-state THz narrow-band sources at the frequency range $\nu > 1$ THz.

The THz imaging applications in use other than astronomical ones have started in the middle of the 1970s (see, *e.g.*, [53]). In that paper, the THz imaging system was based on an HCN laser (operation wavelength $\lambda = 337$ µm, $\nu = 0.89$ THz).

In this decade, it was known that an ultra-short optical pulse colliding a photoconductor can generate THz pulses. The highest frequency of the THz pulses depends on the optical pulse width and on the electron mobility in the photoconductor as well. In the mid-1970s, the technology of generating the optical pulses from a mode-locked Nd:glass laser and high-resistivity Si [54] were used for switching with the photoconductive (PC) structure. This PC switch is often called as the Auston switch. The availability of short optical pulses and the development of the PC ultrafast technology have promoted the THz optoelectronics. In 1983, Auston and Smith [55], using the sampling technique, showed the coherent detection of a short burst of THz radiation.

Ge extrinsic photoconducting (though they are noisy in spite of the cooling to liquid He temperatures) detectors were used up to a wavelength of 220 µm (stressed Ge:Ga photoconductor) [56].

---

[1] The Schottky barrier diode is named after German physicist Walter H. Schottky, who analyzed the metal–vacuum barrier (Schottky–Nordheim) and, later, the metal-semiconductor rectifier junction [46]. A point contact metal–semiconductor rectifier was first patented by J.C. Bose in 1904 (see above).

Materials Research Forum LLC
https://doi.org/10.21741/9781644900758

In spite of the fact that THz radiation has been known for a long time, the technical applications in biomedicine still have not been developed due to a lack of suitable sources and detectors [11]. During the last two to three decades, these problems have been solved to a certain degree.

In the late 1970s, there were performed investigations [57] of hemoglobin and alcohol dehydrogenase that were irradiated. The functional effects were investigated. The authors studied isolated biological systems including enzymes, antibodies, biomolecules, and artificial liposomes. Applied was a sweep irradiation from 0.075 to 0.115 THz. Not explicitly specified exposure time was given, but it can be calculated to be about three hours [58]. The alcohol dehydrogenase activity was measured at the irradiation power densities $W \approx 10...50$ mW/cm$^2$. No changes of more than 0.1% were seen. For hemoglobin, the oxygen-binding capacity was measured after the irradiation with 3...13 mW/cm$^2$, and no changes of more than 0.4% were observed. The experimental investigations looked at all biological levels from isolated biomolecules to animals. But, till 2012, no investigation of effects on human organism is available. The history of THz radiation influence on biological systems till the 2010s was reviewed in [58].

The development in computer technologies (1980s) had chosen Fourier-transform systems for the spectroscopy through the IR and THz regions [9].

In the 1980s, the common today methods of THz radiation generation with the PC antenna (PCA) and optical rectification (OR) were considered [59–61].

In [62], it was shown that PCAs used as an emitter and a detector have frequency spectra that extend from $\nu \sim 100$ GHz to over 1 THz. These are important for THz time-domain spectroscopy (TDS) and imaging. The advent of mode-locked Ti:sapphire femtosecond lasers [63] in the early 1990s greatly expanded the field of THz applications in spectroscopy and imaging. Today, a lot of the research in THz spectroscopy and imaging is carried out using such time-domain spectrometers.

In the 2000s, the THz imaging was applied to the non-destructive quality check of hidden damages in foams after the accident in 2003 with Space Shuttle Columbia.

The generation and coherent detection of sub-picosecond electronic transients conditioned by Auston and Nuss [64] and by Fattinger and Grischkowsky [61] were performed.

The first lasing in an electrostatic accelerator free electron laser (FEL) emitting in THz spectral range was demonstrated at the University of California, Santa Barbara (UCSB) [65]. It covered the spectral range from 0.3 to 0.77 THz with 10-kW peak power. These kind of sources are space-consuming having room-building size.

Detectors and Sources for THz and IR                    Materials Research Forum LLC
Materials Research Foundations **72** (2020)              https://doi.org/10.21741/9781644900758

At the end of the 1980s, NbN superconducting hot electron bolometers (HEBs) were proposed [66] as THz mixers with a time constant of about 40 ps. They allowed one to realize the bandwidth of several GHz with an advantage of ~ 1 µW local oscillator power (LOP) compared to ~1 mW LOP for SBDs (for Refs., see, *e.g.*, [67]). This superconductor film (NbN) HEB is able to work as a thermal detector in a wide spectral range in the IR and THz domains, but the significant advantage is in that it is also able to work as a mixer with wide bandwidth (several GHz). HEBs have been known for quite a while, since the first work on the InSb low-temperature hot-electron heterodyne detector [68, 69]. However, this detector can be used in heterodyne systems only, where the bandwidth lies in the MHz range, because of long recombination times (~$10^{-6}$ s).

In 1993, the PC antenna fabricated on the high-speed low-temperature grown (LTG) GaAs as photomixers to generate CW THz radiation [70] were used. The development of such procedure served as the basis for THZ spectroscopy and imaging.

Since the first demonstration of THz wave TDS in the 1980s, there has been a series of significant advances as more intense THz sources and higher sensitivity detectors provided new opportunities for understanding the basic science in the THz frequency range. As developments will move forward, THz science will not only have an impact on the material characterization and identification, but also will have potential applications in the fields of communications, imaging, medical diagnosis, health monitoring, environmental control, and chemical and biological sensing, as well as security and quality-control applications. The twenty-first century research in the THz band is one of the most promising areas to study the transformational advances in imaging and other interdisciplinary fields [71].

In the late 1990s, the novel near-field probes have been developed, and various applications have been demonstrated (see, *e.g.*, [72]).

In 1996, Dyakonov and Shur [73] considered the THz response of the two-dimensional electron gas in a field-effect transistor (FET) channel. Following the mentioned considerations, the sub-THz/THz detection was demonstrated in III–V HEMTs [74], and more recently, in graphene [75]. Compared to many other uncooled detectors, these devices (*e.g.*, Si-MOSFETs and III-V HEMTs) can be manufactured at the foundry level, as their technology readiness is high.

Recently, the performances acceptable for many THz and IR applications have been obtained with uncooled microbolometer technology. The relatively low-cost arrays with a high number of pixels were demonstrated. The most reliable and good performances has been obtained both in IR and THz with good dynamic range using $VO_x$ and $\alpha$-Si thin film materials with thermally isolated microbridges [76].

In the early 2000s, the demonstrations of THz wireless communications were conducted using both pulsed and continuous waves, which were generated from photoconductors and photodiodes excited by pulse lasers and intensity-modulated lasers (for Refs., see, *e.g.*, [77]. A little bit later, the wireless link employing a 120 GHz band was the first demonstrated commercial THz communication system with an allocated bandwidth of 18 GHz, which offers the 10…20 Gbit/s modulation for transmission distances of over 5 km. In the technique of THz communications, the important progress is connected with the development of active silicon photonics integrated technologies [78, 79].

## 2.3 Brief history of IR technologies

The story of infrared radiation technologies took many years to reach the level of use that is recognized today. The foundation of the IR detectors mainly in the 1960s–1970s led to the growth of thermovision and ground-based IR astronomy in the atmosphere transparency windows and a little bit later to the observations from the low- and high-altitude observatories to diminish or exclude the water vapor absorption.

The people guessed that the imaging in the IR range is a technology for getting additional information from objects that are invisible (*e.g.*, under night conditions) to the human eye, which is only sensitive within the spectral range approximately from 0.4 to 0.75 μm. At first, researchers relied on the radiation from the Sun.

The first recorded account of an infrared experiment appears to be that by Jean Batista Della Porta of Naples in his book "Magiae Naturalis" published in 1589 [11]. He found that heat could be sensed, when locating a candle in front of a silver plate. When the plate was removed, the sensing of heat from the candle flame was reduced.

In the experiments made in 1667 at the Academia del Cimento, it was shown that the mass of ice placed in a vessel radiates. The registered radiation was concentrated by a concave mirror on a long vertical thermometer and ice made a sensible repercussion of cold upon the thermometer [80].

It was found [80] in the note by Petrus van Musschenbroek (1692–1761), probably relating to an experiment made by himself (1755), that the heat from a charcoal fire placed out of an evacuated vessel, but nearby, passes easily through this vessel, and raises indications of the thermometer placed at the center of this vessel.

Temperature is a long established indicator of heat. As soon as the thermometer was invented, S. Sanctorius used it in 1612 to measure the heat from the Sun. Using the thermometer, one can registered the thermal balance of a body to control health conditions and the temperature rising or falling, which enables one to conclude about the human or animal body disfunction. C. Huygens, O. Rømer, and D. Fahrenheit all proposed the need for a calibrated scale of thermometers in the late seventeenth and early eighteenth centuries. A. Celsius proposed a centigrade scale based on ice and boiling water. However, he suggested that boiling water should be zero and melting ice 100 on his scale. It was C. Linnaeus in 1750 who proposed the reversal of this scale, as it is known today. The clinical thermometer, which is used universally in medicine for about 150 years, has been proposed by C. Wunderlich in 1868. In the 1960s, there were proposed liquid crystal sensors for temperature measurments [80, 81].

The IR technologies obviously have started with amateur astronomer William Herschel's experiments with a thermometer in 1800. He placed a prism in the path of a Sun beam, set thermometers after a prism in various regions of the rainbow and showed that different colors registered different temperatures. By shifting a thermometer further after the red part, he observed some warming. He has concluded it could only been caused by a form of light that is invisible for the human eye. In 1840, John Herschel made a simple image by the evaporation of a carbon and alcohol mixture using focused sunlight. He named the image a "thermogram" [81, 82].

Remote sensing of the IR radiation became of practical value at the end of the 1930s – early1940s. It has continued to develop steadily from the middle of the 1950s. Since that time, the radiometric determination of human body's temperature became an important tool, both for the medical diagnostics and monitoring treatment. Human heat is associated with many conditions such as inflammation and infection, and these conditions can be detected/identified by the IR thermography. It is a non-invasive and painless tool for the skin-temperature control over physiological functions [83].

The heat transfer by radiation is of great importance in medicine, as the radiation flow from the body with temperature around $T \approx 310$ K ($T \approx 37$ C) is large. At this temperature, there is situated the maximum radiation power (see Fig. 2.1), and the amount of radiation power in the atmospheric window from $\sim 7.5$ to 14 μm, in which mainly the thermovision cameras are operated, is appreciable. That is why the thermovision is an important instrument to control the thermal balance of the human body. The nude human body emits about 1 kW into the environment in the whole spectral range from $\lambda = 0$ to $\lambda = \infty$ according the Stefan--Boltzmann law $P(T) = \sigma_B \cdot T^4$. Here, $\sigma_B = 5.6686 \times 10^{-12}$ W/(cm$^2$ K$^4$) is the Stefan--Boltzmann constant, and it is assumed that the surface area of human body is $S \approx 2$ m$^2$.

Materials Research Forum LLC

https://doi.org/10.21741/9781644900758

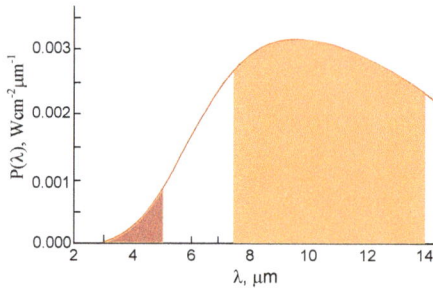

*Figure 2.1. Blackbody spectral radiant exitance at T = 300 K. The colored regions demonstrate the portions of the total power available for MWIR (3–5 μm) and LWIR (7.5–14 μm) intervals, respectively.*

At the same temperature of the environment, the human body is in equilibrium with it and, therefore, does not loose energy. Whereas, *e.g.*, at an environment temperature lowered by $\Delta T \approx 20$ K, the human body heat losses are about 250 W, and an undressed person will quickly chill.

The IR technologies find extensive applications in technical vision systems, since the temperature of the objects heated to mean Earth surface temperature is about 300 K at the maximum emission. Within the transmission windows 3...5 and 8...14 μm, it allows using such instrumentation as thermal passive imagers for remote sensing and vision at large-scale distances at the night or day surveillance, when different kinds of camouflages are used for objects to be hidden in the visible spectral range.

The development of IR detectors in the 19th century and at the beginning of 20th one was basically connected with the thermal uncooled detectors, which response to "heat" changing their properties under the radiation within the whole spectral range. Those were thermocouples and bolometers. In 1830, L. Nobili has proposed a thermocouple as an IR detector [84]. In 1833, the multielement thermopile was applied by M. Melloni to show that a person 10 m away could be detected by focusing his or her thermal energy on that device [7]. In 1880, Langley [85], whose main interest was in the use of the detectors in astronomy, proposed a bolometer as a thermal detector and stated that his bolometer could detect a cow moving across the field at 1/4 mile away.

In 1893, Rubens and Snow [86] presented an investigation on the refraction of rays in the materials pointed out. In 1896, Nichols [87] found that the reflectivity of crystalline quartz rose over a narrow wavelength range near 9 μm from a few % up to almost the reflectivity of a metal. The reflected radiation from a crystalline quartz is due to lattice vibrations and was detected with the new radiometer. The spectral range of the high reflectivity in ionic crystals (the range between the longitudinal and transverse optical phonon frequencies) now is attributed to the restrahlen band.

Rubens and Nichols built a spectrometer using multiple restrahlen plates to isolate a very narrow wavelength band. By changing the reflecting plates, they were able then to produce nearly the monochromatic radiation at a number of different wavelengths beyond 50 µm [11].

At the end of the 1890s, the radiation from hot bodies and its wavelength dependence were of important concern. Unfortunately, the experimental data obtained and compared with the blackbody used as an ideal source did not agree with the theory. In 1900, H. Rubens and K. Kurlbaum [88], using a restrahlen spectrometer, obtained the required experimental data in the IR range. H. Rubens immediately visited Max Planck to give him the results who wrote down the equation the same day. This result was published in 1900 [89] and now is called Planck's Radiation Law.

In 1909, A. Einstein, analyzing the energy and momentum fluctuations in the blackbody radiation, assumed the validity of Planck's law and showed that the expressions for the mean-square energy and momentum fluctuations split into a sum of two terms. The first is a wave term that dominates in the Rayleigh–Jeans (long wavelength) range of the spectrum and the second is a particle term that dominates in the Wien's law (short wavelength) spectral range [90] (cited in Ref. [91]). Both terms were necessary to describe the fluctuations for the complete blackbody spectrum.

In 1911, H. Rubens and coworkers showed that the mercury arc lamp in a quartz envelope was a long-wavelength IR source able to emit very long wavelength radiation (210 and 324 µm, THz range) [92, 93]. This source is still capable to generate radiation at frequencies in the 30…50 THz interval.

In 1913, one of the first examples for security applications was probably presented. It was L. Bellingham, who patented "an IR eye," capable of detecting the distance to "icebergs" [7].

The second kind of detectors, apart of the thermal detectors that are called now the photon detectors, was mainly developed during the 20th century although the photoconductivity effect (observable in a photon detector) was discovered in 1873 (W. Smith) in experiments with selenium as an insulator. In 1917, T.W. Case developed photoconducting detectors on the base of $Tl_2S$. In 1904, verily a photovoltaic detector in galena (natural PbS) – solid-state diode detector to detect EM waves – was patented by J. Bose [20] (see Subsection 2.2 "Brief history of THz technologies").

The period between the World Wars I and II can be characterized by the development period of photon detectors and image converters. In that period, the idea of an image tube was proposed in 1928 [94]. In 1934, the first successful IR converter tube (Holst' cup) was created [95]. This tube consisted of a photocathode in close proximity to a fluorescent screen. Electrons knocked out from the photocathode by IR photons were striking the fluorescent screen thus transferring an IR image into the visible region. These image tubes (now called intensifiers with photocathode, micro-channel plates, and fluorescent screen as the basic elements), are sensitive in the short interval ($\lambda \sim 0.8...1.1$ µm) of the IR spectrum.

In 1928, M. Czerny documented the first infrared image of a human subject [96]. The infrared thermal imaging used for recording the surface temperatures of human body in early trials in medicine started in 1952 in Germany [97]. A single IR bolometer for the thermal measurement of definite regions of the human body's surface for diagnostic purposes was developed [98]. In 1954, the first medical association of thermography was established in Germany.

From that time, the medical IR imaging covers a broad field of applications. Among these applications − female breast cancer, neurology, vascular imaging, forensic medicine, surgery, *etc.* and now are efficient means of a noncontact radiometric technique [99, 100].

In the 1930s–1940s, there were considerable needs to register the radiant emission from objects. In 1934, J. Hardy showed that the human skin surface has the characteristics of a near perfect blackbody radiator, being highly efficient in the irradiative heat exchange [101, 102]. He pointed out that it is important to know the precise value of emissivity, because the emissivity decrease to 0.945 from 0.98 causes the skin temperature determination error to be 0.6 °C. J. Hardy had shown that the human skin, regardless of color, is a highly efficient radiator with an emissivity of 0.98, which is close to that of a black body. In 1960, K. Lloyd-Williams *et al.* showed that many tumors are hotter than adjacent skin sections [103]. Therefore, it was hoped for that the thermosense technique can be used for screening for breast cancer.

In 1933, E.W. Kutzscher (Germany) discovered that lead sulphide (PbS) is photoconductive to about 3 µm wavelength. These detectors were the first practical infrared detectors that have found a variety of applications during the World War II. After World War II, R.J. Cashman in the USA found that other lead salts (PbSe and PbTe) can be used as infrared detectors [104] in the MWIR range.

During and after the World War II, the IR detector technology development was primarily driven by military applications. In the former USSR, there were developed systems for the army and the navy. The progress was gained in night vision devices mainly for military applications (IR vidicons and other electro-optical converters).

The first cooled bolometer was invented in the 1940s [105]. The bolometer was cooled down to the superconducting region using liquid helium. The superconducting transition point of tantalum at 4.4 K was used. A later version was made with niobium nitride, which has a transition at about 15 K, and the proper operation of a bolometer at 14.3 K was achieved, where the cooling system with liquid hydrogen was more easily available [106].

The significant post-World War II event was the invention of room-temperature pneumatic detectors. In 1946, Zahl and Golay published the paper titled "Pneumatic heat detector" [107] referred more to the THz range. One year later, Golay individually published paper [108] and patented the pneumatic detector referred to as the Golay Cell. These inventions were the extension of works made by Zahl and Golay in the 1930s. In 1938, Zahl patented a "pneumatic cell detector". One year later, he and Golay patented a "System for detecting sources of radiant energy" (see [9]).

In 1959, Lawson and co-workers [109] proposed a narrow band-gap mercury-cadmium-telluride (MCT) ($Hg_{1-x}Cd_xTe$) solid solution as the material with variable band-gap to be used for IR detectors with the sensitivity wavelength changeable by the chemical composition "x". This opened a new era in the IR detector technology (a little bit later in 1960, Sneider and Gavrishak in Ukraine also grew $Hg_{1-x}Cd_xTe$ for IR detectors [110]. The development of this material inaugurated a revolutionary step in the broad development of cooled IR detectors. Presumably in the nearest decade, different vision systems with large focal plane arrays with ultimate performance for IR spectra from ~1 μm to ~ 20 μm will be based on HgCdTe semiconductor for both Earth and Space locations.

In the 1970s, the development of a multielement photon detector linear array has started. This formed the basis of a real-time imaging process. Somewhat later, the computer technology made a widespread impact on improvements in thermal imaging cameras, both on the image quality and the image frame rate. However, the early imaging systems were large with very limited facilities for display and temperature measurements [111]. The computer image processing of thermograms resulted in increased possibilities for the quantitation and the archiving of images. Therefore, this enhanced the needs for the standardization of the IR biomedical imaging [112, 113].

The other type of photon detectors appeared after the publication of paper [114] in 1985. In this research, the intersubband optical transitions in quantum wells (QWs) were first observed. By 1987, the basic operation principles for QW infrared photodetectors (now frequently cold QWIPs) demonstrating the sensitive infrared detection were formulated. The QWIP arrays were used in Landsat Data Continuity Mission (2013) in IR bands at 10.8 and 12.0 µm [115].

Photon detectors were dominating in IR technologies till the end of the 20th century. The essential drawback of photon detectors with ultimate performance is the need of cryogenic cooling. This is necessary to prevent the thermal generation of charge carriers, thus raising the noise level.

The second revolution in the thermal imaging began in the last decades of the 20th century with the results of investigations of small-area low-mass uncooled thermal detectors for military and civilian applications. In 1978, Texas Instruments (USA) patented uncooled ferroelectric infrared detectors using barium strontium titanate (BST – (BaSrTiO$_3$). A little bit later in 1982, another uncooled detector technology (resistive microbolometer technology) was developed by Honeywell (USA) under the guidance of R. Andrew (see [116]). Later, it became clear after the realization that the key to the high bolometer performance was not the resistive material but the structure's thermal isolation. The material chosen at Honeywell was vanadium oxide (VO$_x$) which has a high (~2...4 % K$^{-1}$) temperature resistance coefficient at room temperature and which was processed using the silicon technology. It was deposited on a microbridge of Si$_3$N$_4$. In the mid of the 1990s (SOFRADIR + ULIS, France), the third technology, amorphous silicon (α-Si), was also developed. This technology is also compatible with the fabrication in a silicon foundry [104].

Now these detector and thermal camera technologies are well mastered by several companies: Raytheon, Teledyne, BAE Systems, DRS Technologies, FLIR, L–3 Communications, Goodrich Corporation, and some others (USA), NEC, Mitsubishi (Japan), XenICs (Belgium), SCD (Israel), INO (Canada), and so on.

Beginning from the late 1970s, the progress in the number of detectors in arrays, which revolutionized IR technologies and made them much more cost-effective was primarily connected with the application of silicon readout integrated circuits (ROICs). Assembling ROICs with different types of detectors allowed the fabrication of IR focal plane arrays (FPAs), which now can contain up to ~10$^7$ and more IR detectors in a single crystal array. Applications of these technologies made possible a discretization of the process of image creation, as well as its processing by the instrumentality of linear and matrix detector arrays from discrete elements. The arrays, both IR and THz, used in vision systems have a set of advantages over the systems with single detectors. This is connected, first of all, with the rate of getting image, space resolution, sensitivity, and visibility range.

The IR thermography (passive vision) is one of the useful techniques that were used in the non-destructive testing of the human body. Using the IR technology has many advantages such as the fact that it is a non-contact, non-destructive, and fast technique, which does not emit any harmful radiation [117]. Now, the IR thermography strives toward using the equipment offering high performance, acceptable accuracy, and low cost [118, 119].

In most cases, the imaging of objects or the environment that is constructed from the IR or THz detector signals in spectral regions from their "invisible" parts of the spectrum can be substantially different from images in the visible spectrum. This is due to different radiation absorptions, transmissions, and reflections of the objects under observation, their emissivities, background temperatures, *etc.*, in various spectral ranges.

The history of IR technologies itself and in application to various types of activity in different periods is presented in a number of papers and books (see, *e.g.*, [120–127]).

## 2.4  Summary

The brief history of THz and IR science and technology as a learning lesson by the historical evolution is presented and discussed identifying important (from Author's point-of-view) steps for their development. The THz science and technology show a rapid growth nowadays and become wide spread in their use despite the explosion of requirements to the applications especially in security, biomedicine, astrophysics, *etc.* The THz science and technology need a deeper and wider knowledge in many scientific and technological aspects. In addition, in spite of the expected progress, the THz technologies are still delayed in a widespread use because of a lack of reliable, cost-effective sensors, sources, and instrumentations produced on a large scale.

The situation is less related to the IR technologies, especially in the thermovision. A few decades ago, the IR technologies were mainly the domain of military applications. In recent two decades, due to the widespread expansion of thermal uncooled detectors, many IR technology advances were realized. The quickly falling down costs of the IR arrays and instrumentations are observed. Uncooled thermal IR arrays have become an alternative to the cooled ones and now are much more commonly used in many commercial, industrial, biomedical and military fields.

For the last several decades, the successful development of IR detectors for large-format small pixel arrays and cameras on their base has led to a significant progress in monitoring the environmental pollution, surveillance and reconnaissance, security imaging, IR astronomy, car driving, imaging in medical diagnostics, *etc.* During the last two decades, the quick application of civil-oriented IR technologies is due to the rapid development of uncooled microbolometer arrays and cameras on their base, which are produced in larger volumes, as compared to other photosensitive devices.

Although no mass-market applications of real THz technologies can be highlighted, because the technology does not yet meet the users' requirements, especially in the easiness of use and in costs, many of the THz applications are emerging and show their usefulness in astrophysics, security, biomedicine, drug and food inspection, non-destructive testing, *etc*. Summarizing the development of THz technologies, we can conclude that, in spite of the great efforts in the past decades, THz applications are still generally at the early stage of development. Many other potential applications are likely to be added in the near future.

## References to Ch. 2

[1] T. Hochrein, Markets, availability, notice, and technical performance of terahertz systems: Historic development, present, and trends, J. Infrared, Millimeter, Terahertz Waves. 36 (2015) 235–254. https://doi.org/10.1007/s10762-014-0124-6

[2] S.S. Dhillon et al. (32 co-authors), The 2017 terahertz science and technology roadmap, J. Phys. D: App. Phys. 50 (2017) 043001.

[3] D.T. Emerson, The work of Jagadish Chandra Bose: 100 years of millimiter-wave research, IEEE Trans. Microwave Theory Techn. 45 (1997) 2267–2273. https://doi.org/10.1109/22.643830

[4] V. Aggarwal, Jagadish Chandra Bose: The real inventor of Marconi's wireless detector, http://web.mit.edu/varun_ag/www/bose_2006.pdf.

[5] G. Marconi, US Patent, Apparatus for wireless telegraphy, 676,332. Patented June 11, 1901. (Application Feb. 23, 1901).

[6] P.K. Bondyopadhyay, Sir J.C. Bose's diode detector received Marconi's first transatlantic wireless signal of December 1901 (The "Italian Navy coherer" scandal revisited), Proc. IEEE. 86 (1998) 259–285. https://doi.org/10.1109/5.658778

[7] C. Corsi, TeraHertz: Quasioptics or sub-millimeter waves? History, actual limits and future developments for security systems, in: C. Corsi, F. Sizov (Eds.), THz and Security Applications, Springer, Dordrecht, 2014, pp. 1–24. https://doi.org/10.1007/978-94-017-8828-1_1

[8] THz Pioneers: IEEE Trans. THz Sci. Technol. 2 (2012) 265–270; 2 (2012) 477–484; 4 (2014) 137–146; 4 (2014) 645–652.

[9] Bründermann E., Hübers H.-W., Kimmitt M.F., 2011. Terahertz Techniques. Heidelberg: Springer. https://doi.org/10.1007/978-3-642-02592-1

[10] W. Gordy, Early events and some later developments in microwave spectroscopy, J. Mol. Struct. 97 (1983) 17–32. https://doi.org/10.1016/0022-2860(83)90172-2

[11] M.F. Kimmitt, Restrahlen to T-rays: 100 years of terahertz radiation, J. Biolog. Phys. 29 (2003) 77–85. https://doi.org/10.1023/A:1024498003492

[12] J.M. Chamberlain, Where optics meet electronics: recent progress in decreasing the terahertz gap, Phil. Trans, R. Soc. Lond. A. 362 (2004) 199–213. https://doi.org/10.1098/rsta.2003.1312

[13] T.G. Blaney, Signal-to-noise ratio and other characteristics of heterodyne radiation receivers, Space Sc. Rev. 17 (1975) 691–702. https://doi.org/10.1007/BF00727583

[14] P. Richards, Bolometers for infrared and millimeter waves, J. Appl. Phys., 76 (1994) 1–24. https://doi.org/10.1063/1.357128

[15] G. Chattopadhyay, Sensor technology at submillimeter wavelength for Space applications, Int. J. Smart Sensing Intell. Systems. 1 (2008) 1–20. https://doi.org/10.21307/ijssis-2017-275

[16] J.M. Lamarre, F.X. Desert, T. Kirchner, Background limited infrared and sub-millimeter instruments, Space Sci. Rev. 74 (1995) 27–36. https://doi.org/10.1007/978-94-011-0363-3_4

[17] H. Kangro, Early History of Planck's Radiation Law, Taylor & Francis, New York, 1976.

[18] P. Siegel, THz pioneer: Richard S. Saylally – Water, water everywhere..., IEEE Trans. Terahertz Sci. Technol. 2 (2012) 265–270. https://doi.org/10.1109/TTHZ.2012.2190870

[19] Jagadish Chundler Bose, On a self-recovering coherer and the study of the cohering action of different metals, Proc. IEEE. 86 (1998) 244–247 (reprinted from: J.C. Bose, On a self-recovering coherer and the study of the cohering action of different metals, Proceedings of the Royal Society, London. LXV, no. 416 (1899) 166–172). https://doi.org/10.1109/JPROC.1998.658776

[20] J.C. Bose, Detector for electrical disturbances, U.S. Patent 755,840. (1904).

[21] G.L. Pearson and W.H. Brattain, History of semiconductor research, Proc. IRE. 43 (1955) 1794–1806. https://doi.org/10.1109/JRPROC.1955.278042

[22] A. Glagolewa-Arkadiewa, Short electromagnetic waves of wave-length up to 82 microns, Nature. 113 (1924) 640. https://doi.org/10.1038/113640a0

[23] A.A. Kostenko, A.I. Nosich, I.A. Tishchenko, Development of the first Soviet 3-coordinate L-band pulsed radar in Kharkov before WW II, IEEE Antennas Propagat. Mag. 44 (2001) 28-49. https://doi.org/10.1109/74.934901

[24] A.A. Kostenko, A.I. Nosich, P.F. Goldsmith, Historical background and development of Soviet quasioptics at nearmillimeter and submillimiter wavelengths, in: T.K. Sarkar, R.J. Mailloux, A.A. Oliner, M. Salazar-Palma, D.L. Sengupta (Eds.), History of Wireless, John Wiley & Sons, Hoboken, N.J., 2006, pp.473–542. https://doi.org/10.1002/0471783021.ch15

[25] I.A Tishchenko, A.I. Nosich, Early quasioptics of near-millimeter and submillimeter waves in IRE-Kharkov, Ukraine: From ideas to the microwave pioneer award, IEEE Microwave Magazine, (Dec. 2003) 32–44. https://doi.org/10.1109/MMW.2003.1266065

[26] A.I. Nosich, Dramatic history and impact of decimeter-wave radar "Zenit" developed in Kharkiv in the 1930s, 2017 XXIInd International Seminar/Workshop on Direct and Inverse Problems of Electromagnetic and Acoustic Wave Theory (DIPED), Dnipro, Ukraine, 25-28 Sept., 2017, pp. 11–14. https://doi.org/10.1109/DIPED.2017.8100546

[27] W. Gordy, Microwave Spectrocopy, Rev. Modern Phys. 20 (1948) 668–717. https://doi.org/10.1103/RevModPhys.20.668

[28] C.A. Burrus, Jr., W. Gordy, Submillimeter Wave Spectroscopy, Phys. Rev. 93 (1954) 897–898. https://doi.org/10.1103/PhysRev.93.897

[29] V.S. Alpher, Ralph A Alpher, George Antonovich Gamow, and the prediction of the cosmic. Microwave background radiation, Asian J. Phys. 23 (2014) 17–26.

[30] D.J. Fixsen, The temperature of the cosmic microwave background. The Astrophys. J. 707 (2009) 916–920. https://doi.org/10.1088/0004-637X/707/2/916

[31] Ch.H. Lineweaver, Cosmic microwave background, Discoveries in modern science. Exploration, Invention, Technology. (2014) 224–229.

[32] A.A. Penzias, R.W. Wilson, A Measurement of Excess Antenna Temperature at 4080 Mc/s, Astrophys. J. 142 (1965) 419–421. https://doi.org/10.1086/148307

[33] R. Kompfner, N.T. Williams, Backward-wave tubes, Proc. IRE. 41 (1953) 1602–1611. https://doi.org/10.1109/JRPROC.1953.274186

[34] R. Warnecke, P. Guenard, Some recent work in France on new types of valves for the highest radio frequencies, Proc. IEE –Radio Commun. Eng. 100, no. 68, pt. III (1953) 351–362. https://doi.org/10.1049/pi-3.1953.0073

[35] W.T. Read, A proposed high frequency negative resistance diode, Bell. Syst. Tech. J. 37 1958 401. https://doi.org/10.1002/j.1538-7305.1958.tb01527.x

[36] A.S. Tager, The avalanche-transit diode and its use in microwaves, *Sov. Phys. Usp.,* 9 1967 892–912. https://doi.org/10.1070/PU1967v009n06ABEH003231

[37] R.L. Jonston, B.C. De Loach, B.G. Cohen, B.S.T.J. briefs: A silicon diode microwave oscillator, Bell Syst. Techn. J. 44 (1965) 369–372. https://doi.org/10.1002/j.1538-7305.1965.tb01667.x

[38] F.A. Brand, V.I. Higgins, L.I. Baranowski, M.A. Druesne, Microwave generation from avalanching varactor diodes, Proc. IEEE. 53 (1965) 1276–1277. https://doi.org/10.1109/PROC.1965.4225

[39] J.B. Gunn, Microwave oscillation of current in III-V semiconductors, Solid St. Commun. 1 (1963) 88–91. https://doi.org/10.1016/0038-1098(63)90041-3

[40] A. Crocker, H.A. Gebbie, M.F.Kimmitt, L.E.S. Mathias, Stimulated emission in the far infra-red, Nature. 201 (1964) 250–251. https://doi.org/10.1038/201250a0

[41] T.Y. Chang, T.J. Bridges, Laser action at 452, 496, and 541 μm in optically pumped $CH_3F$, Opt. Commun. 1 (1970) 423–426. https://doi.org/10.1016/0030-4018(70)90169-0

[42] P.L. Richards, High-resolution Fourier transform spectroscopy in the far-infrared, J. Opt. Soc. Am. 54 (1964) 1474–1484. https://doi.org/10.1364/JOSA.54.001474

[43] S.J. Webb, D.D. Dodds, Inhibition of bacterial cell growth by 136 gc microwaves, Nature. 218 (1968) 374–375. https://doi.org/10.1038/218374a0

[44] C.F. Blackman, S.G. Benane, C.M. Weil, J.S. Ali, Effects of non-ionizing electromagnetic radiation on single-cell biologic systems, Annals of the New York Acad. Sci. 247 (1975) 352–366. https://doi.org/10.1111/j.1749-6632.1975.tb36010.x

[45] M. Nicolson, Broad-band microwave transmission characteristics from a single measurement of the transient response, Instrumentation and Measurement, IEEE Trans. Instrum. Meas. 17 (1968) 395–402. https://doi.org/10.1109/TIM.1968.4313741

[46] W. Schottky, R. Stormer, F. Waibel, Uber die Gleichrichterwirkungenan der Grenze von Kupferoxydul gegen aufgebrachte Metallelektroden, (On the rectifying action of cuprous oxide in contact with other metals), Z. Hochfrequenz. 37 (1931) 162–167, 175–187

[47] A.J.M. Kreisler, Submillimeter wave applications of submicron Schottky diodes, Proc. SPIE. 666 (1966) 51–63.

[48] P.L. Richards, T.M. Shen, R.E. Harris, F.L. Lloyd, Quasiparticle heterodyne mixing in SIS tunnel junctions, Appl. Phys. Lett. 34 (1979) 345–347. https://doi.org/10.1063/1.90782

[49] W. Faries, K.A. Gehring, P.L. Richards, Y.R. Shen, Tunable far-infrared radiation generated from difference frequency between two ruby lasers, Phys. Rev. 180 (1969) 363–365. https://doi.org/10.1103/PhysRev.180.363

[50] T. Yajima, N. Takeuchi, Far-infrared difference frequency generation by picosecond laser pulses, Jap. J. Appl. Phys. 9 (1970) 1361–1371. https://doi.org/10.1143/JJAP.9.1361

[51] R.F. Kazarinov, R.A. Suris, Possibility of amplification of electromagnetic waves in a semiconductor with a superlattice. Fizika i Tekhnika Poluprovodnikov. 5 (1971) 797–800.

[52] J. Faist, F. Capasso, D. Sivco, C. Cirtori, *et al.*, Quantum cascade laser, Science. 264 (1994) 553–556. https://doi.org/10.1126/science.264.5158.553

[53] D.H. Barker, D.T. Hodges, T.S. Hartwick, Far infrared imagery, Proc. SPIE. 67 (1975) 27–34.

[54] D.H. Auston, Picosecond optoelectronic switching and gating in silicon, Appl. Phys. Lett. 26 (1975) 101–103. https://doi.org/10.1063/1.88079

[55] D.H. Auston, P.R. Smith, Generation and detection of millimeter waves by picosecond photoconductivity, Appl. Phys. Lett. 43 (1983) 631–633. https://doi.org/10.1063/1.94468

[56] A.G. Kazanskii, P.L. Richards, E.E. Haller, Far infrared photoconductivity of uniaxially stressed germanium, Appl. Phys. Lett. 31 (1977) 496–497. https://doi.org/10.1063/1.89755

[57] P. Tuengler, F. Keilmann, L. Genzel, Search for millimeter microwave effects on enzyme or protein functions. Zeitschrift Fur Naturforschung C – J. Biosci. 34 (1979) 60–63. https://doi.org/10.1515/znc-1979-1-214

[58] H. Hintzsce, H. Stopper, Effects of terahertz radiation on biological systems, Critical Rev. Environmental Sci. Technol. 42 (2012) 2408–2434. https://doi.org/10.1080/10643389.2011.574206

[59] G. Mourou, C.V. Stancampiano, A. Antonetti, A. Orszag, Picosecond microwave pulses generated with a subpicosecond laser-driven semiconductor switch, Appl. Phys. Lett. 39 (1981) 295–296. https://doi.org/10.1063/1.92719

[60] D.H. Auston, K.P. Cheung, P.R. Smith, Picosecond photoconducting Hertzian dipoles, Appl. Phys. Lett. 45 (1984) 284–286. https://doi.org/10.1063/1.95174

[61] C. Fattinger, D. Grischkowsky, Terahertz beams, Appl. Phys. Lett. 54 (1989) 490–492. https://doi.org/10.1063/1.100958

[62] P.R. Smith, D.H. Auston, M.C. Nuss, Subpicosecond photoconducting dipole antenna, IEEE J. Quant. Electron. 24 (1988) 255–260. https://doi.org/10.1109/3.121

[63] D.E. Spence, P.N. Kean, W. Sibbett, Sub-100 fs pulse generation from a self-modelocked titanium: sapphire laser, Conference on Lasers and Electro-optics, CLEO,

Technical Digest Series: Optic. Soc. of America. 1990, pp. 619–620.

[64] D.H. Auston, M. C. Nuss, Electrooptic generation and detection of femtosecond electrical transients, IEEE J. Quant. Electron. 24 (1988) 184–197. https://doi.org/10.1109/3.114

[65] L.R. Elias, J. Hu, G. Ramian, The UCSB electrostatic accelerator free electron laser: First operation, Nucl. Instrum. Meth. Phys. Research. A237 (1984) 203–206. https://doi.org/10.1016/0168-9002(85)90349-3

[66] E.M. Gershenzon, M.E. Gershenson, G.N. Goltsman, A.D. Semenov, A.V. Sergeev, On the limiting characteristics of high-speed superconducting bolometers, Sov. Phys. Tech. Phys. 34 (1989) 195–201.

[67] F. Sizov, THz radiation detectors: the state of the art, Semicond. Sci. Techn. 33 (2018) 123001. https://doi.org/10.1088/1361-6641/aae473

[68] E.H. Putley, Indium antimonide submillimeter photoconductive detectors, Appl. Opt. 4 (1965) 649–657. https://doi.org/10.1364/AO.4.000649

[69] F. Arams, C. Allen, B. Peyton, E. Sard, Millimeter mixing and detection in bulk InSb, Proc. IEEE. 54 (1966) 612–622. https://doi.org/10.1109/PROC.1966.4781

[70] E.R. Brown, K.A. McIntosh, F.W. Smith, M.J. Manfra, C.L. Dennis, Measurements of optical-heterodyne conversion in low-temperature-grown GaAs, Appl. Phys. Lett. 62 (1993) 1206–1208. https://doi.org/10.1063/1.108735

[71] X.-C. Zhang, J. Xu, 2010. Introduction to THz Wave Photonics. New York-Dordrecht-Heidelberg-London: Springer. https://doi.org/10.1007/978-1-4419-0978-7

[72] A.J.L. Adam, Review of near-field terahertz measurement methods and their applications, J. Infrared Milli Terahz Waves. 32 (2011) 976-1019. https://doi.org/10.1007/s10762-011-9809-2

[73] M. Dyakonov, M.S. Shur, Plasma wave electronics: Novel terahertz devices using two dimensional electron fluid, IEEE Trans. Electr. Devices. 43 (1996) 1640–1645. https://doi.org/10.1109/16.536809

[74] J.-Q Lü, M.S. Shur, J.L. Hesler, L. Sun, R. Weikle, Terahertz detector utilizing two-dimensional electronic fluid, IEEE Electr. Device Lett. 19 (1998) 373–375. https://doi.org/10.1109/55.720190

[75] L. Vicarelli, M.S. Vitiello, D. Coquillat, A. Lombardo, et al., Graphene field-effect transistors as room-temperature terahertz detectors, Nature Mater. 11 (2012) 865–871. https://doi.org/10.1038/nmat3417

[76] F. Simoens, J. Martyrs, Terahertz real-time imaging uncooled array based on antenna – and cavity-coupled bolometers, Philosoph. Trans. R. Soc. A. 372 (2014)

20130111. https://doi.org/10.1098/rsta.2013.0111

[77] T. Nagatsuma, G. Ducournau, C.C. Renaud, Advances in terahertz communications accelerated by photonics, Nature Photonics. 10 (2016) 371–379. https://doi.org/10.1038/nphoton.2016.65

[78] G.-H. Duan, Ch. Jany, A. Le Liepvre, A. Accard, *et al.*, Hybrid III–V on silicon lasers for photonic integrated circuits on silicon, IEEE J. Select. Topics Quant. Electron. 20 (2014) 6100213. https://doi.org/10.1117/12.2044258

[79] C.R. Doerr, Silicon photonic integration in telecommunications, Front. Phys. 3 (2015) 1–16. https://doi.org/10.3389/fphy.2015.00037

[80] J. Lequeux, Early infrared astronomy, J. Astron. History Heritage, 12 (2009) 125–140.

[81] F.J. Ring, Pioneering progress in infrared imaging in medicine, Quantitative InfraRed Thermography J. 11 (2014) 57–65. https://doi.org/10.1080/17686733.2014.892667

[82] F.J. Ring, B.F. Jones, Historical development of thermometry and thermal imaging in medicine, in: M. Diakides, J. D. Bronzino, and D. R. Peterson (Eds.), Medical Infrared Imaging: Principles and Practices, CRC Press, Boca Raton, 2013, pp. 2.1–2.6.

[83] L.J. Jiang, E.Y. Ng, A.C. Yeo, S. Wu, *et al.*, A perspective on medical infrared imaging, J. Med. Eng. Technol. 29 (2005) 257–267. https://doi.org/10.1080/03091900512331333158

[84] E.S. Barr, Historical survey of the early development of the infrared spectral region, Amer. J. Phys. 28 (1960) 42–54. https://doi.org/10.1119/1.1934975

[85] S.P. Langley, The bolometer, Nature. 25 (1881) 14–16. https://doi.org/10.1038/025014a0

[86] H. Rubens, B.W. Snow, On the refraction of rays of great wavelength in rock salt, sylvine, and fluorite, Phil. Mag. 35 (1893) 35–45. https://doi.org/10.1080/14786449308620376

[87] E.F. Nichols, A method for energy measurements in the infrared spectrum and the properties of the ordinary ray in quartz for waves of great wavelength, Phys. Rev. 4 (1897) 297–313. https://doi.org/10.1103/PhysRevSeriesI.4.297

[88] H. Rubens, F. Kurlbaum, Anwendung der Methode der Restrahlen zur Prufung des Strahlungsgesetzes, Annalen der Physik, 4 (1901) 649–666. https://doi.org/10.1002/andp.19013090402

[89] M. Planck, Über eine Verbesserung der Wienschen Spektralgleichung, Verhandlungen der Deutschen Physikalischen Gesselschaft, 2, (1900) 202–204

Materials Research Forum LLC
https://doi.org/10.21741/9781644900758

[90] A. Einstein, Zum gegenwärtigen Stand des Strahlungsproblems. Physikalische Zeitschrift, 10 (1909) 185–193.

[91] R.H. Stuewer, Einstein's Revolutionary Light-Quantum Hypothesis, Acta Phys. Polon. B. 37 (2006) 543–558.

[92] H. Rubens, O.V. Baeyer, On extremely long waves emitted by the quartz mercury lamp, Phil. Mag. 21 (1911) 689–703. https://doi.org/10.1080/14786440508637081

[93] H. Rubens, R.W. Wood, Focal isolation of long heat-waves, Philosoph. Magazine. 21 (1911) 249–261. https://doi.org/10.1080/14786440208637025

[94] G. Holst, J.H. de Boer, M.C. Teves, C.F. Veenemans. Foto-electrische cel en inrichting waarmede uit een primair, door directe lichtstralen gevormd beeld een geheel ofnagenoed geheel conform secundair optisch beeld kan, Dutch patent 27062 (1928); British Patent 326200; D.R.P. 535208.

[95] G. Holst, J.H. de Boer, M.C. Teves, C.F. Veenemans, An apparatus for the transformation of light of long wavelength into light of short wavelength, Physika. 1 (1934) 297–305. https://doi.org/10.1016/S0031-8914(34)90036-7

[96] M. Czerny, Über photographie im ultraroten, Zeit. f. Physik. 53 (1929) 1–*12.*

[97] R. Berz, H. Sauer, The medical use of infrared-thermography. History and recent applications, Thermografie-Kolloquium 2007, Vortrag 04, 1-12, 2007 (www.ndt.net/search/docs.php3?MainSource=61).

[98] E. Schwamm, J. Reeh, Die Ultrarotstrahlung des Menschen und seine Molekular spektroskopie, Hippokrates. 24 (1953) 737−742.

[99] F.J Ring, E.Y.K. Ng, Infrared thermal imaging standards for human fever detection, in: M. Diakides, J.D. Bronzino, D.R. Peterson (Eds.), Medical Infrared Imaging: Principles and Practices, CRC Press, Boca Raton, 2013, pp. 22.1−22.5.

[100] I. Fernandez-Cuevas, J.C.B. Marins, J.A. Lastras, P.M. Gomez Carmona, et al., Classification of factors influencing the use of infrared thermography in humans: A review, Infrared. Phys. Technol. 71 (2015) 28–55. https://doi.org/10.1016/j.infrared.2015.02.007

[101] J.D. Hardy, The radiation of heat from the human body. J. Clinical Invest. 13 (1934) 615–620. https://doi.org/10.1172/JCI100609

[102] J. Hardy, The radiation power of human skin in the infrared, Am. J. Physiol. 127 (1939) 454–462. https://doi.org/10.1152/ajplegacy.1939.127.3.454

[103] K. Lloyd-Williams, F. Lloyd-Williams, R. Handley, Infrared radiation thermometry *in* clinical practice, Lancet. 2 *(*1960*)* 958–959. https://doi.org/10.1016/S0140-6736(60)92028-6

[104] A. Rogalski, History of infrared detectors, Opto-Electr. Rev. 20 (2012) 279–308. https://doi.org/10.2478/s11772-012-0037-7

[105] D.H. Andrews, W.F. Bruksch, W.T. Zeigler, E.R. Blanchard, Attenuated superconductors for measuring infra-red radiation, Review of Scientific Instruments. 13 (1942) 281–291. https://doi.org/10.1063/1.1770037

[106] D.H. Andrews, R.M. Milton, W. DeSorbo, A fast superconducting bolometer, J. Opt. Soc. Am. 36 (1946) 518–524. https://doi.org/10.1364/JOSA.36.000518

[107] H.A. Zahl, M.J.E. Golay, Pneumatic heat detector, Review Scient. Instrum. 17 (1946) 511–515. https://doi.org/10.1063/1.1770416

[108] M.J.E. Golay, A pneumatic infra-red detector, Rev. Sci. Instr. 18 (1947) 357–362. https://doi.org/10.1063/1.1740949

[109] W.D. Lawson, S. Nielson, E.H. Putley, A.S. Young, Preparation and properties of HgTe and mixed crystals of HgTe−CdTe, J. Phys. Chem. Sol. 9 (1959) 325–329. https://doi.org/10.1016/0022-3697(59)90110-6

[110] A.D. Shneider, I.V. Gavrishak, Structure and properties of HgTe-CdTe system, Solid St. Phys. 2 (1960) 2079−2081 (In Russian).

[111] E.F.J. Ring, K. Ammer, The technique of infrared imaging in medicine, Thermology international. 10 (2000) 7–14.

[112] E.F.J. Ring, Standardization of thermal imaging in medicine: Physical and environmental factors, in: M. Gautherie, E. Albert, L. Keith (Eds.), Thermal Assessment of Breast Health, MTP Press Ltd., Lancaster-Boston-The Hague, 1983, pp.29–36.

[113] K. Ammer, The Glamorgan Protocol for recording and evaluation of thermal images of the human body, Thermology Intern. 18 (2008) 125–129.

[114] L.C. West, S.J. Eglash, First observation of an extremely large-dipole infrared transition within the conduction band of a GaAs quantum well, Appl. Phys. Lett. 46 (1985) 1156–1158. https://doi.org/10.1063/1.95742

[115] https://www.nasa.gov/pdf/723395main_LDCMpresskit2013-final.pdf.

[116] P.W. Kruse, Uncooled Thermal Imaging. Arrays, Systems and Applications, SPIE Press, Bellingham, 2001. https://doi.org/10.1117/3.415351

[117] N. Avdelidis, T.-H. Gan, C. Ibarra-Castanedo, X. Maldague, Infrared thermography as a nondestructive tool for materials characterization and assessment, Proc. SPIE. 8013 (2011) 8013OK. https://doi.org/10.1117/12.887403

[118] F. Khodayar, S. Sojasi, and X. Maldague, Infrared Thermography and NDT: 2050 Horizon, Quantitative InfraRed Thermography J. 13 (2015) 210–231.

https://doi.org/10.1080/17686733.2016.1200265

[119] U. Raghavendra, U.R. Acharya, E.Y.K. Ng, J.-H. Tan, A. Gudigar, An integrated index for breast cancer identification using histogram of oriented gradient and kernel locality preserving projection features extracted from thermograms, Quantitative InfraRed Thermography J. 13 (2016) 195–209. https://doi.org/10.1080/17686733.2016.1176734

[120] R.A. Smith, F.E. Jones, R.P. Chasmar, 1958. The Detection and Measurement of Infrared Radiation. Oxford: Clarendon. https://doi.org/10.1063/1.3062526

[121] R.D. Hudson, 1969. Infrared System Engineering. New Jersey: Wiley-Interscience.

[122] L.M. Biberman, R.L. Sendall, Introduction: A brief history of imaging devices for night vision, in: L.M. Biberman (Ed.), Electro-Optical Imaging: System Performance and Modeling, SPIE Press, Bellingham, 2000, pp. 1-1–1-26 (2000).

[123] Sakai K., 2005. Terahertz Optoelectronics. Berlin: Springer. https://doi.org/10.1007/b80319

[124] A.S. Gilmore, High-definition infrared FPAs, Raytheon Technology Today, issue 1. 4–8 (2008).

[125] C. Corsi, History highlights and future trends of infrared sensors, J. Modern Optics. 57 (2010) 1663–1686. https://doi.org/10.1080/09500341003693011

[126] N. Sclar, Properties of doped silicon and germanium infrared detectors, Progr. Quant. Elect. 9 (1984) 149–257. https://doi.org/10.1016/0079-6727(84)90001-6

[127] V. Vavilov, Thermal NDT: historical milestones, state-of-the-art and trends, Quantitative InfraRed Thermography J. 11 (2014) 66–83. https://doi.org/10.1080/17686733.2014.897016

# Chapter 3. Types of detectors and figures of merit

There is a large variety of IR and THz detectors for different areas of applications. Among them, there are different kinds of thermal detectors (uncooled, cooled, and superconductive), quantum photoconductive, photovoltaic detectors, Schottky barrier diodes for both IR and THz ranges, field-effect and high electron mobility transistors for THz sensing, quantum well detectors, and others.

The operation principles and parameters of some of them will be considered here and in Chapter 4.

## 3.1 Remarks

Today, among IR, THz, and mm technical systems, the most advanced are the IR vision and spectroscopic ones, which are related to the fact that the objects on the Earth surface have the temperature T ≈ 273...320 K with the maximum emissivity at about $\lambda \approx 10$ μm (Fig. 3.1) and a rather high power density. A lot of detectors and arrays operating in the background limited performance (BLIP) regime for this spectral interval were designed and manufactured. The vision in IR is also propped by the well-examined model object that is a black body.

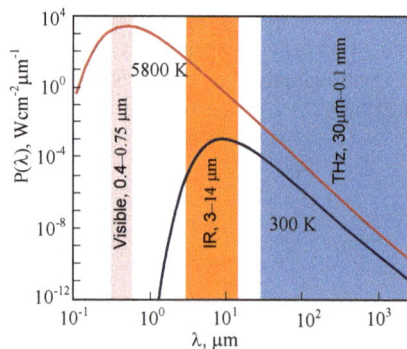

Figure 3.1. The black body radiation law from the UV to THz spectral ranges for two temperatures: about the temperature of the Sun (T = 5800 K, which is close to the effective temperature of the Sun (T = 5777 K [1]) and the background temperature at the Earth surface (T = 300 K).

According to the Wien's law, the maximum emissivity of heated objects is dependent on their temperature T as $\lambda_{max} \times T = 0.2898$ cmK. Therefore, for the residual temperature after the Big-Bang (T = 2.725 K), the maximum emissivity is observed at $\lambda_{max} \approx 1.06$ mm, *i.e.*, in the THz frequency range.

For the equilibrium power spectral density $P(\lambda, T)$ of the radiation of a black body into the $2\pi$ solid angle for the wavelength interval $d\lambda$ (spectral radiant exitance or radiant emittance), the expression (see, *e.g.*, [2, 3]) known as Planck's formula

$$P(\lambda, T) \times d\lambda = \frac{2\pi h c^2}{\lambda^5} \times \frac{d\lambda}{\exp[ch/(k_B \lambda T)] - 1}, \ W \times cm^{-2} \times \mu m^{-1} \tag{3.1}$$

(based on the experiments of H. Rubens [4]) is valid.

From formula (3.1), it follows that $P(\lambda,T)$ is growing with the temperature. Here h is the Planck constant, $k_B$ is the Boltzmann constant, and c is the speed of light in space.

At photon energies $h\nu \ll k_B T$, the spectral radiant exitance into the $2\pi$ solid angle takes the form (Rayleigh–Jeans approximation first experimentally found by H. Rubens and F. Kurlbaum) $P(\lambda,T)$

$$P(\lambda, T) \times d\lambda = \frac{2\pi c}{\lambda^4} \times k_B T \times d\lambda \tag{3.2}$$

In such approximation, $P(\lambda,T)$ is changing linearly with the black body temperature T in the THz frequency range (Fig. 3.1.) for T equal to 5800 and 300 K. For T = 30 K, such linear dependence will be seen at $\lambda > 2$ mm.

The most fundamental limit of the sensitivity of a detector comes from the photon noise due to fluctuations of a detected photon flux. The differences in the radiation power spectral densities of the electromagnetic radiation in different spectral EM ranges lead to different upper performance limits (BLIP regime) of the detectors operating in the different parts of the electromagnetic spectra. For the black body radiation, the BLIP is defined by fluctuations in the photon fluxes (photon noise) whose thermal average variance in the number of photons per mode is

$$\langle (\Delta N_{ph})^2 \rangle = \langle (N_{ph}) \rangle + \langle (N_{ph}) \rangle^2. \tag{3.3}$$

For a photon gas (that follows the Bose–Einstein statistics) in equilibrium, the number of photons per standing wave mode in a box at the temperature T is $N_{ph}$ = $[\exp(h\nu/k_BT)-1]^{-1}$.

At $h\nu \gg k_BT$, this defines the noise level in the photon statistics. Thus, $NEP_{ph}$ of a detector in a BLIP regime mainly in the IR spectral range ($\lambda < 20$ μm, the source T ~ 300 K) is dominated by the Poisson processes (the fluctuation of the number of photons is proportional to the square root of their number, $\langle(\Delta N_{ph})^2\rangle \sim \langle(N_{ph})\rangle$). At temperature T ≈ 30 K the Poisson statistics is valid in the wavelength range $\lambda < 150$ μm.

At $h\nu \ll k_BT$ (the Rayleigh-Jeans approximation), the Gaussian statistics dominates (the fluctuation of the number of photons is proportional to their number, $\langle(\Delta N_{ph})^2\rangle \sim \langle(N_{ph})\rangle^2$) in the photon statistics that defines the noise level in the detector BLIP regime mainly in the THz and sub-THz ranges. In the Rayleigh-Jeans approximation, the $NEP_{ph}$ is decreasing (improving), as the wavelength increases. In the whole wavelength range, both statistics should be taken into account.

As the source temperature decreases, the region of changing the statistics to another one is shifting to lower radiation frequencies. In astrophysical observations from the space platforms (the background and the instrument temperatures are about several Kelvin degrees) at low background temperatures, for estimations of the noise equivalent power limited by photon number fluctuations both statistics should be taken into account

The estimations of the photon $NEP_{ph}$ can be conducted according to the considerations in [5] within the antenna theorem $A·\Omega = \lambda^2$ validity. The antenna theorem states that, for a single spatial mode or diffraction limited beam, the throughput is $A·\Omega = \lambda^2$, where A is the area of a circular aperture of the optical system, $\Omega$ is the solid angle and $\lambda$ is the wavelength. The value of $A·\Omega$ is an invariant of the optical system.

The first term in Eq. (3.3) is the shot noise produced by a Poisson process, when photons are not correlated (the occupation number is small) [6]. In this case, the different sources contributing to $NEP_{ph}$ can be considered as statistically independent and adding the power of a source to the noise.

The second term dominates in the long wavelength limit (microwave and radio waves). In this case, the different sources contributing to $NEP_{ph}$ cannot be considered as statistically independent and adding the power of sources to the noise, since, within the same radiation mode, photons come by groups giving the noise, which is larger compared to the Poisson process. The reason for this is that if several photons occupy the same space mode (the occupation number is large) with the same volume of coherence, they will produce interference phenomena. Therefore, the diffraction phenomena should be taken into account [6] especially in the case of observed sources with angular diameters much smaller than the characteristic size of the diffraction pattern.

These estimations are valid only under specific conditions of photon noise in the one mode beam produced by an incoherent source of black body radiation and are usually not met in instrumentation. It was also supposed that the radiation is uniformly distributed over the geometric distance (the beam throughput $A\Omega = \lambda^2$). Within these estimations, one cannot consider the contributions of different sources at the same wavelength and, in the same space of coherence, as statistically independent. Therefore, one cannot compute the noise from two sources and add their power. This statement is important for long wavelength operation systems working at wavelengths of the Rayleigh-Jeans approximation [6]. For example, it was shown in [7] that, for the wavelength range $\lambda <$ 400 µm, the telescope temperature is the dominant factor defining the BLIP operation conditions. At $\lambda >$ 400 µm, both emissivity and temperature of the parts of a telescope (under low background conditions) are important. Thus, low emissive telescope/mirror designs should be used.

Figure 3.2 illustrates the effect of a cold telescope on the radiation background seen by single-mode detectors pointed at some dark part of the sky with low luminosity with different resolutions $\Delta v/v$. The detector NEP should be lower than the indicated NEP in order that the noise is less than the astronomical background.

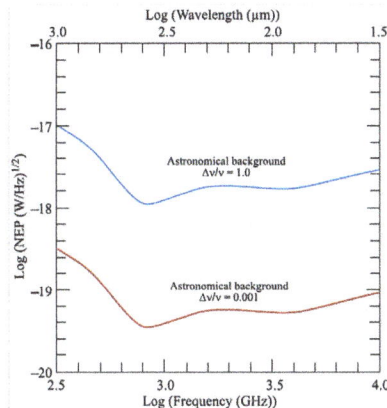

*Figure 3.2. NEP of the dark sky astronomical background as a function of the wavelength for two different resolutions, $\Delta v/v = 1$ for photometry and, $\Delta v/v = 0.001$ for moderate resolution extragalactic spectroscopy. The detector NEP should be well below these values in order that the overall noise is dominated by astronomical background [8].*

The performance of cosmic astronomical facilities is strongly limited by the thermal emission from the telescope optics [9, 10], as is shown in Fig. 3.3.

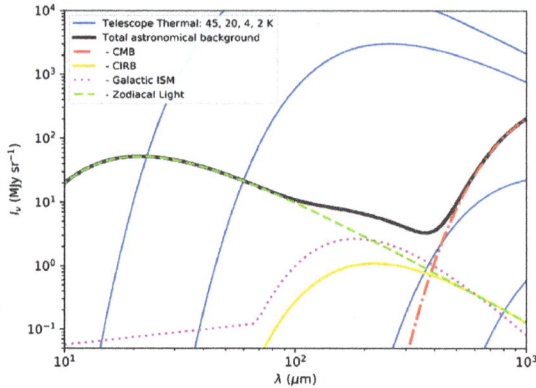

*Figure 3.3. A comparison between the primary astrophysical continuum backgrounds at IR and THz wavelengths (CMB – cosmic microwave background, CIRB – cosmic IR background, ISM – interstellar medium (emission), and Zodiacal emission from the interplanetary dust) and representative thermal emission from the telescope optics at three temperatures, assuming a uniform thermal emissivity of 4%. The astrophysical backgrounds assume observations outside the atmosphere toward high ecliptic and galactic latitudes, and at a distance of 1 AU (astronomical unit) from the Sun [10].*

The use of "cold" telescopes is indispensable in the THz spectral range in cosmic observations as, *e.g.*, at $\lambda \approx 300$ μm the thermal emission from a telescope at T = 4 K is five orders of magnitude lower than for a telescope at T = 45 K. This enables the detection of the cosmic IR or THz background and galactic interstellar medium radiations, and zodiacal light [9, 10].

With Herschel telescope and its temperature T ~ 85 K, the telescope emission was the dominant noise term for both its photo-detector array camera and spectrometer (PACS) and spectral and photometric imaging receivers (SPIRE). Therefore, the ultimate performance operation of the far-infrared (THz) in orbital missions requires deeply cooled telescopes and instruments.

## 3.2   General classification of detectors

Important parameters of many cooled IR and THz detectors–the detectivity D* and the noise equivalent power (NEP)–have reached now the upper limit performance set by background photon fluctuations (radiation shot noise). For uncooled detectors, these parameters are mainly limited by inner noises.

Most of the radiation detectors can be divided into three categories: photon (or quantum), thermal, and rectification detectors. Some of them can be used as fast mixers in coherent systems.

Different types of detectors are described in numerous books and reviews (see, *e.g.*, [11–19]. In dependence on the circuit type registration, the detectors are divided into direct and coherent (heterodyne) detectors.

Many of these detectors can operate both under room-temperature conditions and at low temperatures (T $\leq$ 4.2 K). However, some of them can operate only under room- or low-temperature conditions. That is why they can be divided into uncooled and cooled detectors as well.

Radiation in the waveband $\approx$(0.5…10) THz in Earth's atmosphere is strongly absorbed by water vapor, making this wavelength band not very attractive for Earth remote sensing applications. Operation of semiconducting photon detectors in this region is strongly limited by the thermo-generation rate. Therefore, advances in the research and fabrication of the detectors and arrays in this radiation frequency range is mostly concentrated on cooled systems for space and ground-based astronomy applications (see, *e.g.*, [20–22]). The extrinsic cold THz detectors and arrays also have found a niche in the long wavelength strategic systems, *e.g.*, for defense space vehicles. QW detectors have found some kind of a relatively narrow practical field of applications mostly at v $\geq$ 3 THz.

### 3.2.1   *Photon (quantum) detectors*

The photon (quantum) detectors respond to the incident photon number. The sensitivity of efficient quantum detectors (photoconductors or photodiodes) can approach the limits of photon noise fluctuations provided that the detector temperature is sufficiently low for photon-induced mechanisms to dominate over thermally induced mechanisms in a detector. Quantum detectors generally have sub-microsecond time constants. Their main disadvantage, to be used in the IR and THz, is associated with the cooling requirements for the optimal sensitivity.

The photon detectors (cooled or uncooled) are widely used for the detection of radiation in different spectral ranges–from the UV to THz. Among photon detectors, semiconductor-based ones are used to detect radiation–the intrinsic (interband), extrinsic and intersubband [quantum well (QW), quantum dot (QD)] detectors.

The intrinsic detectors involve the interband optical transitions (Fig. 3.4.a). Thus, the spectral response of an ideal photon detector with involving interband transitions is depending on the wavelength of radiation and for an ideal detector abruptly falls to zero at the cut-off wavelength $\lambda_{co}$, which corresponds to $\lambda_{Eg}$ connected with the band-gap $E_g$ of a semiconductor. For a non-ideal detector, the long wavelength part of sensitivity is not abrupt. Generally, $\lambda_{co}$ is taken at 50% of sensitivity (Fig. 3.5). These intrinsic detectors are generally used in the spectral range $\lambda < 30$ μm ($\nu > 10$ THz) and have high quantum efficiency from the visible region out to $\nu \approx 10$ THz.

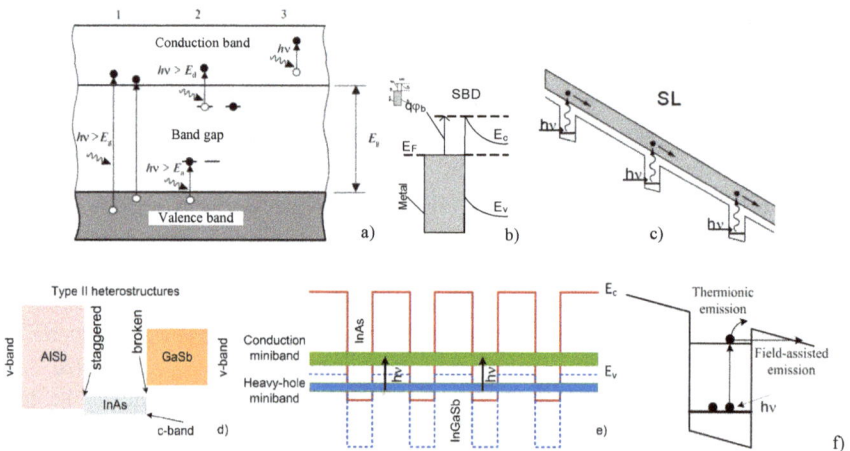

*Figure 3.4. Photon mechanisms of excitation of the electron subsystem in semiconductor detectors. a) From right to left: intrinsic (interband) excitation, extrinsic excitation, intraband excitation (free carrier absorption). $E_g$, $E_d$, $E_a$ are the band gap and the energies of donor and acceptor states, respectively. b) Excitation of carriers at the metal-semiconductor contact (photoemission detectors or SBD detectors (Schottky-Mott model). $E_F$ is the Fermi level, $q\varphi_b$ is the potential barrier height and $\varphi_b$ is the built-in potential. c) Electron excitation out of a stack in type I SLs or multiple QWs (MQWs) with applied bias. In this example of SLs, optical transitions take place from the bound ground states to quasi-bound states. d) Type II staggered or broken heterostructures. e) Type II "staggered" SL and optical transitions between the "minibands". f) Schematic of a QD structure and possible current transport mechanisms.*

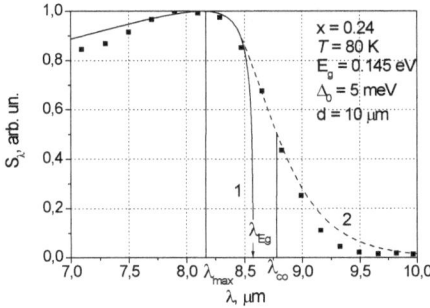

*Figure 3.5. Typical spectral dependence of the sensitivity in $Hg_{1-x}Cd_xTe$ photodiodes with chemical composition $x = 0.24$. 1 is the calculated sensitivity with regard for the non-parabolic of c- and v-bands, 2 is the calculated curve involving the static and dynamic disorders, squares are the experimental data, d is the sample thickness. $\lambda_{Eg}$ is the wavelength, corresponding to the band-gap energy of an ideal photodetector, and $\lambda_{co}$ is the cut-off sensitivity wavelength of a real detector [23]. (Reproduced with permission from Elsevier)*

The extrinsic detectors involve the optical excitation from the impurity donor or acceptor levels in the gap to conduction (valence) bands (Fig. 3.4.a.). They are used in the spectral range $\lambda \approx 20...220$ µm ($\nu \approx 15...1.3$ THz).

The SBD detectors (Fig. 3.4.b) involve the carrier transition under the optical excitation over the potential barrier height $q\varphi_b$. They are used in the IR range (as a rule at $\lambda < 10$ µm) because of a high thermal generation rate in the devices with lower built-in potential $\varphi_b$.

The QW, SL and QD detectors involve the intersubband optical transitions between the quantized energy levels in QW (SL) (Fig. 3.4.c-f) or in QDs. As a rule, they can be applied for the radiation detection in the spectral range $\lambda \approx 6...100$ µm ($\nu \approx 3...50$ THz).

In the low part of the THz spectral range, the superconducting pair breaking detectors, in which the photon energy overcomes the binding gap, are used.

Simple schematics of some photon detectors are shown in Fig. 3.6. The physical processes in these devices were considered elsewhere (see, *e.g.*, [13, 14] and will be briefly discussed in Ch. 5.

The angle of incidence of the light for type I MQW detectors is related to the specific selection rules for optical dipole intersubband transitions between the energy sublevels in the same conduction band, which occur between the envelope states. This situation differs for the interband optical dipole transitions between the Bloch states from the valence band to the conduction one. Intersubband transitions are only possible for the component of the electric field in the EM wave, which is perpendicular to the type I QW or SL plane. That is why one should use the tilted radiation configuration or some grating, which changes the direction of the illumination in diffracted waves (see, *e.g.*, [24]).

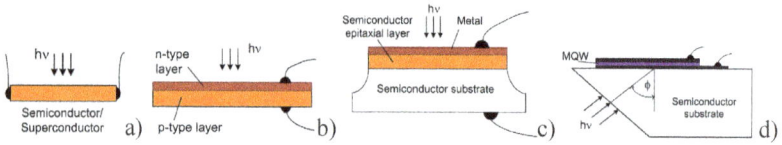

*Figure 3.6. Simple schematics of typical photon (quantum) detectors: a) photoconductor (intrinsic or extrinsic) or superconductor, b) p-n-photodiode or p-n-heterojunction, c) Schottky barrier diode, d) multiple quantum well (MQW) detector, where $\phi$ is the light incidence angle (about the Brewster angle).*

In Fig. 3.7, is shown the typical spectral dependences of the intrinsic and extrinsic photon detectors. One can see the rather abrupt of the long wavelength response as the photo-excited carriers can only appear in the conduction (valence) band overcoming the band-gap or the energy of an impurity level in the gap. The quantum efficiency for extrinsic semiconductors decreases, as the cut-off wavelength increases. For intrinsic IR detectors with anti-reflecting coatings, the quantum efficiency can be close to 100%. For HgCdTe detectors, it can attain ~ 96 % [25, 26].

The IR SBD detectors (also called photo-emissive detectors) on the base of silicides, using the electron transitions over barriers and cooled down to T ~ 70 K, are typically sensitive in the IR region $\lambda \leq 11$ μm (see, *e.g.*, Fig. 3.8). Now, they are relatively rarely used as single detectors and arrays because of low quantum efficiencies that are at least an order less than in intrinsic detectors (see, *e.g.*, Fig. 3.7). That is a cause for the small sensitivity falling down with the wavelength. Nevertheless, the low sensitivity together with the low noise level allow one to get the low-noise equivalent difference temperature NEDT ≈ 33 mK [27] at the expense of a relatively low frame rate in matrix arrays.

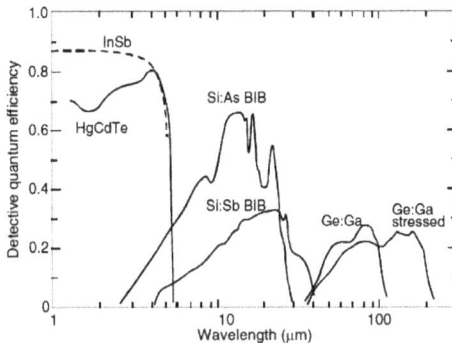

*Figure 3.7. Quantum efficiencies, of semiconductors used in the wavelength interval between 1 μm and 200 μm. InSb 1024×1024 pixel test array, HgCdTe 1024×1024 pixel MBE grown array without anti-reflection coating, Si:As and Si:Sb: 256×256 pixel BIB arrays, Ge:Ga and stressed Ge:Ga 16×25 pixel arrays [26]. (Reproduce with permission from Springer Nature).*

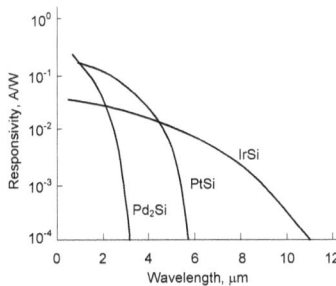

*Figure 3.8. Spectral dependences of the sensitivity of IR detectors on the base of Schottky barriers Pd₂Si, PtSi, and IrSi [28]. (Reproduce with permission from SPIE).*

The SBDs are majority carrier devices, and semiconductor p-n-junctions or heterojunctions and intrinsic photoconductors are minority carrier devices. Since there are no minority carrier storage effects, SBDs are potentially capable for the operation up to frequencies approaching the reciprocal of the dielectric relaxation time $\tau_d \sim 10^{-11}$... $10^{-12}$ s (Maxwell relaxation time). In the semiconductor minority carrier devices, the operation frequency is determined by the recombination time of free carriers, $\tau$, which, in dependence on the semiconductor, its quality, and operation temperature, can be within $\tau \sim 10^{-2}...10^{-8}$ s. For superconductor direct detectors, the operation frequency can be about $10^{-11}$ s (see Ch. 5).

Another photon technologies for IR or THz detectors involve the creation of QW (or SL) and QD structures in III-V materials (Ga(As,N)/(Al,Ga)As/N), AlInSb/InSb or II-VI (HgTe/CdTe) materials [15]. In these detectors, the relatively low-temperature operation is needed and, as a rule, is limited by the spectral range $\nu > 2$ THz. The QW and QD detectors, because of the optical transitions between the quantized energy levels, have a narrow wavelength response. In Fig. 3.9, the typical spectral responses of III-V QWs are shown.

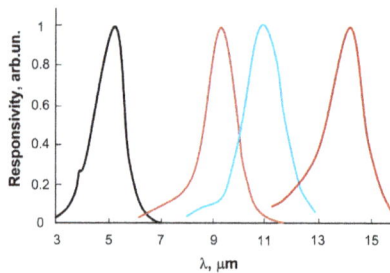

*Figure 3.9. A four-color 640 × 512 imaging focal plane array (3 to 15 microns) for a hyperspectral QWIP imaging system [29]. (Reproduce with permission from SPIE).*

The specific geometry (Fig. 3.6.d) of the radiation input relatively to the surface of type I SL or QW detectors is caused by the specific selection rules in this case.

Within the simple model concerning the interband and intraband dipole optical transitions (see, *e.g.*, [24]), these transitions depend on the dipole matrix element

$$\boldsymbol{\varepsilon} \cdot P_{ij} = \left\langle F_i \mid \boldsymbol{\varepsilon} \cdot \mathbf{p} \mid F_j \right\rangle ,$$

(3.4)

where $\boldsymbol{\varepsilon}$ is the polarization vector, which is perpendicular to the propagation vector $\mathbf{q}$ of an electromagnetic wave with frequency $\omega$, and $\mathbf{p}$ is the electrical dipole moment. Here, $F_{i(j)}$ is the electron wave function in a semiconductor well, which is a product of the Bloch function $U_{ik}(\mathbf{r})$ for "i" band and X(z), which is the envelope function describing the motion of the electron in the z-direction.

Because there are a rapid variation of the periodic part of the Bloch functions $U_{ik}(r)$ on the elementary cell length and a slow change in the envelope function $X_i(z)$, one can represent Eq. (3.4) in the form

$$\varepsilon \times P_{ij} \sim \varepsilon \int_{\Omega} U_{ik}^* \mathbf{p} U_{jk} \times d\tau \times \int_V X_{in}^* X_{jm} \times d\tau + \varepsilon \int_{\Omega} U_{ik}^* U_{jk} \times d\tau \times \int_V X_{in}^* \mathbf{p} X_{jm} \times d\tau \tag{3.5}$$

where $\Omega$ is the volume of the elementary cell, and $V = S \cdot L$ is the sample volume.

From the analysis of Eq. (3.5), it is seen that the allowed optical transitions are split into two classes. There are: (i) interband dipole optical transitions which take part between QW subbands originating from different band extrema ($i \neq j$) and defined by atomic-like dipole matrix elements $\langle U_{ik}^* | \mathbf{p} | U_{jk} \rangle$; and (ii) intersubband ($i = j$) optical transitions which are defined by dipole matrix elements between envelope functions of the same band.

The interband optical transitions are defined by the first term on the right side of the Eq. (3.5), as in the second term for interband transitions $\langle U_{ik}^* | U_{jk} \rangle = 0$ due to the different parity of Bloch functions. For additional selection rules in SLs and MQWs, one should evaluate also the integral $\langle X_{in}^* | X_{jm} \rangle$, which is non-zero, if $\Delta n = n - m = 0$. This means that the intersubband transitions (between the states in the conduction and valence bands) will take place between the quantum states of the same number $n = m$ at the normal radiation incidence.

The intraband dipole optical transitions in bulk semiconductors within the same band are forbidden and may only be induced by phonons or impurities to provide the momentum conservation. However, in SLs or MQWs [the second part on the right side of Eq. (3.5)], $\langle U_{ik}^* | U_{jk} \rangle \neq 0$ for envelope functions for the same band, and one needs to estimate the integral $\langle X_{in}^* | \mathbf{p} | X_{jm} \rangle$, which can be non-zero for $n \neq m$. Here $n$ and $m$ are the numbers of subbands.

The interband dipole optical transitions for a polarization $E \| x, y$ ($x, y$ is the sample surface) at the radiation normal incidence are allowed, if only the initial and final states coincide. For these polarizations, the intersubband transitions are only possible for the free carrier absorption mechanism that is an analog for the free carrier absorption in bulk semiconductors. However, for the in-plane or oblique radiation propagation in SLs or

MQWs, when the electric vector of an EM wave has a component perpendicular to the sample surface (E||z), the optical transitions between different subbands (of the same band) arise. This explains the tilted configuration of light incidence for SLs or QWs of type I (Fig. 3.6.d).

For type II superlattices, the light incidence perpendicular to the surface is not forbidden, as the dipole transitions take place between the states of different bands, and the intersubband optical transitions within the same band in these devices are not involved.

### 3.2.2 Thermal (power) detectors

Thermal detectors absorb radiation power and convert it into heat. In the ideal case, the response in the thermal detectors is wavelength-independent, as the signal depends on the radiation power only. Among these detectors, the uncooled ones seem to be the most widely used devices because of their relatively simple design and lower cost. For many applications, their sensitivity and speed are satisfactory.

The thermal detectors can be subdivided into several types. Among such types of detectors are pneumatic, thermistor, metal and composite bolometers, pyroelectrics, thermocouples, free carrier hot electron bolometers (HEBs), semiconducting and superconducting HEBs low-temperature bolometers, and mechanical cantilevers. The latter requires relatively complex readout systems and has worse detectivity compared to other uncooled thermal detectors because of the additional mechanical noise. Therefore, they are of limited use as IR and THz detectors.

Free carrier HEBs are based on the free-carrier absorption. Radiation heats a free electron gas. The term "hot electrons" was introduced [30] to describe nonequilibrium electrons in semiconductors. In semiconducting HEB, the incident radiation power is absorbed directly by free carriers, the crystal lattice temperature remaining essentially constant [31]. The electron distributions could be formally described by the Fermi function distribution, but with an effective elevated temperature. In metals, the mobility changes are much less pronounced, and the electron heating does not affect the metal resistance unless the change in the effective temperature is comparable with the temperature at the Fermi level. The typical spectral dependence of the sensitivity (detectivity) for InSb HEB is shown in Fig. 3.10.

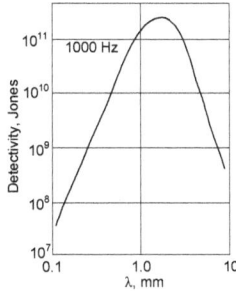

*Figure 3.10. Spectral dependences of detectivity D\* of an InSb free carrier hot electron bolometer [8].*

The change of the temperature in a thermal detector (lattice temperature) can be calculated from the heat balance equation

$$C_{th} \frac{d\Delta T}{dt} + G_{th}\Delta T = \varepsilon \cdot W ,$$
(3.6)

where $\varepsilon$ is the emissivity of a detector, $C_{th}$ is the heat capacity, and $G_{th}$ is the heat conductance. If the incident rtadiation power varies periodically with time as $W = W_0 \cdot \sin(i\omega t)$, then Eq. (3.6) for the stationary case has the solution

$$\Delta T = \frac{\varepsilon W_0}{G_{th}\sqrt{1+\omega^2 \tau_{th}^2}} ,$$
(3.7)

where $\tau_{th} = C_{th}/G_{th}$ is a thermal constant (time constant in "thermal" circuits), which is an analog of the time constant $\tau = R \cdot C$ in electrical circuits (R and C are, respectively, a resistor and a capacitance). Typical value $\tau_{th}$ is about 10 ms for uncooled thermal detectors. Then the voltage sensitivity, which is equal to the ratio of the output signal voltage to the input radiation power, is

$$S_V = \frac{\Delta V}{\Delta T} \times \frac{\varepsilon W_0}{G_{th}\sqrt{1+\omega^2 \tau_{th}^2}} .$$
(3.8)

Here, the coefficient $\Delta V/\Delta T$ characterizes how the temperature changes influence the detector voltage output.

Thermal detectors were the first detectors for the registration of radiation in the IR range. Today, their different modifications designed from a large number of materials for different temperatures from $T \approx 0.03$ K to $T \approx 300$ K are among the basic detectors and arrays operating in the IR and THz spectral ranges.

To get a high sensitivity, the thermal detectors should have low thermal mass (low heat capacity $C_{th}$, J/K) and good thermal isolation (low thermal conductance $G_{th}$, W/K). The design of these detectors should minimize both the heat capacity $C_{th}$ of the sensitive element and the heat losses from either conductive ($G_{th}$) or radiative mechanisms. The schematic of some of thermal detectors is presented in Fig. 3.11.

Typical thermal uncooled detectors are $VO_x$ or $\alpha$-Si microbolometers. In these materials, the temperature coefficient of resistance (TCR) $\alpha_{th}$ is of several percent that provides their high responsivity.

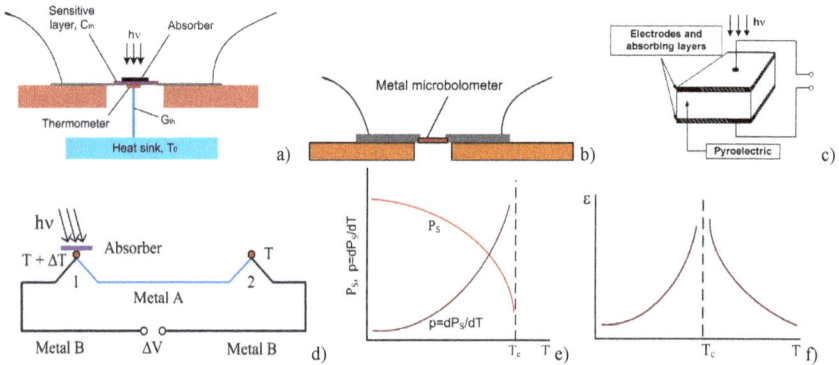

*Figure 3.11. a) Schematic diagram of the composite bolometer, absorber + thermometer. $C_{th}$ is the specific heat and $G_{th}$ is the thermal conductance. b) Schematic design of a metallic bolometer, c) Schematic design of a pyroelectric detector. d) Thermocouple detector schematic, e) Temperature dependences of the spontaneous polarization P and the pyroelectric coefficient p in a vicinity of the Curie temperature $T_c$. f) temperature dependence of the electrical permittivity $\varepsilon$ in a vicinity of the Curie temperature $T_c$.*

Microblometer (or bolometer) is a resistance element with a high temperature coefficient of resistance $\alpha_{th} \sim 3\% \times °C^{-1}$ (*e.g.*, in $VO_x$ or $\alpha$-Si microbolometers) and has a small heat capacity $C_{th}$. TCR can be positive or negative. The absorbed radiation changes the temperature of a bolometer and its resistance R. The coefficient $\alpha_{th}$ is defined by the expression

$$a_{th} = \frac{1}{R} \frac{dR}{dT}.$$
(3.9)

The change of a bolometer temperature by $\Delta T$, which is defined in the case of composite bolometer by a thermometer (Fig. 3.11.a), causes a change of a bolometer resistance R by $\Delta R$ that leads to the output signal

$$\Delta V = I_B \times \Delta R = I_B R \, \alpha_{th} \, \Delta T,$$
(3.10)

where $I_B$ is the bolometer current. For the voltage responsivity $S_V$ of a bolometer, it follows

$$S_V = \frac{I_B R \alpha_{th} \varepsilon}{G_{th} \sqrt{1 + \omega^2 \tau_{th}^2}}.$$
(3.11)

In a metallic bolometer (microbolometer) (Fig. 3.11.b), the TCRs are positive and are about an order lower ($\sim 0.3\% \times {}^\circ C^{-1}$) compared to $VO_x$ or $\alpha$-Si microbolometers, but this type of bolometers are featured by a lower level of noises and have close NEP values at room temperature.

In a pyroelectric detector (Fig. 3.11.c), the operation principle is a combination of the capacitor and the generator. Both the dependences of the spontaneous polarization $P_S$ (Fig. 3.11.e) and the permittivity $\varepsilon$ on the temperature (Fig. 3.11.f, dielectric bolometer) can be employed in the temperature range below the transition temperature $T_c$.

When the IR or THz radiation is falling down onto a pyroelectric crystal, it heats the detector surface of area S leading to a temperature change $\Delta T$ and, therefore, to a change of the polarization by $\Delta P$, and the electric charge with density $\Delta Q/S$ is generated. In the external circuit, the current $I = dQ/dt = p \times S \times d(\Delta T)/dt = p \times S \times (dT/dt)$ will arise. Here, dT/dt is the rate of change of the temperature with time. When the radiation is turned off, the crystal cools down, and an opposite charge is generated. Because of the existence of the multiplier dT/dt, at a constant radiation flow the responsivity of the pyroelectric detector $S_V = 0$ at this condition.

Setting $\Delta T$ from the heat balance equation solution for a periodic radiation signal with frequency $\omega$, the generated current I under the assumption of the typical modulation frequency $\omega$ dependence $\sim (1 + \omega^2 \tau_{th}^2)^{-1/2}$ for thermal (pyroelectric) detectors is

$$I = \frac{\varepsilon p S W_0 \omega}{G_{th}\sqrt{1+\omega^2\tau_{th}^2}}$$

(3.12)

with the typical thermal time $\tau_{th} \sim 10^{-2}$ s.

In addition, because of the high resistance of a pyroelectric detector, a high-value load resistor should be used to measure the signal. That is why the voltage generated in such circuit is also dependent on the electrical time constant $\tau$, that is dependent on the load shunt conductance G and the capacitor C of the detector and can differ substantially from $\tau_{th}$. The generated voltage is given by

$$V = \frac{I}{G\sqrt{1+\omega^2\tau_e^2}}.$$

(3.13)

In summary, the voltage responsivity is given both by the thermal constant $\tau_{th}$ and time constant $\tau = G \times C$:

$$S_V = \frac{V}{W_0} = \frac{\varepsilon p S W_0 \omega}{G_{th} G \sqrt{1+\omega^2\tau_{th}^2} \times \sqrt{1+\omega^2\tau_e^2}}.$$

(3.14)

Wavelength-independent pyroelectric detectors are manufactured from ferroelectric (pyroelectric) crystals. These crystals, also called polar crystals, are spontaneously polarized, since each unit cell of the crystal has a permanent electric dipole moment aligned with a specific crystal axis. Actually, the spontaneous polarization is a fundamental property of many crystalline systems. Ten of 32 crystal classes are pyroelectric, they do not have the center of symmetry. If an external electric field can reverse the dipole, the material is said to be ferroelectric. All ferroelectric materials are pyroelectric, but not the other way around [12].

In a thermocouple (Fig. 3.11.d, differential thermocouple), the voltage signal is proportional to the temperature difference at contacts 1 and 2:

$$\Delta V = \alpha_S \times \Delta T \text{ ,} \tag{3.15}$$

where $\alpha_S$ is the Seebeck coefficient. For the voltage sensitivity, we have

$$S_V = \frac{\Delta V}{W_o} = \frac{\alpha_S \varepsilon}{G_{th}\sqrt{1+\omega^2\tau_{th}^2}} \text{ .} \tag{3.16}$$

For a series connected N thermocouples, called a thermopile, the sensitivity increases N times.

### 3.2.3 *Rectification detectors*

Any detector with the nonlinear current-voltage characteristics can be used directly as a rectification direct detector or as a mixer in coherent systems. Frequently used are devices having a strong electric field quadratic nonlinearity. Examples of devices with nonlinear current-voltage characteristics are forward biased SBDs, SIS tunnel junctions, HEBs, SLs, FETs, *etc.* (Fig. 3.12). However, to achieve the efficient application in coherent receivers and a low noise in THz and sub-THz spectral ranges, only several types of detectors can be used.

The typical rectification detector used in the THz range is the SBD. The SBD in its simple form consists of a metal layer that contacts a semiconductor (Fig. 3.6.c). The SBD detector consists of the sensitive non-linear element itself (SBD junction), parasitic elements, and antenna. Any antenna-type area is much larger than the sensitive element (junction) area of a detector.

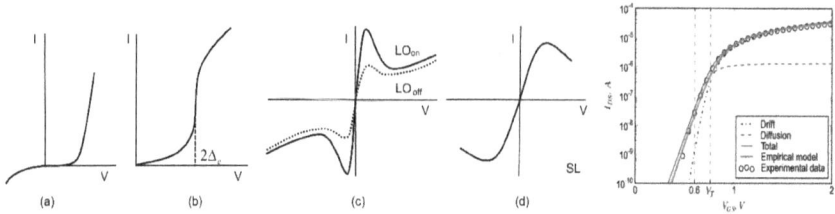

*Figure 3.12. I–V characteristics of detector sensitive nonlinear elements. (a) SBD, (b) SIS, (c) HEB (LO$_{on}$, LO$_{off}$ means at radiation and without radiation, respectively), (d) SL, (e) FET.*

The metal/semiconductor junction exhibits rectifying behavior (Fig. 3.12.a). As direct (incoherent) rectification detectors, SBDs (and other type of rectification detectors) are used at low-signal level regimes, where they operate as quadratic detectors. In this case, the signal is proportional to the square of the electric field ($E^2$) in the electromagnetic wave (proportionally to the first degree of the radiation power). In the low-signal regime (square law detection) for SBD biased by $V_0$ and under the microwave radiation with frequency $\omega$, the current-voltage characteristics [32] with the voltage $V_j = V_0 + V_P \cdot \cos(\omega t)$ is

$$I = I_{sat} \cdot [\exp(V_j/V_t) - 1] = I_{sat} \cdot [V_j/V_t + 1/2(V_j/V_t)2 + \ldots] =$$

$$(I_{sat} \cdot V_P/V_t) \cdot \cos(\omega t) + (I_{sat}/4) \cdot (V_P/V_t)^2 \cdot [1 + \cos(2\omega t)]. \tag{3.17}$$

It was assumed $V_0 \ll V_P = V_A$, $V_P < V_t$, $\cos^2(\omega t) = \frac{1}{2}(1 + \cos 2\omega)$. Here, $V_t$ is the thermal voltage ($V_t = nk_B T/q \approx 0.026$ V at T = 300 K), $V_P$ is the amplitude of the signal, and $I_{sat}$ is the saturation current, n is the ideality factor (n $\approx$ 1), $k_B$ is the Boltzmann constant, q is the elementary charge, and $V_j$ is the diode junction voltage (see Fig. 3.13).

To describe the radiation frequency dependences of SBD detectors, we should consider the additional parasitic components of such devices (see, *e.g.*, [33]). In Fig. 3.13, we show some of them, which define, to a great degree, the frequency and temperature dependences of SBD detectors.

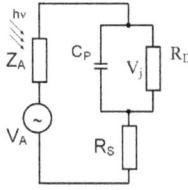

*Figure 3.13. One of the SBD's simplified schematic representations with regard for the basic parasitic components. Here, $Z_A$ is the antenna impedance, $V_A$ is the radiation voltage amplitude, which is produced in the antenna, $R_S$ is the series parasitic active resistance; $R_D = R_{SBD}$ is the SBD differential active resistance; $C_P$ is the parasitic reactance (usually capacitive), $R_S$ is the active serial resistance.*

The parameters $C_P$, $Z_A$, and $R_D$ are dependent on the radiation frequency. Because of it, they largely condition the frequency and temperature dependences, which are observed in the SBD detectors (see, *e.g.*, Fig. 4.4 for SBD strong radiation frequency response dependence). Similar frequency and temperature dependences are typical of FET THz detectors too.

For the rectified current $I_{dc}$ which can be used for estimations in the low-signal regime, one has

$$I_{dc} = (I_{sat}/4) \times (V_P/V_t)^2, \ (V_P < V_t, \ V_0 \ll V_P). \tag{3.18}$$

For low-barrier SBDs with the barrier height $\varphi \sim 0.2$ eV of area $A \sim 1 \ \mu m^2$ to satisfy the low-signal regime (square law region), it should be $I_{SBD} < 1 \ \mu A$.

In the square law region, the waveform of the SBD and other non-linear structures, the detected voltage and the current are proportional to the square of the electric field E in the input signal. Under larger signal levels, the diode current is a function of the input signal level. This occurs, when the junction voltage $V_j$ exceeds the thermal voltage $V_t$.

The SBDs are majority carrier devices and, *e.g.*, semiconductor p-n-junctions are minority carrier devices. Since there are no minority carrier storage effects, SBD devices are potentially capable for the operation up to frequencies approaching the reciprocal of the dielectric relaxation time $\tau_d$ of the semiconductor crystal. These frequencies can be of the order of $10^{11}...10^{12}$ GHz. However, other considerations, particularly the series resistance and junction capacitance, will be more important in the determination of the frequency response for practical planar SBDs, placing their upper frequency limit at lower values [32]. Up to now, the SBDs remain one of the widely used direct detection THz devices and active elements in cooled and uncooled terahertz coherent receivers.

Similar to SBDs, diode-like nonlinear current-voltage characteristics are demonstrated by the self-switching diodes (SSDs) which can also be used as THz detectors [34]. The SSDs are based, on the geometric symmetry breaking of a semiconducting nanochannel, but they are not based on the use of any doped p-n junctions or Schottky barrier structures and are, hence, completely different from conventional diodes. As the SSDs possess a much lower parasitic capacitance as compared with conventional diodes because of their planar architecture, the electrical signal rectification in them has ultrahigh speed, and they give the potential possibility to use them as local oscillators in coherent systems.

For using rectification devices as direct detectors with nonlinear characteristics in arrays, it is desirable to use them at zero biases. The condition of zero bias reduces the noise level and simplify the design of arrays. However, *e.g.*, FET (HEMT) and SSD devices to be effectively used need some bias.

### 3.3   Basic parameters of detectors - figures of merit

Radiation detectors (in dependence on the spectral region of applications, there are used equivalent terms such as photodetectors, photoelectric detectors, radiation sensors, receivers) represent detectors of electromagnetic signals. In them, the absorbed electromagnetic radiation results in changes of physical parameters in the detector sensitive element. The absorbed electromagnetic radiation modifies the motion of electrical charges or heats the detector lattice or electron gas in the sensitive elements, which, as a rule, are fixing by changes of their electrical parameters.

The tie line of the IR or THz detectors of their main idea is the acquisition of information from the objects under observation, as every photon in any spectral range is a carrier of information. In detectors, the transformation of absorbed electromagnetic radiation, in most cases, is ending in the appearance or changing the electrical signals. The absorbed radiation heats the electron or lattice subsystems or changes the electron energy distribution, thus modifying the motion of charged carriers. Such alterations are fixing by measuring the changes of physical parameters of a detector.

The principles of radiation registration in the THz and IR ranges and the physics of detectors can differ substantially. But all of them can be characterized by the same set of parameters, though the parameters that characterize both the IR and THz detectors need the account for a large number of experimental characteristics (*e.g.*, electrical, radiometric, environmental ones) to be involved. For linear or two-dimensional arrays,

Detectors and Sources for THz and IR                    Materials Research Forum LLC
Materials Research Foundations **72** (2020)          https://doi.org/10.21741/9781644900758

the procedure of characterizing devices by the figures of merit is more complicated as compared to that for a single detector.

A lot of books and papers devoted to the description of figures of merit of detectors can be cited (see, *e.g.*, [13, 35–38]), though they are mainly consider the determination of IR detector parameters. Many of them are also applicable to THz detectors.

For a single IR detector, the next figures of merit are generally used: the responsivity (R) (or sensitivity), the detectivity (D* or D-star), the noise equivalent power (NEP), the noise equivalent temperature difference (NETD) (or the noise equivalent difference temperature NEDT), which are the most important performance parameters. For characterization of the THz detectors, the responsivity, the noise temperature $T_n$, and the NEP are usually used. The detectivity (in which the detector sensitive area is included) generally is not used for sub-THz/THz detectors, since these detectors, as a rule, are operating with antennas, and, in many cases, it is difficult to estimate their antenna effective area. Therefore, their effective area for collecting a radiation commonly can be estimated only approximately. For IR single element detectors, the most important performance parameter for sensing applications is the detectivity, which represents the signal-to-noise ratio and the area of a sensing element.

*3.3.1   Responsivity*

Absorbed electromagnetic radiation modifies the motion of electrical charges or heats the detector lattice or electron gas in the sensitive element, which, as a rule, is fixed by changes of their electrical parameters. An electrical signal (current or voltage) is dependent on the incident radiation intensity–the photon flow. The number of electrons $N_e$ excited in a detector or heat it its lattice or free electrons is relevant to the number of photons or their energy. This relevance is the quantum efficiency (or coupling efficiency) $\eta$.

The number of photons absorbed in quantum detectors and the number of generated electrons are high in the IR and THz under the general background conditions (the number of photons $N \sim 10^{12}...10^{18}$ $cm^{-2}{\cdot}s^{-1}$). Therefore, except for low photon flows, *e.g.*, at very low background temperatures allowing the photon counting [39], finding the number of photons and electrons is practically almost impossible. Instead, there can be used the ratio of the output electrical current per incident optical power which is called the current responsivity parameter $R_I(\lambda) = <I_s(\lambda)>/<P(\lambda)> = \eta{\times}q{\times}\lambda/h{\cdot}c = \eta{\times}q/h\nu$. Here, $<I_s(\lambda)>$ and $<P(\lambda)>$ are the root mean square (rms) values and q, h, c, and $\lambda$ are the elementary charge, Planck's constant, speed of light in vacuum, and wavelength of the incident light, respectively.

The responsivity R assumes the certain minimal signal level that can be detected and used as the characteristic of the effectiveness of conversion (steepness of conversion) of the electromagnetic radiation power <P(λ)> (measured in W) on a detector at the monochromatic wavelength λ to detector's electrical output signal <U(λ)> (measured in V)

$$R_U(\lambda, f) = \frac{\langle U(\lambda) \rangle}{\langle P(\lambda) \rangle}, \text{ V/W},$$ 

(3.19)

for the voltage responsivity or $R_I(\lambda, f) = \langle I(\lambda) \rangle / \langle P(\lambda) \rangle$ (A/W) for the current responsivity. Here again, $\langle U(\lambda) \rangle$, $\langle I(\lambda) \rangle$, and $\langle P(\lambda) \rangle$ are the root mean square (rms) values. Further, it is assumed that the rms symbol $\langle \ldots \rangle$ is omitted.

The blackbody total voltage responsivity $R_U^{int}(T)$ (the detector responsivity to the blackbody radiation) is defined by

$$R_U^{int}(T) = \frac{U}{\int_0^\infty P(\lambda(T))d\lambda} = \frac{U}{\frac{2\pi^5 k_B^4}{15c^2h^3} \times T^4} = \frac{U}{\sigma_B T^4},$$

(3.20)

where P(λ,T) is defined by Planck's equation (3.1). Here, $k_B = 1.38048 \times 10^{-23}$ J/K and $\sigma_B = 5.670367 \times 10^{-8}$ Wm$^{-2}$K$^{-4}$ are the Boltzmann and Stefan-Boltzmann constants, respectively. For the current responsivity, instead of the signal voltage $U_s$, there should be used the signal current $I_s$.

For an ideal quantum photodetector, the spectral responsivity $R_I^{\lambda_{co}}$ in the wavelength range $0 \le \lambda \le \lambda_{co}$ can be introduced by the expression

$$R_I^{\lambda_{co}} = \frac{U}{\int_0^{\lambda_{co}} \frac{\lambda}{\lambda_{co}} P(\lambda(T))d\lambda} = \frac{U}{\frac{ch}{\lambda_{co}} \int_0^{\lambda_{co}} N(\lambda(T))d\lambda},$$

(3.21)

where $\lambda_{co}$ is the cut-off wavelength of the ideal quantum detector responsivity, N(λ,T) is the number of photons at a certain wavelength λ, and T is the background temperature.

For an ideal quantum photodetector, the dependence of the responsivity on the wavelength is

$$R_\lambda = \frac{\lambda}{\lambda_{max}} R_\lambda^{max} \, , \qquad\qquad (3.22)$$

where $R_\lambda^{max}$ is the maximal responsivity of the ideal detector at the wavelength $\lambda_{co}$.

The signal current in the quantum detector at a certain radiation wavelength $\lambda$ is

$$I_{s,\lambda} = \eta q N_\lambda \times A_d = \eta q \times (P_\lambda / h\nu) \times A_d, \qquad\qquad (3.23)$$

where the rate of signal-carrier generation $\eta N_\lambda A_d = \eta(P_\lambda/h\nu)A_d$, $N_\lambda$ (photons·s$^{-1}$·cm$^{-2}$) is the average number of signal photons arriving at the detector, and $A_d N_\lambda$ is the average rate of arrival of signal photons, $P_\lambda = [W/cm^2]$.

For the current responsivity at the wavelength $\lambda$, we get

$$R_I(\lambda) = I_{s,\lambda}/P_\lambda = \eta \times q \times (\lambda/h) \times c = \eta q/h\nu = 0.806 \times \eta\lambda, \, (A/W). \qquad\qquad (3.24)$$

Here, $\lambda$ on the right part of the expression is in $\mu m$. For example, for $\lambda = 10.5$ $\mu m$ and $\eta = 0.65$ (the typical value for uncoated HgCdTe photodetectors), $R_I(\lambda) = 5.1$ A/W.

### 3.3.2   Noise equivalent power (NEP)

The noise equivalent power (NEP) is a convenient parameter with which detectors can be compared. It is defined as the incident rms radiation power $P = (<P^2>)^{1/2}$ falling down on the detector generating a rms output signal $S = (<S^2>)^{1/2}$, which is equal to the rms noise level $N = (<N^2>)^{1/2}$ (N is a noise current $I_n$ or a noise voltage $U_n$). The NEP is the signal level S that produces the signal-to-noise ratio $S/N = 1$:

$$NEP = \frac{P \times N}{S} = P \times \frac{I_n}{I_s} = P \times \frac{U_n}{V_s}, \, W. \qquad\qquad (3.25)$$

The NEP can be written via the detector responsivity ($R_I = I_s/P$ or $R_U = U_s/P$) as

$$NEP = \frac{U_n}{R_U} = \frac{I_n}{R_I}, \; W, \qquad\qquad (3.26)$$

where $U_n$ and $I_n$ are the noise voltage and the noise current, respectively.

The NEP is a criterion of the detector in measuring weak optical signals. The smaller the NEP, the better the detector.

The NEP is also quoted for a fixed reference electronic bandwidth. When comparing different detectors, it is convenient to normalize such parameter to the unit bandwidth. As for the direct detectors in BLIP regime (limited by the background radiation fluctuations), the noise electrical signal is proportional to the square root of the pass bandwidth ($I_n$, $U_n$ ~ $(\Delta f)^{1/2}$ [35–37, 40]. Then, for the normalization to the bandwidth $\Delta f$, the expression for noise equivalent power is

$$NEP = \frac{I_n}{R_I \times (\Delta f)^{1/2}} = \frac{U_n}{R_U \times (\Delta f)^{1/2}}, \; W/Hz^{1/2}. \qquad\qquad (3.27)$$

For the BLIP detection for direct detectors it follows (see below Ch. 4) that

$$NEP = \left( \frac{2h\nu}{\eta} \times P_B \right)^{1/2}, \; W/Hz^{1/2}. \qquad\qquad (3.28)$$

From radiometric considerations, the background power registered by a detector varies proportionally to the area of a receiver optics $A_r$ ($P_B$ ~ $A_r$). Then the direct-detection detector noise equivalent power for the BLIP detection is

$$NEP \sim A_r^{1/2}. \qquad\qquad (3.29)$$

This dependence is important for imaging in the IR and THz wavelength ranges, where a background radiation level is high and is the main source of a detector noise level. It is seen that the more sensitive detector is (smaller NEP), the smaller receiver optics can be, and, thus, the weaker background radiant power falling down on the detector be promoting a realization of the better ultimate performance of a direct detector. With the same optics, the better NEP allows one to get larger distances for the detection and recognition of objects.

For heterodyne detectors, the heterodyne power level $P_0$, as a rule, can be made large to make the quantum noise (shot noise) to be the dominant one. Therefore, for heterodyne non-photoconductive detectors for intermediate frequency signals, when the radiation heterodyne power $P_0 \gg P_s$, where $P_s$ is the signal radiation power, the current at an intermediate frequency is governed by the expression $< I(t)_{IF}^2 >= 2R_I^2 P_s P_0$ and the noise is $< I_n^2 >= 2q < I_0 > \Delta f = 2qR_I P_0 \Delta f_{IF}$ (shot noise). Then the signal-to-noise ratio is

$$SNR = \eta P_s/(h\nu\Delta f) \tag{3.30}$$

and, for a photoconductive detector,

$$SNR = \eta P_s/(2h\nu\Delta f). \tag{3.31}$$

The noise equivalent power NEP normalized to $\Delta f$ in the noise limit for the heterodyne detection for a non-photoconductive detector is

$$NEP = P_{s,het}^{min}/\Delta f_{IF} = \frac{h\nu \times \Delta f_{IF}}{\eta \Delta f_{IF}} = \frac{h\nu}{\eta}, \text{ W/Hz.} \tag{3.32}$$

For the heterodyne detection with a photoconductive detector

$$NEP = P_{s,het}^{min}/\Delta f_{IF} = \frac{2h\nu \times \Delta f_{IF}}{\eta \Delta f_{IF}} = \frac{2h\nu}{\eta}, \text{ W/Hz.} \tag{3.33}$$

In contrast to direct-detection detectors, when $[NEP] = W/Hz^{1/2}$, $[NEP] = W/Hz$ in heterodyne detectors.

In the SWIR and visible spectral ranges, there is a difficulty in the provision of a mutual coherence detection of $P_s$ and $P_0$ over the detector area, because large IF conversion losses can take place. This is a cause why, as a rule, the heterodyne detection is applicable in the sub-THz and radio frequency spectral ranges, as the wavelengths at these frequencies are comparable or greater than the detector dimensions ($\geq 100$ μm), and a spatial coherence needed over the detector area usually takes place. However, it is possible to realize the heterodyne regime and the coherent detection even in the visible wavelength region.

### 3.3.3 Detectivity

The figure of merit – the detectivity D* – was defined by R.C. Jones in 1958 [31]. It is used to compare the sensitivity of detectors and is called D* (D star) or Jones. D* is the signal-to-noise (S/N) ratio of a detector measured under the specified test conditions, referenced to a 1-Hz bandwidth and normalized to the square root of the detector area. The specified test conditions usually consist of the blackbody source temperature that, for infrared detectors, is often taken as 500 K or 800 K, and the signal chopping frequency. If the background temperature is other than the room temperature (T = 295 or 300 K), then this should be noted.

In many cases, it is found that the NEP for direct detectors scales with the square root of their area A (as the photon fluctuation noise ($\langle(\Delta N_{ph})^2\rangle \sim \langle(N_{ph})\rangle$ in the spectral region, where the Poisson statistics is valid), while the signal increases in proportion to the detector area. Then the figure of merit–the specific detectivity D* = $(A)^{1/2}$/NEP–is approximately constant.

By normalizing the measured S/N ratio by the square root of the detector area, the D* figure of merit recognizes that the statistical fluctuations of the background photon flux incident on a detector (photon noise) are dependent on the square root of the number of photons. Therefore, this figure of merit provides a comparison of detectors, which may be produced and tested with different areas.

The detectivity is the figure of merit that combines the current responsivity $R_I$ with the detector noise current $I_n$ (or the voltage responsivity $R_U$ with the voltage noise $U_n$) (see, e.g., [2, 31, 36, 38]). When normalized to the bandwidth $\Delta f$, it takes the form

$$D^* = \frac{\sqrt{A_d \times \Delta f}}{NEP} = \frac{R_I \sqrt{A_d \times \Delta f}}{I_n}, \text{ cm} \cdot \text{Hz}^{1/2}/\text{W}, \qquad (3.34)$$

The specific detectivity D* (D-star) is a normalized parameter mainly of IR single detectors and characterizes the quality of detector. D* involves the detector area and, therefore, is not frequently applied to the characterization of sub-THz/THz detectors, since those detectors have, as a rule, antennas whose effective area frequently is not well determined. When the dominant detector noise is due to the random arrival of photons (background noise), the detector is operating in the BLIP regime.

The detectivity D* is used for the comparison of the detector threshold responsivities. The larger D* is, the better the detector, and, *vice versa*, the smaller the NEP. The value of D* can be interpreted as the signal-to-noise ratio at the detector output with area of 1 cm$^2$ at the noise equivalent bandwidth of 1 Hz, when the radiation power of 1 W is falling down on a detector. Therefore, for comparison of different detectors with different areas and different noise equivalent bandwidths, the detectivity is adduced to the unit area and the unit noise equivalent bandwidth. The detectivity D*, thereby, is defined as the inverse average threshold power $P_{thr}$ of radiation which is needed for getting the signal-to-noise ratio S/N = 1 for a detector with unit area and unit noise equivalent bandwidth:

$$D^* = \frac{1}{P_{thr}} = \frac{\sqrt{A_d \times \Delta f}}{NEP}, \ cm \cdot Hz^{1/2}/W \qquad (3.35)$$

For the ideal quantum photodetector, the detectivity for monochromatic radiation D*($\lambda$) is

$$D^*(\lambda) = \frac{\lambda}{\lambda_{max}} \times D^*(\lambda_{max}), \qquad (3.36)$$

where D*($\lambda_{max}$) is the detectivity at the maximal responsivity of the quantum ideal photodetector, and $\lambda_{max} = \lambda_{co}$ for the ideal photodetector.

The detectivity D* for a number of IR detectors is presented in Fig. 3.14.

### 3.3.4 Noise equivalent temperature difference (NETD)

For the characterization of IR passive imaging systems, such parameter as NETD or NEDT (noise equivalent difference temperature) is introduced (see, *e.g.*, [2, 31, 36, 38]).

NETD is the incremental temperature above the background that produces a signal-to-noise ratio equal to unity. Two equivalent definitions of NEDT are as follows [42]:

1. NEDT is a change in the temperature of a black body of the infinite lateral extent, which causes, when is viewed by an imaging system, a change in the signal-to-noise ratio. This results in the change of the electrical output of the pixels of a focal plane array

or else of the read-out electronics, which receives an input signal from the pixels of the array.

2. NEDT is the difference in temperatures between two side-by-side black bodies of large lateral extents which gives rise, when are viewed by an imaging system, to a difference in the signal-to-noise ratios in electrical outputs of two halves of the array viewing the two black bodies. These can be measured either at the output of the focal plane array or at the output of the read-out electronics.

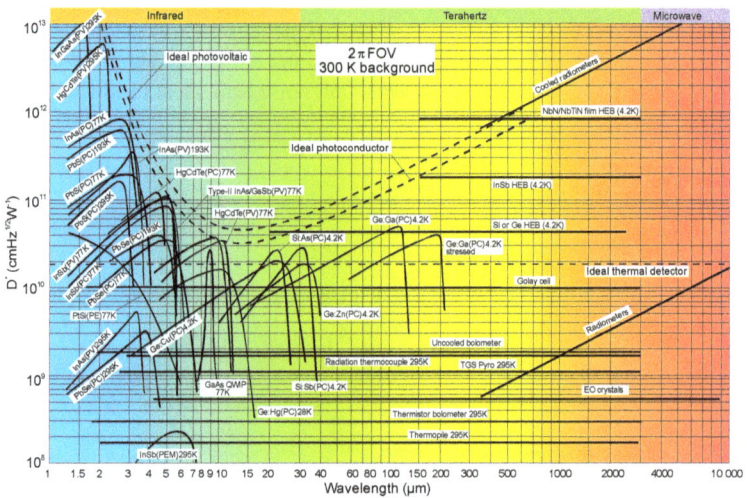

*Fig. 3.14. Comparison of the detectivity D\* of various available detectors operating at the indicated temperature [41]. (Reproduce with permission from SPIE).*

In either definition, NETD can be expressed as that of a single pixel or it can be the average of the pixels of an array. NEDT is one of the most important parameters that defines the quality either a single detector or the focal plane arrays (FPAs) in the technical vision systems.

The classical expression for NETD (see, *e.g.*, [2]) for the axial point of an axisymmetric optical system with detectors sensitive in the spectral range between $\lambda_1$ and $\lambda_2$ wavelengths for signal-to-noise ratio $I_s/I_n = 1$ and "white noise" is:

$$NEDT = \Delta T = \frac{4(F/\#)^2 \times (\Delta f)^{1/2}}{A_d^{1/2} \times \int_{\lambda_1}^{\lambda_2} \frac{\partial P(\lambda, T)}{\partial T} \tau_{op} \tau_{atm} \tau_f \sqrt{\eta} D*(\lambda, T) d\lambda}, \ K, \tag{3.37}$$

Here $F/\#$ is the optics f-number, $\tau_{op}$, $\tau_{atm}$, $\tau_f$ are transmission coefficients of optics, atmosphere, and filter, respectively, $\eta$ is the detector quantum (coupling) efficiency, and $A_d$ is the detector sensitive area.

NETD and NEP can be connected with each other. In the case of $D* = (A_d \cdot \Delta f)^{1/2}/NEP =$ const (*e.g.*, thermal detectors) for $F/\# \approx 1$, NEP can be estimated from the classical expression for NEDT [43]

$$NEP \approx \frac{NETD \times A_d \times \sqrt{\eta} \times \int_{\lambda_{co}}^{\lambda_u} \frac{\partial P(\lambda, T)}{\partial T} \cdot d\lambda}{4}, \ W, \tag{3.38}$$

where $\tau_{op}$, $\tau_{atm}$, $\tau_f$, and $\eta$ are taken to be equal to unity.

The parameter NEDT is a system parameter. Optimizing the parameters of an objective, electronics (*e.g.*, changing an accumulation time in ROICs), detector geometry, *etc.*, one can change NEDT over a wide range. Therefore, NEDT is used as a figure of merit not for the comparison of radiation detectors, but essentially for the comparison of system parameters both in the IR range [2, 38] and at the THz frequency range [44]. For passive imaging systems, NEDT is a more proper parameter for the system characterization and is suitable for direct measurements for staring imagers compared to NEP. NEDT and NEP are connected with each other. In the case of $D* \approx$ const (thermal detectors), this relationship is simple.

NEDT (for determining the ultimate performance) and the modulation transfer function (MTF) for IR and THz arrays and systems are the relevant figures of merit, which primarily define the performance metrics of imaging systems. MTF is connected with the spatial resolution.

Important characteristics for photon and thermal IR detectors are spectral dependences of the detectivity $D*$ and $\partial P(\lambda, T)/\partial T$. For photon detectors, the detectivity $D*$ increases from $\lambda_{co} \approx 14$ µm to shorter cut-off wavelengths at the background temperature $T \approx 300$ K (see Fig. 3.14), and the experimental values for detectors with different $\lambda_{co}$ are close to the theoretical values for the background-limited detector performance. At longer wavelengths ($\lambda_{co}$ > 14 μm), the experimental values of D* should also grow with the wavelength. But, because of additional intrinsic noises in these long-wavelength detectors (at $\lambda_{co} \sim 25...200$ μm as a rule extrinsic photoconductors), the values of D* are lower, as compared to the ones predicted by calculations for the background-limited performance.

The function $\partial P(\lambda,T)/\partial T$ at the background temperature T = 300 K has a maximum at $\lambda$ = 8.035 μm and decreases relatively slow in the longer wavelength region. At shorter wavelengths ($\lambda$ < 8 μm), it decreases faster (see Fig. 3.15). The relatively fast decrease of the function $\partial P(\lambda,T)/\partial T$, as the wavelength decreases, results in increasing (degradation) the NEP and NEDT parameters (see Eq. (3.37)) that is partly compensated by an increase of the detectivity D* in quantum detectors. However, this is a cause for that the uncooled IR thermal detectors (with $D^*(\lambda) \approx$ const) in the spectral region $\lambda$ < 5 μm are basically less sensitive, as compared to photon detectors, and are less used in this part of IR spectrum.

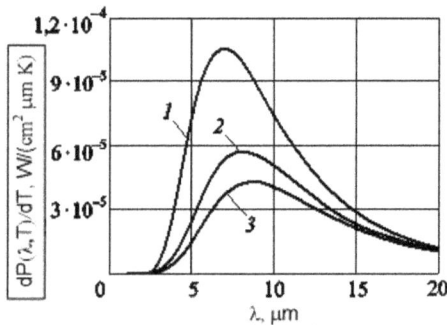

*Figure 3.15. Derivative $\partial P(\lambda,T)/\partial T$ of Planck's law for the black body radiation as a function of the wavelength. T, K: 1 – 320, 2 – 300, 3 – 280.*

The radiation emission maximum under the conditions on the Earth's surface ($T_s \approx$ 293 K) is at the wavelength of about 10 microns (30 THz). Most radiation in the Universe is emitted at wavelengths longer than 10 microns, and this peaks at about 100 microns (3 THz), if we exclude contributions from the cosmic microwave background. Radiation at those wavelengths highlights warm phenomena and the processes of changes such as the star formation, formation of planetary systems, galaxy evolution, atmospheric constituents, dynamics of planets, comets, and tracers for the global monitoring of the Earth. Sensors in the IR ($\lambda \approx 1...30$ μm) and THz ($\lambda \approx$ 30 μm...3 mm) ranges in many cases demonstrate the unprecedented sensitivity of cooled and superconducting detectors (see below) for the Earth observation, security applications, medical diagnostics, astrophysical, planetary, and ground-based imaging instruments, *etc.*

For space-based platforms, where the instruments are not limited by atmospheric losses and absorption, the overall instrument sensitivity is dictated by the sensitivity of the sensors themselves. Moreover, some of the cryogenic heterodyne receivers at sub-millimeter wavelengths provide the almost quantum-limited $h\nu/k_B$ sensitivity (see Fig. 3.16, where the limiting noise temperatures of some detectors are shown). However, the frequency sources at sub-millimeter wavelengths with adequate output power for transmitters and local oscillators (coherent or heterodyne detection) are not easily available, and pose the greatest challenge for the advancement of this field [20].

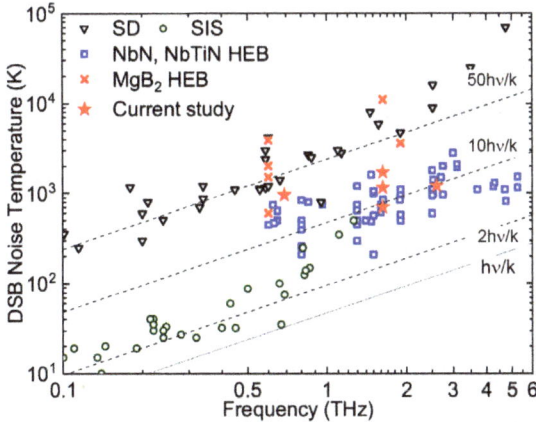

*Figure 3.16. DSB noise temperature performance of SIS, HEB, and SBD mixers. For comparison, the 1-, 2-, 10-, and 50-times quantum noise limit lines (after [45]) are shown. At $\nu > 0.7$ THz ($\sim eV_{gap}/h$), the photons have not enough energy to effectively break the Cooper pairs and the operation characteristics of SIS detectors go down. (Reproduced with permission from E. Novoselov)*

### 3.3.5   Modulation transfer function and system's resolution

For arrays, the figures of merit for estimations of the vision system performance is not the detectivity D* (applied, as a rule, to IR or THz single detectors), but the noise equivalent power, noise equivalent difference temperature, the modulation transfer function (MTF), and the phase transfer function (PTF) parameters [2, 38, 40]. MTF is one of the primary parameters used in the vision system design, analysis, and specification as a whole, which mostly defines the image quality from the resolution point of view.

MTF is the magnitude, and PTF (it determines the image position and orientation, rather than the size of details [38, 40]) is the phase of the complex-valued optical transfer function (OTF), which plays a key role in the evaluation and optimization of an optical system.

As usual, for technical vision systems, the OTF is maximum at the spatial frequency $f = 1/d$, where d is the extent of one cycle at the target modulation. For optical systems, d is measured in the angular space. The human visual system exhibits the maximum response at spatial frequencies that are low, but are non-zero [40].

The optical system resolution depends on the scattering spot of the lens. A variety of an optical system designs are possible, each of which allows achieve a certain resolution measure, and can be implemented [2, 38].

To combine an optical resolution and a detector pitch, the function $\lambda F/\# \, / \, V_d$ is introduced. Here, F/# is the f-number (focal ratio), $\lambda$ is the average wavelength, and $V_d$ is the detector pitch. In the spatial domain, the important parameter is the ratio of the Airy disk diameter to the detector size $A_{Airy} / V_d$.

The important characteristics of the IR or THz vision system in the imaging are determined by the spatial resolution and contrast, which can be characterized by the Rayleigh criterion, Nyquist frequency, and spatial frequency at a specific value of the MTF. The imager quality effectiveness (efficient spatial bandwidth), can also be determined by the product of the Nyquist frequency and the MTF [46].

For a diffraction limited circular aperture, the Airy disk diameter is:

$$A_{Airy} = \theta \times fl = 2.44 \times (\lambda/D) \times fl. \qquad (3.39)$$

Here, the f-number $F/\# = fl/D$, $\theta = 2.44 \times (\lambda/D)$ is the detector angular parameter, fl is the effective focal length.

In the frequency domain, the important parameter is the ratio of the optics cut-off frequency to the detector cut-off frequency, which is defined for a corrected optics by $v_0 = 1/(\lambda F/\#)$ (cycles or lines per mm).

NEDT and MTF are considered as the performance metrics of the IR and THz imaging systems: the sensitivity and spatial resolution, respectively. NEDT is connected with the temperature resolution (the minimum temperature difference that can be registered), and MTF concerns the spatial resolution. MTF and PTF are measures of how the system responds to spatial frequencies. They do not contain any signal and intensity information [2]. MTF defines how the image can be resolved by the system.

$MTF_{sys}$ is the parameter used for the design and characterization of a system, which accounts for the diameter of a scattering circle (or the Airy disk diameter $A_{Airy}$), detector dimensions, electronic system, display, *etc.* and is characterized by the product of optics $MTF_{opt}$, electronics $MTF_{el}$, detector $MTF_{det}$, *etc.* Here, it is accepted that MTF is mainly determined to be the product of $MTF_{opt}$ and $MTF_{det}$, as it is frequently taken for IR systems (see, *e.g.*, [2, 46, 47]). Then, for one-dimensional approach (linear array) along the x-axis for any optical or quasi-optical system, we have

$$MTF_{sys} (\nu_x) = MTF_{opt}(\nu_x) \times MTF_{det}(\nu_x), \tag{3.40}$$

where $\nu_x$ is the spatial frequency (pairs of lines per mm) along the x-axis, $MTF_{opt}(\nu_x)$ is the MTF of an objective, and $MTF_{det}(\nu_x)$ is the spatial MTF of a linear array detector.

The spatial $MTF_{det}$ is the module of the normalized Fourier transform of geometric sizes of the sensitive element. For one-dimensional approach and square-law detectors,

$$MTF_{det}(\nu_x) = \frac{\sin(\pi \times V_d \times \nu_x)}{\pi \times V_d \times \nu_x}, \tag{3.41}$$

where $V_d$ is the detector period (detector pitch).

$MTF_{det}$ has at first a zero value at $V_d \times \nu_x = 1$. At larger spatial frequencies, $MTF_{det}$ is negative, which means that the contrast reverses. In a design of the vision system, the detector size should satisfy $V_d \leq 1/\nu_x(max)$, where $\nu_x(max)$ is the maximum space frequency of a formed image in a detector plane along the x-axis.

$MTF_{opt}$ is dependent on the spatial cut-off frequency, which is defined by $\nu_0 = (\lambda F/\#)^{-1}$, pairs of lines per mm, for a corrected optics. For square pixels, the detector $MTF_{det}$ is the periodic function of the detector instantaneous FOV and the spatial frequency.

$MTF_{opt}$ for circular aperture with the diffraction-limited optics [2] can be represented as

$$\mathrm{MTF}_{\mathrm{opt}}(\nu_x) = \begin{cases} \dfrac{2}{\pi} \cdot (\arccos x - x\sqrt{1-x^2}), & at \quad 0 \le x \le 1, \\ 0, & at \quad x > 1 \end{cases} \qquad (3.42)$$

where $x = \nu_x/\nu_c$, and $\nu_c = (\lambda F/\#)^{-1}$ is the cut-off spatial frequency. The image quality is mainly a function of $\lambda F/\# \, / \, V_d$.

The ideal $\mathrm{MTF}_{\mathrm{det}}$ is dependent on the detector size, while the diffraction limited $\mathrm{MTF}_{\mathrm{opt}}$ is dependent on the F/# and $\lambda$. The relationship between the optics and a detector can be described by the ratio $\lambda F/\# \, / \, V_d$. This figure of merit can then be used to describe the regions, where a system is described as diffraction-limited (optics limited) and is considered as detector-limited [2, 48].

For the spatial resolution, the ratio of the Airy disk diameter to the detector size is the most significant. This parameter is important for the determination of the detection/identification range and NEDT [49].

The optical system (*e.g.*, IR or THz camera, human eye, or telescope) resolution depends on the diameter of the minimum spot size (diffraction-limited Airy disk diameter $A_{\mathrm{Airy}} = 2.44 \times \lambda F/\#$) at the focal plane and the detector sensitive pixel size $V_d$. The Airy disk diameter increases with F/# and, as a rule, can surpass the pixel size in the optical system. For large detector's sensitive size, the Airy disk diameter can be much less than the detector size.

To describe and to design the IR or THz system operation by spatial resolution and contrast, the $\mathrm{MTF}_{\mathrm{sys}}$ dependences are usually used. They can be presented as the relationships between the optical $\mathrm{MTF}_{\mathrm{opt}}$ and the detector $\mathrm{MTF}_{\mathrm{det}}$, where the ratio $\lambda(F/\#)/V_d$ is involved.

Using the MTF, we can consider several ratios of "objective–detector" relations at frequencies matching: the spatial optical cut-off frequency $\nu_{\mathrm{co}}$, $\nu_{\mathrm{co}} = \dfrac{1}{\lambda F/\#}$; the spatial detector cut-off frequency $1/V_d$ and the Nyquist frequency $1/(2V_d)$. These relationships are presented by the graphs, which are plotted below in Figs. 3.17 and 3.18.

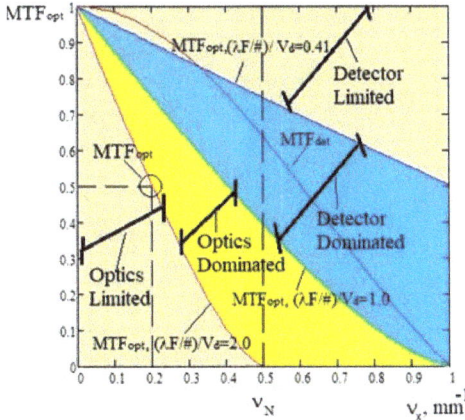

*Fig. 3.17. The dependences of $MTF_{opt}$ for different $\dfrac{\lambda \cdot F/\#}{V_d}$ ratios. At the spatial frequency $v_x = 0.2\ mm^{-1}$, the contrast is $\approx 50\ \%$.*

As $\lambda F/\#\ /\ V_d$ decreases, it is going to a regime of detector-limited operation. The operation in the detector-limited region ($\lambda F/\#\ /\ V_d \ll 1$, the detector pitch is much larger the Airy disk diameter), changing the aperture diameter does almost not influence the spatial resolution.

For large values of $\lambda F/\#\ /\ V_d$, the system becomes optics-limited. In the optics-limited region ($\lambda F/\#\cdot/\ V_d \gg 1$), the spatial resolution is poor (the image is unclear). Changing the detector pitch in this case does not influence the spatial resolution.

The verge of these approximations can be found from the parity of the diameter of the circle of confusion $2r_{Airy}$ of the optical system and the detector pitch $V_d$:

$$2.44 \times \lambda \times F/\# = V_d\ . \tag{3.43}$$

From this ratio, we have

$$\frac{\lambda \times F/\#}{V_d} = 0.41\ . \tag{3.44}$$

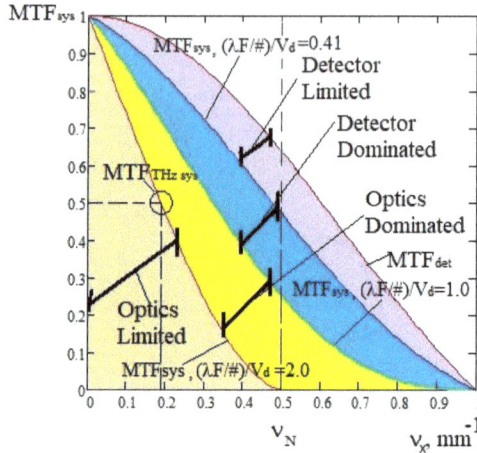

*Fig. 3.18. The dependences of MTF$_{sys}$ for different $\dfrac{\lambda \cdot F/\#}{V_d}$ ratios. A spatial frequency$t$ $v_x$ = 0.2 mm$^{-1}$, the contrast ≈50 % can be provided (see the experimental data on the spatial resolution at Fig. 3.20). (By courtesy from A. Shevchik-Shekera).*

In this case, the image resolution is slightly unclear, as compared to the detector-limited approximation. According to this condition, there is valid the restriction to match the "objective–detector" system to the detector period, matching the Airy disk diameter to the detector pixel pitch, $d_{dif} = 2.44 \times \lambda F/\# = V_d$. In the optics-limited region, the image appears more unclear, as compared to the detector-limited image, and smaller detectors have no influence on system's spatial resolution.

Nowadays, the trends in technical vision systems are directed to decrease the detector pitch as it allows one to increase the number of detectors in an array of the same area. The pixel pitch reduction leads to a more compact system (for a given field of view) due to the array size reduction. The focal length will decrease proportionally to the pixel pitch reduction. For example, going in the IR systems from a 25-μm to 12-μm pitch, the gain is more than a factor of 2 [50]. The weight of the system will follow accordingly. This is a cause for the system cost reduction, because of a reduction of the optics sizes and, as a whole, the technical vision system.

It has a benefit when obtaining the images with the diffraction-limited optics with low f-numbers, *e.g.*, in short- or middle IR wavelength range, where the small size detector pitches now are designed [48, 51, 52].

At system's level, a possibility will consist of keeping the same system size (same focal length and f-number) to improve the detection range, as is shown in Fig. 3.19.

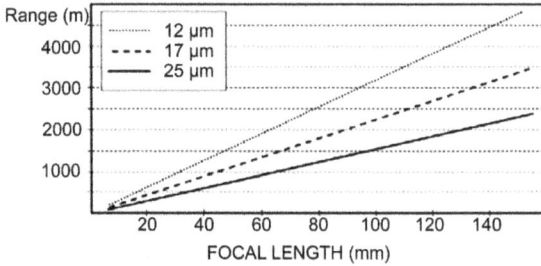

*Figure 3.19. Recognition range performance for ¼ VGA with different pixel pitches (NATO target 2.3 m × 2.3 m) [50].*

For $\mathrm{MTF}_{opt}(v_x)$ from (3.42) with the use of the relation $\lambda F/\# = 0.41 V_d$, we obtain

$$M_{opt}(v_x) = \frac{2}{\pi} \cdot (\arccos(0.41 \cdot V_d \cdot v_x) - 0.41 \cdot V_d \cdot v_x \sqrt{1 - (0.41 \cdot V_d \cdot v_x)^2}.$$

(3.45)

In view of the limit of the optics at zero for $\mathrm{MTF}_{opt}$ [the Nyiquist frequency $v_N = 1/(2V_d)$], Eq. (3.42) yields

$$\frac{2}{\pi} \cdot (\arccos(\lambda \cdot F/\# \cdot \frac{1}{2 \cdot V_d}) - \lambda \cdot F/\# \cdot \frac{1}{2 \cdot V_d} \sqrt{1 - (\lambda \cdot F/\# \cdot \frac{1}{2 \cdot V_d})^2} = 0$$

(3.46)

and $\dfrac{\lambda F/\#}{V_d} = 2$.

This condition shows the limits of the optical system. Using $\lambda F/\# = 2V_d$, we get

$$M_{opt}(v_x) = \frac{2}{\pi} \cdot (\arccos(2 \cdot V_d \cdot v_x) - 2 \cdot V_d \cdot v_x \sqrt{1 - (2 \cdot V_d \cdot v_x)^2}$$

(3.47)

The equality of the optical cut-off frequency $\frac{1}{\lambda F/\#}$ to the cut-off spatial frequency of a detector $(1/V_d)$ gives $\frac{\lambda \cdot F/\#}{V_d} = 1$. Therefore, using the condition $\lambda F/\# = V_d$ and (3.42), we arrive at the relation

$$M_{opt}(v_x) = \frac{2}{\pi} \cdot (\arccos(V_d \cdot v_x) - V_d \cdot v_x \sqrt{1 - (V_d \cdot v_x)^2}$$

(3.48)

The curve shown in Fig. 3.17, when calculating Eq. (3.48), separates the gap between the cut-off curve showing the limit area, where the detector prevails, and the cut-off curve presenting the limit for the optical system to the bands with the dominating optics and the detector.

From estimations of MTF$_{sys}$, one can make conclusions about the quality of a THz vision system, which is qualified by the spatial resolution and contrast. For such THz system, we have

-   The curve describing MTF$_{sys}$ almost coincides with the cut-off curve, which shows restrictions of the optics due to restrictions of the detector operation at frequencies that are higher the Nyquist frequency.
-   As the designed optical system is close to the diffraction-limited one, it is possible to get a higher contrast owing to a radiation wavelength decrease (radiation frequency increase).

There was designed the THz vision system with the parameters $\lambda \approx 2.14$ mm, $V_d = 1$ mm, and $F/\# = 1$. The system allows one [see (3.48)] to get the contrast $\approx 0.5$ at a spatial frequency of 0.2 mm$^{-1}$, which corresponds to the spatial resolution $\Delta \approx 5$ mm, Fig. 3.20.

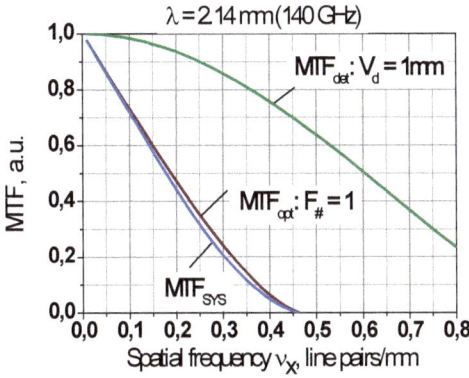

Fig. 3.20. Evaluated dependences for $MTF_{sys}$ of a 140 GHz imaging system at the $V_d = 1$ mm and $F_\# = 1$.

In this case, a spatial frequency of 0.2 mm$^{-1}$ provides a spatial resolution of 5 mm (see Fig. 3.21.c) and allows one to resolve different separately situated smaller objects in imageries, *e.g.*, a clip fabricated from a wire 0.8 mm in diameter.

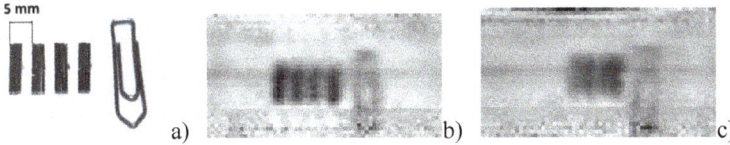

Fig. 3.21. 5 mm pattern period (spatial frequency $v_x = 0.2$ mm$^{-1}$), (a) the imaging of the pattern and clip in the visible, (b) 5 mm pattern period ($v_x = 0.25$ mm$^{-1}$), (c) 4 mm pattern period ($v_x = 0.2$ mm$^{-1}$). $v = 140$ GHz. Imaging of the pattern and clip through 5 mm rubber layer. It was used the 40 element FET linear array, the rate 200 mm/s, the scan time $t = 1$ s). (By courtesy of A. Golenkov).

The 4-mm pattern period practically cannot be resolved (Fig. 3.21.c) in this designed vision system, which is operating at the radiation frequency $v = 140$ GHz with $V_d = 1$ mm and $F_\# = 1$. However, the image of a clip is clearly seen.

## 3.4   Summary

The figures of merit of main detectors and arrays have been briefly discussed. There exist now a large variety of traditional cooled and uncooled THz and IR detectors and arrays, as well as new propositions based on novel materials and structures. They will be considered in Ch. 5. It can be a rather difficult task to measure and to compare the performance of THz and IR detectors and arrays. The cause for this is that one should consider a variety of parameters, which depend on different conditions such as the power level of radiation, detector output level, operational temperature, background conditions, detector dimensions, electronics, *etc.*, which should be kept under control.

For the last two decades, the THz and IR technologies were mainly the domain of space astronomy, security, and surveillance technologies. During that time period, they have shifted to an increasing number of everyday-live new applications (*e.g.*, biomedicine, drug control, material science, food control, art analysis, *etc.*). As concerning the sensors in the recent two-three decades, the thermal detectors both deeply cooled and superconducting ones have realized many significant advances and applications reaching a background limiting performance even under very low background temperature conditions, as well as the uncooled ones used, *e.g.*, in microbolometer arrays and cameras for the thermovision, environmental pollution, surveillance and reconnaissance, security imaging.

Large-format multispectral THz and IR detectors and arrays for the real-time imaging and spectroscopy will play an important role in many applications–astrophysics, security, spectroscopy, communications biomedicine, non-destructive testing, drug and food control, *etc.* To be effectively used in these applications, the figures of merit of detectors and arrays should be applied, and the conditions of their usage should be carefully controlled.

Much longer THz wavelengths in the far-field approximation severely degrade the resolution of images compared with imaging at visible or IR wavelength ranges. To get a higher contrast and resolution in the THz band is possible owing to a radiation wavelength decrease or when restoring the THz imaging by the deconvolution technique [53, 54]. Filters in time and frequency domains can be used to filter out noise, low-frequency spectrum, and diffraction distortions. By application of this procedure one can achieve an enhancement on the quality and resolution of their THz images. E.g., the resolution 0.32 times the physical size of the focused beam was gained [54] (periodical stripes with a period of 0.8 mm are resolved at radiation frequency $\nu$ = 300 GHz (wavelength $\lambda$ = 1 mm)).

**References to Ch. 3.**

[1] Information on https://en.wikipedia.org/wiki/Effective_temperature.

[2] R.D. Hudson, Infrared System Engineering, Wiley-Interscience, New York, 1969.

[3] G.C. Holst, Electro-Optical Imaging System Performance, SPIE Optical Eng. Press, Bellingham, 2003.

[4] H. Kangro, Early History of Planck's Radiation Law, Taylor & Francis, New York, 1976.

[5] P.L. Richards, Bolometers for infrared and millimeter waves, J. Appl. Phys. 76 (1994) 1–24. https://doi.org/10.1063/1.357128.

[6] J.M. Lamarre, Photon noise in photometric instruments at far-infrared and submillimeter wavelengths, Appl. Opt. 25 (1986) 870–876. https://doi.org/10.1364/AO.25.000870

[7] J.M. Lamarre, F.X. Desert, T. Kirchner, Background limited infrared and submillimeter instruments, Space Sci. Rev. 74 (1995) 27–36. https://doi.org/10.1007/978-94-011-0363-3_4

[8] A. Rogalski, F. Sizov, THz detectors and focal plane arrays, Opto-Electr, Rev. 19 (2011) 346–404. (Open Access). https://doi.org/10.2478/s11772-011-0033-3

[9] B.S. Karasik, A.V. Sergeev, Daniel E. Prober, Nanobolometers for THz photon detection, IEEE Transactions on THz Science and Technology, 1 (2011) 97–111. https://doi.org/10.1109/TTHZ.2011.2159560

[10] D. Farrah, K.E. Smith, D. Ardila, Ch.M. Bradford, *et al.*, Review: Far-infrared instrumentation and technological development for the next decade, J. Astronom. Telescopes, Instrum. Systems. 5 (2019) 020901. (Open Access). https://doi.org/10.1117/1.JATIS.5.2.020901

[11] G.H. Rieke, Detection of Light. From the Ultraviolet to the Submillimeter, Cambridge University Press, Cambridge, 2003. https://doi.org/10.1017/CBO9780511606496

[12] Y.-S. Lee, Principles of Terahertz Science and Technology, Springer, New York, 2009.

[13] A. Rogalski, Infrared detectors, Boca Raton: CRC Press, 2011.

[14] R.J. Keyes, Optical and Infrared Detectors, Springer, Berlin, 2009.

[15] A. Rostami, H. Rasooli, H. Baghban, Terahertz Technology: Fundamentals and Applications, Spinger, Berlin, 2011. https://doi.org/10.1007/978-3-642-15793-6

[16] C. O'Sullivan, J.A. Murphy, Field Guide to Terahertz Sources, Detectors, and Optics, SPIE Press, Bellingham, 2012. https://doi.org/10.1117/3.952851

[17] K.-E. Peiponen, J.A. Zeitler, M. Kuwata-Gonokami (Eds.), Terahertz Spectroscopy and Imaging, Springer, Heidelberg–New York–Dordrecht–London, 2013. https://doi.org/10.1007/978-3-642-29564-5

[18] F. Sizov, Terahertz radiation detectors: State-of-the art, Semicond. Sci. Technol. 33 (2018) 123001. https://doi.org/10.1088/1361-6641/aae473

[19] D.L. Woolard, W.L. Loerop, M.S. Shur, Terahertz Sensing Technology, Vol. 2: Emerging Scientific Applications and Novel Device Concepts, World Scientific, Singapore, 2004. https://doi.org/10.1142/5396

[20] G. Chattopadhyay, Submillimeter-wave coherent and incoherent sensors for space applications, in: S.C. Mukhopadhyay, R.Y.M. Huang (Eds.), Sensors, Springer, Berlin–Heidelberg, 2008, pp. 387–414. https://doi.org/10.1007/978-3-540-69033-7_19

[21] M.C.E. Huber, A. Pauluhn, J.L. Culhane, J.G. Timothy, K. Wilhelm, A. Zehnder, Observing Photons in Space. A Guide to Experimental Space Astronomy, Springer, New York, 2013. https://doi.org/10.1007/978-1-4614-7804-1

[22] U.U. Graf, C.E. Honingh, K. Jacobs, J. Stutzki, Terahertz heterodyne array receivers for astronomy, J. Infrared Millimeter, and Terahertz Waves, 36 (2015) 896–921. https://doi.org/10.1007/s10762-015-0171-7

[23] A.G. Golenkov, F.F. Sizov, Z.F. Tsybrii, L.A. Darchuk, Spectral sensitivity dependences of backside illuminated planar MCT photodiodes, Infr. Phys. Technol. 47 (2006) 213–218. https://doi.org/10.1016/j.infrared.2004.12.001

[24] F. Sizov, A. Rogalski, Semiconductor superlattices and quantum wells for infrared optoelectronics, Progr. Quant. Electr. 17 (1993) 93–164. https://doi.org/10.1016/0079-6727(93)90005-T

[25] D. Figer, J. Lee, E. Corrales, J. Getty, L. Mears, HgCdTe detectors grown on silicon substrates for observational, Proc. SPIE. 10709 (2018) 1070926.

[26] W. Raab, Semiconductors for low energies: incoherent infrared/sub-millimetre detectors, in: M.C.E. Huber, A. Pauluhn, J.L. Culhane, J.G. Timothy, K. Wilhelm, A. Zehnder (Eds.), Observing Photons in Space. A Guide to Experimental Space Astronomy, Springer, New York, 2013, pp. 525–542. (Permission Springer Nature). https://doi.org/10.1007/978-1-4614-7804-1_30

[27] T. Shiraishi, H. Yagi, K. Endo, M. Kimata, *et al.*, PtSi FPA with improved CSD operation, Proc. SPIE. 2744 (1996) 33–43. https://doi.org/10.1117/12.243489

[28] M. Kimata, N. Tsubouchi, Schottky barrier photoemissive detectors, in: A. Rogalski (Ed.), Infrared Photon Detectors, SPIE Optical Eng. Press, Bellingham, 1995, pp. 299–349.

[29] M.N. Abedin, T.F. Refaat, J. Zawodny, S.P. Sandford, *et al.*, Multicolor focal plane

array detector technology: A Review, Proc. SPIE. 5152 (2003) 279–299.
https://doi.org/10.1117/12.505887

[30] E.M. Conwell, High Field Transport in Semiconductors, Academic Press, New York, 1967.

[31] P.R. Norton, Photodetectors, in: Handbook of Optics, 3d Edition, Volume II: Design, Fabrication and Testing, Sources and Detectors, Radiometry and Photometry, part 5, Ch. 24, M. Bass (Ed.), McGraw Hill, New York, 2010, pp. 15.3–15.100.

[32] A.M. Cowley, H.O. Sorensen, Quantitative comparison of solid state microwave devices, IEEE Trans. Microwave Theory Techn. MTT-14 (1966) 588–602.
https://doi.org/10.1109/TMTT.1966.1126337

[33] M. Sakhno, A. Golenkov, F. Sizov, Uncooled detector challenges: Millimeter-wave and terahertz long channel field effect transistor and Schottky barrier diode detectors, J. Appl. Phys. 114 (2013) 164503. https://doi.org/10.1063/1.4826364

[34] Sh.R. Kasjoo, A.M. Song, Terahertz detection using nanorectifiers, IEEE Electr. Device Lett. 34 (2013) 1554–1556. https://doi.org/10.1109/LED.2013.2285162

[35] W.L. Wolfe, G.J. Zissis (Eds.), The Infrared Handbook, Office of Naval Research, Washington, 1989.

[36] J.D. Vincent, Fundamentals of Infrared Detector Operation and Testing, Wiley, New York, 1990.

[37] M. Schlessinger, I.J. Spiro, Infrared Technology Fundamentals, second ed., Marcel Dekker, New York, 1995.

[38] G.C. Holst, Testing and Evaluation of Infrared Imaging Systems, SPIE Press, Bellingham, 1998.

[39] J. Wei, D. Olaya, B.S. Karasik, S.V. Pereverzev, *et al.*, Ultrasensitive hot-electron nanobolometers for terahertz astrophysics, Nature Nanotechn. 3 (2008) 496–500.
https://doi.org/10.1038/nnano.2008.173

[40] N. Kopeika, A System Engineering Approach to Imaging, SPIE Optical Eng. Press, Bellingham, 1998. https://doi.org/10.1117/3.2265069

[41] A. Rogalski, Next decade in infrared detectors, Proc. SPIE. 10433 (2017) 104330L-1, in: Electro-Optical and Infrared Systems: Technology and Applications XIV, Eds. D.A. Huckridge, R. Ebert, H. Bursing. https://doi.org/10.1117/12.2300779

[42] P.W. Kruse, Uncooled Thermal Imaging, SPIE Press, Bellingham, 2001.

[43] F. Sizov, V. Reva, A. Golenkov, V. Zabudsky. Uncooled detector challenges for THz/sub-THz arrays imaging. J. Infrared, Millimeter, and Terahertz Waves, 32 (2011) 1192–1206. https://doi.org/10.1007/s10762-011-9789-2

[44] J.A. Cox, R. Higashi, F. Nusseibeh, C. Zins, MEMS-based uncooled THz detectors, Proc. SPIE, 8031 (2011) 8031OD. https://doi.org/10.1117/12.884066

[45] E. Novoselov, MgB$_2$ hot-electron bolometer mixers for sub-mm wave astronomy (Thesis), Chalmers Reproservice, Göteborg, 2017. https://doi.org/10.1117/12.2233402

[46] G.C. Holst, Imaging system performance based upon Fλ/d, Opt. Eng. 46 (2007) 103204. https://doi.org/10.1117/1.2790066

[47] D. Lohrmann, R. Littleton, C, Reese, D, Murphy, J. Vizgaitis, Uncooled long-wave infrared small pixel focal plane array and system challenges, Opt. Eng. 52 (2013) 061305. https://doi.org/10.1117/1.OE.52.6.061305

[48] G.C. Holst, H. Driggers, Small detector in infrared system design, Opt. Eng. 51 (2012) 096401. https://doi.org/10.1117/1.OE.51.9.096401

[49] M.A. Kinch, State-of-the-Art Infrared Detector Technology, SPIE Press, Bellingham, 2014. https://doi.org/10.1117/3.1002766

[50] L. Tissot, P. Robert, A. Durand, S. Tinnes *et al.*, Status of uncooled infrared detector technology at ULIS, France, Defence Sci. J. 63 (2013) 545–549. (Open Access). https://doi.org/10.14429/dsj.63.5753

[51] W.E. Tennanat, D.J. Gulbransen, A. Roll, M. Carmody, *et al.*, Small-pitch HgCdTe photodetectors, J. Electron. Mater. 43 (2014) 3041–3046. https://doi.org/10.1007/s11664-014-3192-4

[52] R.K. McEven, D. Jeckells, S. Bains, H. Weller, Developments in reduced pixel geometries with MOCVD grown MCT arrays, Proc. SPIE. 9451 (2015) 94512D. https://doi.org/10.1117/12.2176546

[53] K. Ahi, A method and system for enhancing the resolution of terahertz imaging, Measurement. 138 (2019) 614-619, doi.org/10.1016/j.measurement.2018.06.044.

[54] W. Ning, F, Qi, Z. Liu, Y. Wang, et.al., Resolution enhancement in terahertz imaging via deconvolution, IEEE Access. 7 (2019) 65116-65121 doi:10.1109/ACCESS.2019.2917531.

Materials Research Forum LLC
https://doi.org/10.21741/9781644900758

# Chapter 4. IR and THz detection principles - direct and coherent detections

This chapter includes the comparison of direct detection and coherent (heterodyne) detection schemes, which can be used in passive or active imaging systems. As usual, the direct (incoherent) detectors are operating in the mode of a linear response to the radiation power (square-law detectors, proportional to the square of the electric field strength in an EM wave). The coherent detectors respond to the product of a local oscillator and the signal input. Here, an attempt is undertaken to select detector devices settled by a brief comparison of detector types.

## 4.1 Information capacity

Today, the IR and THz technologies find broad use in the imaging, information and telecommunication technologies, *etc.* employing the photons flux coming onto a detector. Each photon is a carrier of information. The qualitative measure of any vision system is the ratio of the amount of information perceived by the vision system to the amount of information contained in the radiation flux entering the imaging or communication system.

In the case of only noise connected with the photon flux fluctuations dispersion $\langle \Delta N_{ph} \rangle \sim \langle (N_{ph}) \rangle^{1/2}$ (Poisson statistics ($h\nu \gg k_B T$), where $N_{ph}$ is the number of photons in photon's flux), the system information capacity $C_M$ (with M sensitive elements in the array or M-number of decomposition elements in the image) is defined [1] by

$$C_M = \frac{M}{8} \log_2 \left[ \frac{1}{k} \sin\left(\frac{\theta}{2}\right) \sqrt{\eta A_d \tau_{acc} N_{ph}} \right], \text{ bit.} \tag{4.1}$$

This expression determines the upper limit of the information capacity of a vision system in one spectral region in the case of only the noise connected with the photon flux fluctuations that, as a rule, settle the upper limit performance of most quantum and thermal cooled detectors. Here, $k = U_{thr} / \sqrt{\langle U_{noise}^2 \rangle}$ is the signal-to-noise threshold ratio, $A_d$ is the detector area, $\eta$ is the detector quantum (coupling) efficiency, $\tau_{acc}$ is the accumulation time, $N_{ph}$ is the number of photons falling down on the detector, and $\theta$ is the plane angle of view. It should be $k > 2$, as, at $k = U_{thr} / \sqrt{\langle U_{noise}^2 \rangle} = 1$, the probability of a false signal is equal to $P_{fs} = 0.159$, *i.e.*, it is a relatively large quantity. At k = 2, the probability of a false signal is $P_{fs} = 0.023$ and rapidly decreases, as k increases [2, 3]. Here, $U_{thr}$ is the threshold signal, and $U_{noise}$ is the noise level. In the case of a multispectral system, its information capacity is proportional to the number of spectral ranges [4].

From Eq. (4.1), it is seen that the number of detectors M or the number of decomposition elements M in the image is a key parameter that determines the information capabilities of the system, as the other parameters are under the logarithm.

It is to be noted that the quantum efficiency η of a detector in the array plays a not very important role, as it is under the square root and logarithm. This is a cause for why the low quantum efficiencies of platinum silicide (η ~ 0.001) and QW or superlattice (SL) (η ~ 0.1...0.2) detectors will give similar results for the information capacity at longer integration times, as compared with arrays composed with detectors with high quantum efficiencies. For most of the THz and IR detectors, η ~ 0.1...0.8, and the information capacities differ by at most 10 % and are close to those for the systems operating in the visible spectral range [1].

Since the information capacity does not depend significantly on the number of photons falling onto the detector, because it is under the square root and logarithm, the dependence of $C_M$ on the spectral range is not very determinative. However, for IR systems on the base of platinum silicide detector arrays (η ~ 0.001), this discrepancy is more pronounced. The same is true for THz vision systems, as the number of photons, even for many active vision systems, is much lower in this spectral range as compared to the IR range (*e.g.*, at a temperature of objects under observation). Therefore, longer accumulation times are needed.

The advances in IR sensor technologies have enabled increased array sizes and decreased pixel sizes to get megapixel arrays at the same time keeping a good resolution [5–7] (see also Ch. 3). Figure 4.1 shows the timeline for the HgCdTe FPA developed at the Raytheon Vision Systems (RVS, formerly Santa Barbara Research Center, SBRC).

Infrared systems require a high resolution for many applications such as the wide-area surveillance, astronomy, medicine, *etc.* to capture more information. Array sizes increase with an exponential rate, following the growth path by Moore's law with the number of pixels doubling every 18 months (Fig. 4.1).

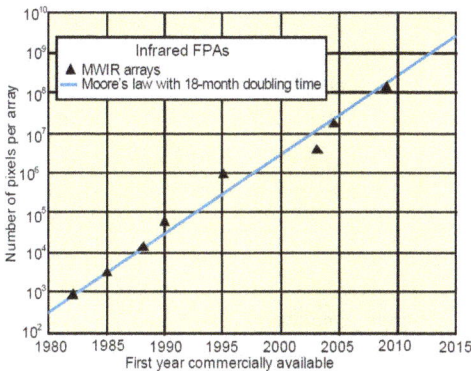

*Figure 4.1. The number of pixels per mid-wavelength ($\lambda \approx 3...5 \ \mu m$) infrared array grows exponentially, in accordance with Moore's law for 35 years with a doubling time of approximately 18 months [8]. (Reproduce with permission from SPIE).*

Today, the number of pixels in IR arrays exceeds $10^8$ (see also Fig. 4.2) and is close to the number of sensitive elements in the human eye ($\sim 2\times10^8$) which is a high-level optical system.

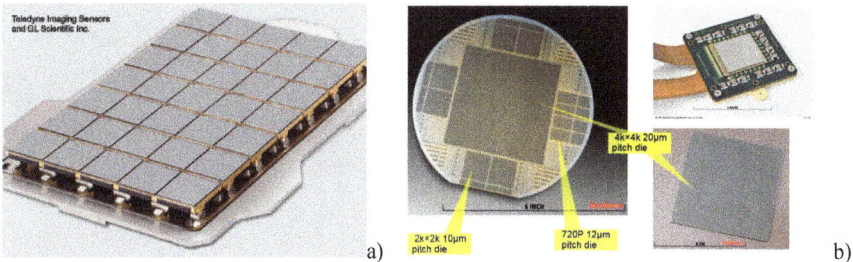

*Figure 4.2. a) Large IR mosaic prototype array with 35 H2RG arrays. The array has a total of nearly 1.47 $10^8$ pixels. Each of the H2RG arrays has 2048 × 2048 pixels [9]; b) Attainable now are single IR arrays 4K × 4K [10] (Reproduced with permission from SPIE).*

The information capacity $C_M$ of the matrix arrays with the readout devices to a great degree is limited by the capacity of accumulative cells, *e.g.*, for silicon readouts, which can accumulate only n* photo-excited electrons [11] a counterweight to possible larger numbers of $n_e = \eta \times A_d \times \tau \times N_{ph} \times \sin^2(\theta/2)$ electrons ($n_e > n^*$) detected by a sensitive element ($n^* \sim 10^6 \ldots 10^7$). These $n_e$ electrons are born by the photons falling down onto the detector. Therefore, in Eq. (1), one should use $n^*$ instead of $\eta \times A_d \times \tau \times N_{ph} \cdot \sin^2(\theta/2)$. This can deteriorate the information capacity $C_M$.

When comparing the detector properties in the visible or IR and THz spectral regions, one can conclude that THz detectors are typically different by operational principles. For example, one of these differences, as a rule, lies in the sizes of a detector. For example, the size of IR detectors (the size of sensitive elements in arrays) is about the wavelength (diffraction limited case connected with resolution $\approx 1.22 \times (F/\#)$, where F/# is the optics f-number, in the far-field approximation). Whereas the THz ones, though having similar proportions, are compared to the wavelength, but only with regard to the antenna dimensions, which are about the wavelength length.

Therefore, there exists a possibility to form large IR matrix arrays with reasonable dimensions and a great number of sensitive elements in them ($\geq 10^7$ pixels; in human eye, there are $\sim 2 \times 10^8$ receptors). In the THz single detectors and arrays, the pitch is also d $\sim \lambda$; however, $\lambda \sim 0.3 \ldots 1$ mm. The difference of the dimensions of sensitive elements in the IR and THz arrays is a cause for the much lesser (by several orders) number of pixels in the THz arrays (Fig. 4.3). This results in different constraints, while designing the imaging systems [12].

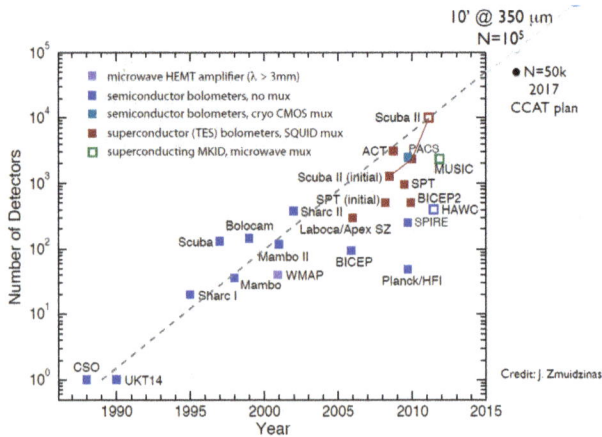

*Figure 4.3. Growth of incoherent superconducting detector number sizes with time for different missions (after [13]).*

Another difference between the IR and THz detectors is related, as a rule, to different physics of input radiation entering a sensitive element. In the IR detectors, it is a direct absorption of radiation by a sensitive element, while the radiation power in THz detectors is typically introduced in a sensitive element by an antenna.

As well, for low-frequency THz range (Rayleigh–Jeans approximation), one cannot consider the contributions from different sources to the detector at the same wavelength as statistically independent, and one cannot compute the noise from two sources and add their power [14].

Among the THz imaging and spectroscopy systems, there are frequently used systems based on the THz pulse imaging – TDS and imaging systems, and also CW photomixer systems (see, *e.g.*, [15, 16]), which as a rule, are not used in the infrared range.

Because the THz detectors are used frequently with antennas, their sensitivity R, which is associated with the input power, is also changing with an antenna effective area as $R \sim v^{-2}$ or a little bit steeply (*e.g.*, in SBD or FET THz detectors, $R \sim v^{-2}-v^{-4}$ in view of the antenna type, frequency range, and the measurement procedure [17–20]). The example of such dependence is shown in Fig. 4.4 for SBDs. Such steep radiation frequency dependences are also observed for all semiconductor-based sources (see Ch. 6).

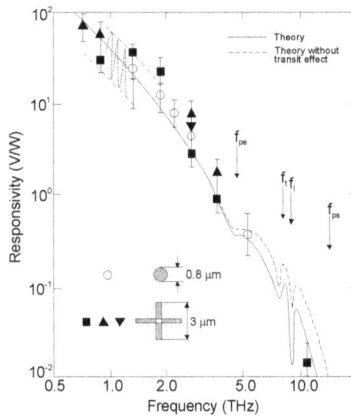

*Figure 4.4. a) Dependence of the SBD voltage sensitivity on the radiation frequency [21].*
*(Reproduced with permission from SPIE)*

## 4.2   General classification of IR and THz detectors

The tie-line of the IR and THz detectors of their main idea is the acquisition of information from the objects under observation, as each photon in any spectral range is the carrier of information. In detectors, the transformation of the absorbed electromagnetic radiation, in most cases, ends in the appearance or changing of electrical signals. The absorbed radiation heats the electron or lattice subsystems or changes the electron energy distribution, thus by modifying the motion of charged carriers. Such alterations are fixed by measuring the changes of detector physical parameters.

Most of the radiation detectors can be divided into three categories: photon (or quantum), thermal, and rectification detectors (see Ch. 3). Some fast detectors can be used as mixers in coherent systems. Different types of IR or THz detectors are described in numerous books and reviews (see, *e.g.*, [22–29]).

In dependence on the circuit type, the registration of the detectors can be divided into direct (incoherent) and coherent. Many of these detectors can operate both under room-temperature conditions and at low temperatures (T ≤ 4.2 K). However, some of them can operate only under room- or low-temperature conditions. That is why they can be divided also into uncooled and cooled detectors.

## 4.3   THz detector challenges - photon, thermal (power), or rectification detectors?

The critical differences between the detection at the THz frequencies and in the IR range lie in the low photon energies (at $\lambda \approx 300$ μm, $\nu \approx 1$ THz, hv $\approx 4$ meV, as compared to the room-temperature thermal energy 26 meV). The minimum spot size (the "waist" or Airy disk) for THz waves is large, which determines the relatively poor spatial resolution of THz vision systems.

The long wavelength limits of currently available IR and THz photon detectors can include the narrow-gap solid solutions ($Hg_{1-x}Cd_xTe$) with tunable band-gap. In the photon detectors, the doped semiconductors with small excitations energies (Si:A, Ge:B), and QWs (SLs) are also used. Here, A and B are dopants. The operation of the IR and THz detectors on their base depends mainly on the available operation temperatures. The relevant noises are connected with the lifetimes of minority carriers (in quantum detectors with $Hg_{1-x}Cd_xTe$, InSb, InAs, PbSe, *etc.*) and majority carriers (extrinsic dopants, SLs, SBDs). The energy gap (or sub-bands in SLs) and the position of dopants in the semiconductor band gap define the thermo-generation rate.

For a photon detector to be effective, the photo-generation rate of carriers, $g$ should be considerably larger than the thermo-generation rate $g_{th}$: $g = \eta \alpha N_{ph} \gg g_{th} = n_{th}/\tau$. The latter one defines the dark current. Here, $\alpha$ is the absorption coefficient, $\eta$ is the quantum efficiency, $N_{ph}$ is the number of photons falling onto a detector, $n_{th}$ is the number of thermo-generated carriers, and $\tau$ is their recombination time. For a typical photon [*e.g.*, narrow-gap HgCdTe, extrinsic Ge, Si [30] or QW(SL) or QD THz] detectors, $g_{th} \gg g$ at $\nu \leq 1.5$ THz and $T \geq 4$ K [31]. That is why these types of THz detectors can operate properly only in high-frequency THz range ($\nu > 1.5$ THz) at $T \geq 4$ K [32–35].

Operation of a photon THz detector is questionable in this temperature range. One specific exception is a photoconductor PbSnTe:In, in which the lifetime $\tau$ is extremely long at $T < 20$ K due to the presence of the energy barrier between the impurity and conductive states [36]).

Therefore, as a rule, only low-temperature superconductor bolometers, Cooper-pair breaking detectors with NEP $\sim 10^{-19}$–$10^{-11}$ W/Hz$^{1/2}$, or rectification ones (*e.g.*, SBD, FET or heterojunction detectors) seem to be taken as THz sensitive devices at $\nu < 1.5$ THz. In the THz spectral range, when choosing direct uncooled or slightly cooled detectors for vision purposes as arrays, the thermal (power) or rectification detectors can be taken.

## 4.4 Incoherent and coherent detection

All radiation detection systems depending on their applications and on the registration circuit type of electromagnetic radiation detection can be roughly divided into two groups [25, 37–40]:

- Incoherent detection systems (with direct detectors), which allow only the signal amplitude detection and are, as a rule, broadband detection systems;
- Coherent detection systems, which allow detecting not only the amplitude of a signal, but also its phase. The coherent detection employs heterodyne techniques.

The choice between the coherent or incoherent detection for a given application at particular THz or IR spectral range is not always obvious, as it depends on the THz or IR spectral band, application purposes, detectors sensitivity, technological possibilities, spectral resolution needs, *etc.*

In the coherent systems, the heterodyne concept of detection is used in which a high-frequency radiation signal ($\nu \sim 10^{11} \ldots 10^{14}$ Hz) is transformed to a signal with much less frequency (intermediate frequency $\nu \sim 1 \ldots 10$ GHz). Further, it can be processed by, *e.g.*, InP-based monolithic microwave integrated circuits (MMICs) low-noise amplifiers and electronics. These systems are narrow-band (selective) systems.

Though these systems are different in operation principles, it is important to make a comparison between the coherent and incoherent systems. However, the caution is warranted when making such a comparison. Direct-detection systems are primarily designed to measure the broadband spectra, and heterodyne systems are directed to the measurements of narrow-band vibration and rotation atomic and molecular fine line structures.

The simplified representation of a direct-detection system is shown in Fig. 4.5. It can be applied in a case where radiation fluxes are not weak. When the levels of signals are low, *e.g.*, in the sub-THz spectral range, the other more complicated facilities should be applied (see, *e.g.*, [41].

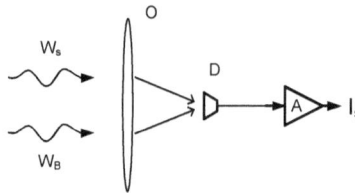

*Figure 4.5. Schematic representation of a direct-detection system, $W_s$ is the signal power, $W_B$ is the background radiation power, O is the focusing optics, D is the detector, A is the amplifier, $I_s$ is the output signal.*

In the coherent (heterodyne) receiver the first operation to be performed is the bi-directional communication over a single path ($W_s$ and $W_O$) in a diplexer. Then the frequency down-conversion or heterodyning in a mixer from the sub-millimeter-wave frequencies to much lower GHz frequencies (intermediate frequency $v_{IF}$) should be carried out. This is accomplished with the use of a mixer and a local oscillator (LO). The mixer noise sets system's responsivity and should be as low as possible. Figure 4.6 schematically shows the typical simplified block diagram of a sub-THz heterodyne receiver.

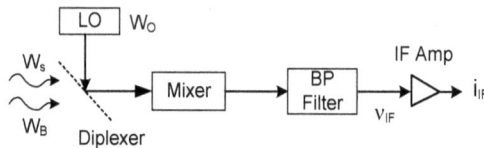

*Figure 4.6. Simplified schematic representation of a heterodyne receiver. $W_s$ is the signal power with the frequency $v_s$, $W_B$ is the background radiation power, $W_O$ is the local oscillator radiation power with the frequency $v_{LO}$, BP Filter is a band-pass filter, and $v_{IF}$ is the intermediate frequency, IF Amp is an intermediate frequency amplifier, and $i_{IF}$ is the output intermediate frequency signal.*

The IF spectrum of a heterodyne receiver is an exact replica of the original radio-frequency (RF), THz or IR spectrum with both the phase and amplitude information. This allows the heterodyne systems to have a high spectral resolution for the study of, *e.g.*, the spectral line spectra and Doppler shift, since $\nu_{IF} \ll \nu_S$. Comparing with direct-detection systems, the coherent ones can be more complicated, but the principle of radiation detection remains the same.

In the heterodyne-detection systems in addition to the signal power $W_s$ and the background radiant one $W_B$, the radiant power $W_{LO}$ from a local oscillator (*e.g.*, laser or any kind of other narrow-band source) is added (Fig. 4.6). LO is required to drive the mixing process. The basic element of a heterodyne receiver is a mixer, which is needed to align $W_s$ and $W_{LO}$ for generating a copy of a signal at the intermediate frequency $\nu_{IF} = |\nu_s - \nu_{LO}|$. It is a key component in the nonlinear mixing element (detector), where the signal and LO radiant powers are coupled using some kind of a diplexer (or a beam-splitter). The latter one spatially combines a signal beam and a LO-beam.

A mixer is the most important component of the heterodyne receiver input stage that is responsible for its responsivity. Its conversion loss contributes to the noise temperature of the heterodyne receiver and a consequent intermediate-frequency amplifier. The signal power losses occur in a diplexer and a detector, but it is a mixer and its distributing circuits, which contribute the most noise to a heterodyne receiver [42]. For a mixer to be efficiently used in the THz or IR arrays, its choice should be dictated by the available LO power in this spectral range, mixer operating temperature, and the sensitivity of a detector.

### 4.4.1 Incoherent (direct) detectors

To evaluate the potential characteristics of detectors, we need to know the characteristics of radiation fluxes. The power flow in a region containing an alternating electromagnetic field such as a plane linear polarized wave that is perpendicular to the wave propagation per unit time (instantaneous power density) per unit area is defined by the Poynting vector: $\mathbf{S} = [\mathbf{E} \times \mathbf{H}]$, $W/m^2$. Here, the magnitudes $\mathbf{E}$ and $\mathbf{H}$ are the values of the sinusoids, and the phase angle between them assumed to be zero. Here $\mathbf{E}$ is the electric field vector and $\mathbf{H}$ is the magnetic field vector.

The time-averaged power flow, of the Poynting vector over a full cycle for the real part of $\mathbf{S}$ than reads

$$\mathbf{S} = \frac{1}{2} \cdot \text{Re}[\mathbf{E} \times \mathbf{H}^*] = \frac{|\mathbf{E}|^2}{2 \cdot \eta_0},$$

(4.2)

Materials Research Forum LLC
https://doi.org/10.21741/9781644900758

the asterisk means the time-domain complex conjugate.

The direction of the flow will be normal to the plane in which the **E** and **H** vectors lie. Here, $\eta_0 = (\mu_0/\varepsilon_0)^{1/2} \approx 377\ \Omega$ is the intrinsic impedance of the free space. In dielectric media, $\delta = 377/n$, where $n = (\mu_r \cdot \varepsilon_r)^{1/2}$ is the refractive index, $\mu_r$ and $\varepsilon_r$ are the magnetic and dielectric constants of the media, respectively. In the free space, $\mu_0 = 4\pi \cdot 10^{-7}$ H/m is the permeability (magnetic constant), and $\varepsilon_0 = (4\pi \cdot 9 \cdot 10^9)^{-1}$ F/m is the vacuum dielectric constant, respectively. The speed of light in the free space $c = (\mu_0 \cdot \varepsilon_0)^{-1/2} = 3 \cdot 10^8$ m/s.

The time average of the power density of a signal $W_s$ that falls on the detector with area $A_d$ is

$$W_s = \frac{|E_s|^2}{2 \cdot \eta_0} \times A_d \text{, W.} \tag{4.3}$$

For the plane waves with frequency $\omega = 2\pi\nu$,

$$E(r,t) = a(t) \times \text{expi}(\omega t + \varphi - kr),$$
$$H(r,t) = b(t) \times \text{expi}(\omega t + \varphi - kr). \tag{.4.4}$$

Here, $a = a(t)$ and $b = b(t)$ are the amplitudes of electric and magnetic fields, respectively (which can be time-varying), **k** is the propagation vector ($k = 2\pi/\lambda$), **r** is the coordinate, and $\varphi$ is the initial phase of a wave.

Assume that radiation is propagating along the z direction, for the real part of the wave

$$E_x = a \times \cos(\omega_s t + \varphi_s - k_z \times z). \tag{4.5}$$

Then, for the instantaneous signal received by a detector with area $A_d$, the radiometric considerations yield

$$W_s(t) = \frac{|\langle E \rangle|^2}{\eta_0} = [a \times \cos(\omega_s t + \varphi_s - k_z z)]^2 \times \frac{A_d}{\eta_0}. \tag{4.6}$$

The instantaneous signal current in the detector is

$$I_s(t) = R_1 \times G \times W_s(t) =$$

$$R_1 \times G \times a^2 \times \cos^2(\omega_s t + \varphi_s - k_z t) \times \frac{A_d}{\eta_0} = \frac{\eta q}{hv} \times G \times \frac{A_d}{\eta_0} \times a^2 \times \left[ \frac{1 + \cos 2(\omega_s t + \varphi_s - k_z z)}{2} \right], \quad (4.7)$$

where $R_I = \eta q/hv$ is the current responsivity, G is the detector internal signal gain (internal detector gain), and $\eta$ is the detector quantum efficiency. For photodiodes, $G = 1$, but, for photoconductive sensors, it is possible that $G \leq 1$ and $G > 1$.

It is assumed that the output detector signals, both for direct and heterodyne detectors, are proportional to the power density of the incident radiation, which means that their response is proportional to the square of the electric field strength of a wave (square-law detector). This assumption is valid, as a rule, in the region of relatively weak radiation power densities (electric field strength of the electromagnetic wave) $\leq k_B \cdot T/q$ where $k_B$ is the Boltzmann constant, T is the temperature, and q is the electron charge.

Since the frequencies $\omega_s$ in the visible, IR, THz, and mm-wave ranges are of the order of $\sim 10^{11}$ to $10^{15}$ rad/s, they are large to be followed by the currents and voltages of electrical signals at the detector output. In dependence of the visible, IR, THz, or mm-wave spectral ranges of applications, the real detector is a time domain integrator with integration time $\tau \gg 2\pi/\omega_s$. The cosine term in relation (4.7) can be neglected, since it is well beyond the detector electronic bandwidth by frequency. Therefore, the average value of a registered signal is

$$I_s = \frac{\eta q}{2hv} \cdot \frac{GA_d}{\eta_0} \cdot a^2. \quad (4.8)$$

For the average power of a signal, it follows from (4.3) that $W_s = a^2 \cdot (A_d/2\eta_0)$, and the average current of signals

$$I_s = \frac{\eta q}{hv} \cdot G \cdot W_s. \quad (4.9)$$

The detectors following this relation are called direct detectors. An important implication of this relation is that, due to the absence of a term in $I_s(t)$ that includes the carrier frequency, the direct-detection detectors are not capable to detect the frequency modulation or phase modulation signals. This is significant for the active imaging and communications in THz or sub-THz regions of the electromagnetic spectrum.

The amplitude of the wave electric field a = a(t) is dependent in the general case on the time representing the amplitude modulation of signal (4.8). If this coefficient changes proportionally to the modulation power (*e.g.*, chopper), it will be an intensity modulation.

The current of an ideal barrier detector (not a photoconductor) due to the action of the background radiation with power density $W_B$ is defined by the expression like (4.7)

$$I_B(t) = R_I \times G \times W_B = \frac{\eta q}{h\nu} \times G \times W_B$$ (4.10)

where $R_I = \eta q/h\nu = \eta q\lambda/hc = 0.806 \times \eta\lambda$ is the current responsivity, and $\lambda$ is the wavelength in microns.

Assume that the basic noise is a background noise caused by fluctuations of the photon flux from the background radiation (quantum noise resulting from the incident radiant power). This defines the background limited performance regime or photon-limited detection of the non-photoconductive detector noise current (see, *e.g.*, [24, 40, 43, 44])

$$<I_{n,B}^2> = 2q \times <I_B> \times \Delta f = 2q \times R_I \times G \times W_B \times \Delta f = 2q \times \frac{\eta q}{h\nu} \times G \times W_B \times \Delta f$$ (4.11)

where $R_I$ is the detector current responsivity.

Such kind of noise (Schottky shot noise) is present in the potential barrier detectors (in this case, instead of $<I_B>$ from the background arising in non-photoconductive detector, one should use the current through a detector $<I_0>$ arising with the bias applied). This expression will be valid for "white noise" not dependent on the frequency, at which the measurements of noise current take place, as these electronic frequencies $f \le 10^7$ Hz, as a rule, are less than that of the current impulses ($\sim 10^{11}$ Hz) at the random passage of electrons through the barrier or the emission from the cathode.

Assuming that other noises are small, as compared with the background noise that defines the ultimate performance of barrier detectors ($G = 1$), for the average signal-to-noise ratio we get

$$\frac{I_S}{I_{n,B}} = \frac{\eta \times q^{1/2} \times W_s}{2^{1/2} h\nu \times R_I^{1/2} \times W_B^{1/2} \times (\Delta f)^{1/2}},$$

(4.12)

and the signal-to-noise ratio is inversely proportional to the square root of the bandwidth $\Delta f$.

Since the current responsivity of the barrier detector $R_I = \eta q/h\nu$ and the signal-to-noise ratio $I_S/I_n = 1$ for BLIP detection, we have (see, *e.g.*, [40])

$$W_{s,dir}^{min} = \left( \frac{2h\nu}{\eta} \times W_B \times \Delta f \right)^{1/2}, W.$$

(4.13)

It is clear that $W_{s,dir}^{min} \sim \Delta f^{1/2}$. Moreover, the minimum detectable signal power $W_{s,dir}^{min}$ can be weaker than the power from the background $W_B$ itself.

The result can also be interpreted as that indicating the equivalent noise power NEP. While comparing different detectors, it is useful to normalize this expression to the square root of the unit bandwidth (noise equivalent power at the bandwidth $\Delta f = 1$ Hz is NEP $= W_{s,dir}^{min} \times I_n/I_S \times (1/\Delta f)^{1/2}$, W/Hz$^{1/2}$).

It is worth to note that, in the direct detectors, there is no connection between the radiation wave bandwidth at the detector input and the electronic bandwidth at the output of the detector circuit. The former is defined by the spectral transmission of the optics used in a system with the detector. The latter is defined by the time response of a detector, its circuit elements, and the output amplifier.

Under conditions when the shot noise is characterized by the photon flow fluctuations in the signal radiation flux ($W_{min} = W_B$) and at signal-to-noise ratio $I_S/I_n = 1$ for the minimum detectable power of non-photoconductive direct detector, we have

$$W_{s,dir}^{min} = \frac{2h\nu}{\eta} \times \Delta f, W.$$

(4.14)

Materials Research Forum LLC
https://doi.org/10.21741/9781644900758

This formula indicates the radiation quantum nature. For the spectral range $\lambda \approx 10$ μm, the minimum detectable power is $W_{s,dir}^{min} \approx 4 \cdot 10^{-20}$ W. For the spectral wavelengths at $\lambda \approx$ 300 μm (radiation frequency $\nu \approx 1$ THz), the minimum detectable power is $W_{s,dir}^{min} \approx 1.3 \cdot 10^{-21}$ W with an electronic bandwidth of 1 Hz and the quantum efficiency $\eta = 1$. For $\nu \approx 1$ THz, the minimum noise temperature is $T_{s,dir}^{min} = 2\,h\nu/k_B \approx 96$ K. The ability to detect such small signals by the direct detectors is limited by the irreducible background photon noise not vanishingly small even for the cosmic background and the telescopes, which are at non-zero temperatures.

For photoconductive detectors, the noticeable generation-recombination noise due to the fluctuations of the average concentration of electrons in a detector appears. In the case of a semiconductor with one level in the gap or an intrinsic semiconductor, the average generation-recombination noise current is [43, 44]

$$< I_{n,g\text{-}r}^2 >= 4q \times I_0 \times G \frac{1}{1+(\omega \tau_0)^2} \times \Delta f$$

.

$$(4.15)$$

Here, $I_0$ is the average current through the detector, $\tau_0$ is the average current carrier lifetime, $G = \tau_0/\tau$ is the device internal gain, where $\tau$ is the drift time for current carriers over the distance between electrodes of the detector, and $\omega = 2\pi f$ is the circular electronic frequency. At the low electronic frequencies $f \ll \tau_0^{-1}$ (as a rule, in semiconductors, $\tau_0 < 10^{-5}$ s), the noise is "white," and it is twice the value of the background shot noise or the Schottky shot noise [see relation (4.11)]. The appearance of the factor 2 is connected with the Shottky shot noise with only the carrier generation process, and the g-r noise is additionally related to the statistical character of the carrier recombination processes.

Thus, for the photoconductive detectors, relation (4.14) valid for barrier detectors, becomes

$$W_{s,dir}^{min} = \frac{4h\nu}{\eta} \times \Delta f$$

.

$$(4.16).$$

In contrast to the heterodyne sensors, whose responsivity is limited by the quantum-noise level (see below), the sensitivity of a direct detector is fundamentally limited only by fluctuations of the energy in the radiation background flux seen by a detector. The lowest-intensity thermal radiation backgrounds are found in the cosmos. In the THz and

IR spectral regions, the backgrounds correspond to the photon remnants of the Big Bang, cosmic infrared background, or interstellar dust emission. The combination of these continua defines the irreducible noise associated with the fluctuation of the number of photons impinging an ideal detector with an optical bandwidth [45].

The background flux fluctuation noise is a fundamental noise adjusting the upper limit performance of direct detectors. The lower this noise (conditioned by the background temperature or by the temperature of the facility and optics), the better the parameter (low NEP) can be attained. Figure 3.3 illustrates the effect of a cold telescope on the radiation background seen by a single-mode detector pointed at some dark part of the sky with very low luminosity.

The associated photon noise NEP is low ($\sim 10^{-19}\ldots 10^{-20}$ W/Hz$^{1/2}$) throughout the entire THz and IR parts of the spectrum. A cooled telescope ($T_{tel} \sim 30\ldots 40$ K) has NEP that is several orders of magnitude higher (worth) above 1 THz than a telescope cooled to 5 K.

Interestingly, the low-background condition affects the way a detector might (or should) operate given the very low arrival rate of THz photons, $N_{Ph}$. This rate remains under $1000$ s$^{-1}$ from 1 to 30 THz. In this situation, the ideal detector should either have a time constant of the order of 100 ms or longer or be able to detect individual THz photons with high fidelity. The situation is complicated by the fact that the background conditions vary across the sky. Therefore, a detector capable of counting single photons under a low optical loading may not be able to count the photons corresponding to a much more intense background, if the detector maximum rate is insufficient.

In this case, some gradual transition to the power detection mode would be useful. The deeply cooled hot electron nano-bolometers can operate in both photon-counting and power detection modes with high power and energy resolutions, which make it an interesting candidate for these future space THz applications [45].

The progress in the sensitivity of THz direct detectors has been impressive in the period of a more than half-century, what is shown in Fig. 4.7 in the case of bolometers used in the THz and sub-mm wave astrophysics.

In astronomy, the driving force for the development of detectors for getting information in the THz wave ranges is largely influenced by the overall instrumentation sensitivity. The possibility to observe changes in the formation of star and planetary systems, galaxy evolution, *etc.*, is determined by the sensitivity of detectors [39, 47]. Such detectors should operate at almost the quantum limit approximation, when the astronomical instrumentation and detectors are cooled, respectively, to about 4 K and ~50 mK. The detectors should have the noise equivalent power NEP $\leq 10^{-20}$ W/Hz$^{1/2}$ [48, 49] to meet the requirements to photon background levels (see Fig. 4.8).

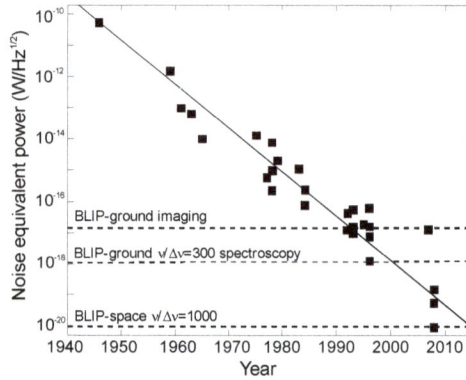

*Fig. 4.7. Improvement in the detector sensitivity over the past 70 years. "Richard's law" predicts the average doubling of sensitivity every two years. The horizontal lines show the photon background-limited noise equivalent powers entailed for ground-based and space-based detectors. Currently available detectors satisfy the background limit for ground-based imaging and spectroscopy. A grating spectrometer on a cold telescope in space requires detectors two orders of magnitude more sensitive than for a similar instrument on the ground or on a suborbital observatory [46].*

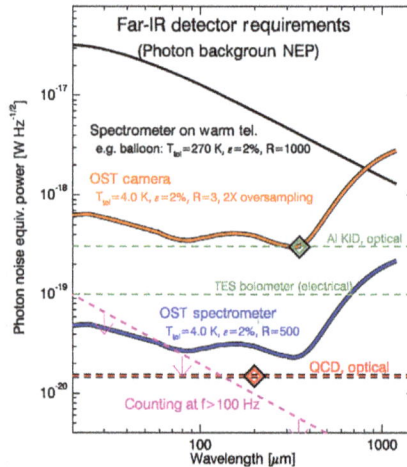

*Figure 4.8. Detector sensitivity requirements to the photon background levels in the far-infrared range. With a cryogenic space telescope, the fundamental limits are the zodiacal dust and galactic cirrus emission [48]. OST – origin space telescope, KID – kinetic inductance detector, TES – transition edge sensor, and QCD – quantum capacitance detector.*

Recent cosmology experiments have discovered that the Universe consists mainly of dark energy (~73 %) and dark matter (~23 %), and the ordinary matter content is only ~4 %. Experiments to resolve the nature of these dark components require ultra-sensitive THz detectors. These requirements can be achieved at THz radiation frequencies [45].

In the direct resistor detectors (*e.g.*, bolometers), another of the basic noises can be the thermal or the Johnson–Nyquist noise (see, *e.g.*, [24, 44, 50, 51]) which appears at sample electrodes due to the statistically distributed potential difference connected with the thermal motion of carriers. For the average thermal noise voltage, the following expression is valid:

$$< U_{J-N}^2 >= \frac{4R \times hf}{\exp(\frac{hf}{k_B T}) - 1} \times \Delta f \quad , \tag{4.17}$$

where R is the sample resistance, and f is the electronic (baseband) frequency. For f << $k_B T/h$, the detector Johnson–Nyquist noise is given by the relation

$$< U_{J-N}^2 >= 4Rk_B T \times \Delta f \quad . \tag{4.18}$$

The presence of the Bose–Einstein factor $[\exp(hf/kT)-1]^{-1}$ argues that the Johnson–Nyquist noise is governed by the Gaussian statistics (the thermal noise is connected with the equilibrium thermal emission into the environment and its back flow into a resistor), whereas the quantum noise (shot noise) is governed by the Poisson statistics.

The performance of the incoherent detectors is limited by the background noise in contrast to coherent detectors, in which the performance is limited by the quantum noise. For coherent detectors, the noise photons would emerge from the amplifier output, even if no signal is present at the input [39].

An advantage of the direct-detection systems lies in their relative simplicity. Moreover, it is possible to design large-format arrays (see, *e.g.*, [52–56]). Most high sensitive imaging systems used the passive direct detection. In the active systems, for operation of which the scene is illuminated, the heterodyne detection can also be used in order to increase the sensitivity to low radiant levels or to image through the scattering media.

### 4.4.2   Coherent receivers

For the coherent detection, the signal received in IR, THz, or mm wave frequency ranges is converted to a GHz band with limited width. The frequency conversion relies on a nonlinear element – the mixer – and an LO. The output IF signal can then be further processed using additional electronics, usually a spectrometer, correlator, or a total power detector. In the process of coherent or heterodyne detection, the amplitudes and phases of incoming waves are converted to a low frequency region. The coherent processes are the base ones in the photo-mixing instrumentations giving the opportunity to obtain the CW THz sources in a wide range of THz frequencies.

The principle of heterodyne detection was first demonstrated in 1901 by R.A. Fessenden in a heterodyne radio receiver circuit [57], in which the incoming radio frequency and the local oscillator frequency were mixed in the crystal diode detector. This receiver did not meet much application because of its local oscillator's stability problem. At the optical frequencies, the heterodyne detection between the incoherent light sources seems have been obtained first in [58]. The optical heterodyne detection began to be studied in 1962, within two years after the construction of the first laser [59].

At the radiation frequencies below 100 GHz, the coherent (heterodyne) detection is extensively used in mobile phones, satellite communications, radar applications, *etc.* At these radiation frequencies, the Earth atmosphere is transparent at long distances. In the cosmos, there are almost no losses.

The advantage of the coherent detection is that a nearly arbitrarily high spectroscopic resolution ($R = \nu/\Delta\nu$) can be obtained by utilizing an intermediate frequency. For a coherent detection, high-resolution requirements $R \geq 10^6$–$10^7$ are achievable. It is a very high resolution, as at a radiation frequency of 1 THz, a resolution of $R = 10^6 \ldots 10^7$ corresponds to a spectral resolution bandwidth of 100 KHz…1 MHz.

The coherent detection is mostly done using heterodyne techniques. The principle of heterodyne detection is shown in Fig. 4.6, where a schematic for heterodyne signal detection is presented. For simplicity, the collection optics is omitted. Again, it is assumed that the detectors in a heterodyne system have linear response to the input radiation power, which means they are square-law detectors, because they respond to the square of the electric field strength of a radiation wave.

Except for the signal radiation power $W_s$ (which may be single-frequency or broadband thermal radiation), the background radiation power $W_B$ and the radiation single-frequency local oscillator power $W_{LO}$ from a local oscillator as the reference (with radiation frequency $\nu_{LO}$) are shown. Because of it, the important advantages offered in coherent systems with heterodyne signal detection are their selectivity and tunability.

Therefore, the heterodyne detectors are almost ideal signal filters. The filtering function, which corresponds to the narrow pass-band of a receiver due to the LO single-frequency, which depends on the presence of tuned LO sources with relatively high power radiation to provide the coherent system operation.

For a simple derivation, it can be assumed that the detector surface with area A has a constant quantum efficiency η over the surface. Let two plane waves be incident normally to this surface, and their **E** vectors lie in the plane of the surface and are parallel to each other. A beam-splitter (diplexer) is needed to align $W_s$ and $W_{LO}$ signals so they be coincident at the mixer (operating at the radiation frequency, in which two signals are mixed) to arrange the usable signal of the intermediate frequency $v_{IF} = |v_s - v_{LO}|$ that is the difference between the signal radiation frequency $v_s$ and $v_{LO}$ from the LO. In the mixer, two frequencies mix to give an output at the intermediate frequency, and its key element is the nonlinear mixing detector. The frequency $v_{IF}$ is of a much lower frequency, as compared to $v_S$ or $v_{LO}$, to be amplified with the electronic instrumentation. The response time of a mixer $\tau_R$ can be as fast as the period of the IF signal, $\tau_R < 1/v_{IF}$ (at $v_{IF}$ ~ 5 GHz $\tau_R < 2 \cdot 10^{-10}$ s). At the same time, the input power is averaged for a time $>> 1/v_S$ which is shorter than $\tau_R$ [60]. The response time may also be limited by the time response of a mixer or a detector. The IF Amp (Fig. 4.6) is an intermediate frequency amplifier, after which a detector that rectifies the IF to give an output signal proportional to the IF power, is situated.

The primary distinction between the coherent and incoherent (direct) detection is the presence or absence of a quantum noise, which leads to preferable spectral ranges of their usage. Coherent receivers preserve the information about both the amplitude and phase of the electromagnetic field while providing the large photon number gain. Coherent receivers are subject to the quantum noise, which can be expressed in terms of the minimum noise temperature $T_n = hv/k_B$ or 48 K/THz (see Fig. 3.16, in which the lowest figure for the DSB noise temperature is presented). Quantum noise is equivalent to the shot noise produced by a background radiation flux connected with photon number fluctuations.

At the radio wavelength range, the background shot noise level is much larger than this value for the quantum noise (especially for SIS or HEB detectors) and, in any case, never falls below the 2.7-K cosmic microwave background. Thus, the use of coherent receivers at radio wavelengths is preferable over direct-detection detectors. In contrast, at optical or infrared wavelengths, the quantum noise of coherent receivers is intolerably large (Fig. 3.16), as compared to the typical backgrounds. Therefore, in this spectral range, the direct detection is preferred [61].

One can neglect the background signal, as heterodyne detectors are weakly sensitive to the background radiation, because of the narrowness of the spectral band, where they operate. With regard for the weak coherence of signals that fall on a detector with area $A_d$ from the object under observation and from LO, we get the following expression for the time-averaged power, which is similar to Eq. (4.3) as for direct-detection detectors:

$$W = \frac{|E|^2}{2\eta_0} \times A_d = \frac{|E_s + E_0|^2}{2\eta_0} \times A_d,$$

(4.19)

where the electric field strength of an electromagnetic wave from LO

$$E_0(r,t) = a_0(t) \times \exp i(\omega_0 t + \varphi_0 - k_0 r).$$

(4.20)

Here, $v_0$ and $\varphi_0$ are the radiation frequency and initial phase of the LO (when one does not neglect the electric field $E_b$ from the background, it does not change the final result, but only complicates a little the intermediate expressions).

For the wave propagation direction along the z-axis, we have

$$E_0 = a_0 \times \cos(\omega_0 t + \varphi_0 - k_{0z}z).$$

(4.21)

When the plane at $z = 0$ is perpendicular to the z-axis and is located at the detector entry, the signal from these two sources will be

$$I(t) = S_1 \times G \times W = S_1 \times G \times [a \times \cos(\omega_s t + \varphi_s) + a_0 \times \cos(\omega_0 t + \varphi_0)] \times \frac{A_d}{\eta_0} = \frac{S_1 \times G \times A_d}{\eta_0} \times$$
$$\left\{ \frac{a^2}{2} \times [1 + \cos2(\omega_s t + \varphi_s)] + \frac{a_0^2}{2} \times [1 + \cos2(\omega_0 t + \varphi_0)] + 2 \times a \times a_0 \times \cos(\omega_s t + \varphi_s) \times \cos(\omega_0 t + \varphi_0) \right\}.$$

(4.22)

In the spectral range from the visible light up to the cm-wavelength band, the frequencies $\omega_s$ and $\omega_0$ are much higher to be followed by the currents and voltages from electrical signals at the detector output. Therefore, the terms of Eq. (4.22) with $\cos2(\omega_s t + \varphi_s)$ and $\cos2(\omega_0 t + \varphi_0)$ can be neglected. The average signal value of a heterodyne detector will be defined by the last term of Eq. (4.22). Taking into account that $(\cos\alpha \times \cos\beta) = 1/2[\cos(\alpha -$

$\beta$)–cos($\alpha$+$\beta$)], the output signal will be determined by the sum and difference of frequencies $\omega_s$ and $\omega_0$ from the signal and LO, respectively. If the signal frequency $\omega_s >$ $\omega_0$, than the corresponding intermediate frequency is positive, and such signal is in the upper side band (USB). If the signal frequency $\omega_s < \omega_0$, then the corresponding intermediate frequency is negative, and such signal is in the lower side band (LSB). For a double side-band receiver, the upper and lower sidebands fold into the IF. In the telecommunication applications, it is undesirable. So, the mixers, in which the upper and lower sidebands are separated, are designed.

For the THz or IR spectral range, the sum of frequencies ($\omega_s$+$\omega_0$) can be neglected. Therefore, the average signal value of the heterodyne detector

$$I(t) = \frac{S_I \times G \times A_d}{\eta_0} \times \left[ \frac{a^2}{2} + \frac{a_0^2}{2} + a \times a_0 \times \cos(\omega_{IF} t + \varphi_{IF}) \right], \qquad (4.23)$$

where the intermediate frequency $\omega_{IF}$ = ($\omega_s$–$\omega_0$) and intermediate phase $\phi_{IF}$ = ($\varphi_s$–$\varphi_0$). If $\omega_s$ and $\varphi_s$ vary with time, so do $\omega_{IF}$ and $\phi_{IF}$. Consequently, unlike direct detectors, heterodyne receivers can detect the frequency and phase modulations, as well as the amplitude modulation $a^2/2$ or $a_0^2/2$ provided the frequency difference ($\omega_s$–$\omega_0$) is relatively small [as a rule, ($\omega_s$–$\omega_0$) < 10 GHz] to pass through the detector electronic and amplifier (IF Amplifier) circuits.

In Eq. (4.23), the IF current is

$$I(t)_{IF} = \frac{S_I \times G \times A_d}{\eta_0} \times a \times a_0 \times \cos(\omega_{IF} t + \varphi_{IF}), \qquad (4.24)$$

where it is assumed that both the signal and LO fields are of constant amplitude.

If $\omega_s = \omega_0$, then $\omega_{IF}$ = 0, and the heterodyne detector is operating in the homodyne regime and is used mainly in the phase modulation schemes.

If $\omega_s \neq \omega_0$, the IF term is time-varying, and the average value of the square signal current of the intermediate frequency

$$< I(t)_{IF}^2 >= \left( \frac{S_I \times G \times A_d}{\eta_0} \right)^2 \times \frac{(a \times a_0)^2}{2}.$$

(4.25)

For the average power of a signal, from (4.3) it follows that $a^2 = 2\eta_0 \times W_s/A_d$ and $a_0^2 = 2\eta_0 \times W_0/A_d$. For the average value of the square signal current with intermediate frequency, we get

$$< I(t)_{IF}^2 >= 2R_I^2 \times G^2 \times W_s \times W_0,$$

(4.26)

where $R_I = \eta q/h\nu = \eta q\lambda/hc$ is the current sensitivity.

From Eq. (4.26), one can conclude that, at large LO power values $W_0$, it is potentially possible to define weak signals at small signal radiation power. It is contrary to Eq. (4.13) for direct-detection detectors, where the signal and background powers can only be close by values.

For the noise equivalent power and the threshold power of heterodyne detectors, these dependences on the noise bandwidth have another appearance in the case of a significant excess of the LO power $W_0$ over the signal radiation power $W_s$ ($W_0 >> W_s$). In this case, the current noise is mainly conditioned by the shot noise. For a barrier detector within the bandwidth $\Delta f$, the noise current $< I_n^2 >=< I(t)^2 >= 2q \times < I_0 > \times \Delta f = 2q \times R_I \times W_0 \times \Delta f$. Comparing with Eq. (4.26) at the signal-to-noise ratio equal to 1, it follows (at G = 1) that

$$W_{s, het}^{min} = \frac{h\nu \times \Delta f}{\eta}, W.$$

(4.27)

For a photoconductor, the minimum registered power $W_{s, het}^{min}$ will be twice larger due to the presence of an additional generation-recombination noise in photoconductors.

It is seen that, for coherent detection and output IF spectrum, a signal is a replica (see Eq. (4.24)) of the input optical spectral distribution, provided that the local oscillator signal is a single-frequency with non-fluctuating phase.

To compare different detectors, it is useful to normalize this expression to unit bandwidth and then, in the case of heterodyne detection, the noise equivalent power NEP = $W_{s, het}^{min}$ $\times N/S \times 1/\Delta f$ (W/Hz), where the signal-to-noise ratio S/N = 1. Therefore, the NEP for heterodyne detectors is measured in "W/Hz" units contrary to the direct detectors, in which NEP is measured in units $W/Hz^{1/2}$.

Coherent signal detection systems use mostly the heterodyne circuit design, since, so far, the proper amplifiers do not exist for high radiation frequencies. In a mixer (consisting a non-linear device), a mixed signal is amplitude-modulated at the intermediate (beat) frequency $v_{IF}= |v_s-v_{LO}|$, where $v_s$ and $v_{LO}$ are signal radiation and local oscillator radiation frequencies, respectively.

There always are two sidebands (Fig. 4.9). After a mixer, the output signal has the same result at $v_{IF}$, no matter if $v_s < v_{LO}$ or $v_s > v_{LO}$. Then the IF signals can be converted or demodulated. Basically, these systems are selective (narrow-band) detection systems. As $v_{IF} \ll v_s$, heterodyne systems have high spectral resolution ($\lambda/\Delta\lambda \sim 10^3-10^7$) that is important, *e.g.*, for spectroscopic systems.

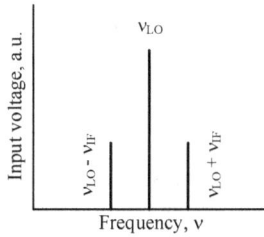

*Fig. 4.9. Sidebands of coherent detection system.*

A heterodyne system will respond to radiation signals in both IF sidebands of the LO frequency. In practice, because of amplifier noise performance considerations, the bandwidth of IF amplifiers does not extend usually to very low frequencies (*e.g.*, below about 10 kHz) for amplifiers operating to several hundred MHz. Therefore, up to several GHz in the THz spectral region that is connected with the bandwidth of a detector used, the receiving spectrum of the heterodyne has a "hole" in it.

As a rule, semiconducting photon detectors are used as direct detectors. If they are used as coherent (heterodyne) ones as HEBs, their central frequency $v$ is about $0.3 \cdot 10^{12}$ Hz, and double sideband is only several or tenth MHz which is controlled by the response time $\tau \sim 10^{-6} \dots 10^{-8}$ s (*e.g.*, InSb, HgCdTe based detectors) that makes them practically unusable.

A mixer is the most important component of the heterodyne receiver input stage that is responsible for its responsivity. Mostly, the choice of a mixer for the particular coherent system is connected with the available local oscillator power and its spectral range applicability, system's sensitivity, and operation temperature.

Although, many CW lines in the THz spectral region are available from the gas lasers to be used as LOs, these are all at specific wavelengths, and the heterodyne detection is restricted to a relatively narrow range on either side of each available wavelength. Here, e.g., gas lasers at $\lambda = 433$ µm (HCOOH), at $\lambda = 184$, 214, and 288 µm ($CH_2F_2$), $\lambda = 337$ µm (HCN), $\lambda = 118$ µm ($H_2O$), and others.

Any nonlinear electronic device can be used as a mixer. However, to achieve efficient conversion and a low noise in the mm and sub-mm wavelength bands, only several types of detectors are acceptable. Frequently, the used mixers are devices having a strong electric field quadratic nonlinearity. Examples are given by the forward biased SBD, SIS tunnel junctions, semiconducting and superconducting HEBs, SLs, FETs, etc. Schematically, current-voltage characteristics of some of these devices are shown in Fig. 3.11. Simultaneously, with a reasonable conversion efficiency and a low noise, these nonlinear devices should possess high conversion operation speeds. This is a need to assure of a wide band-pass for the consequent amplification of signals at much lower frequencies ~1...10 GHz. For example, the ordinary InSb detector HEB is not very suitable for these purposes. Indeed, because of the relatively long relaxation times $\tau \sim 10^{-6}...10^{-7}$ s, the achievable bandwidth at the intermediate frequency of a heterodyne receiver will be only within ~1...10 MHz.

Due to the complexity and relatively high LO powers needed for the conversion of frequencies down to $\nu \leq 10$ GHz, heterodyne receivers are mostly used as single pixel instruments. For astronomy applications with the need for high spectral resolution of $\nu/\Delta\nu > 10^6$, they are the only instruments capable of resolving the velocity structure of spectral lines of atoms and molecules in space. To increase not only the productivity of astronomical high-resolution spectrometers (e.g., to increase the data output), the multielement arrays are needed. Only during the last two decades, the THz array (though with moderate number of elements, $N < 20$) receivers were deployed for astronomical instrumentation (for Refs., see [62]). It is expected that the number of heterodyne detectors in arrays will be increased up to 100 pixels.

SIS mixers are the most sensitive mixers available today at radiation frequencies $\nu < 1.2$ THz (Fig. 3.16) with the quantum noise temperature limit $T_n \approx 2h\nu/k_B$ at $eV_{gap}/h$ (frequencies $\nu < 0.67$ THz). However, for the optimal performance, they require magnets to suppress Josephson currents. SIS mixers are characterized by a low LO pump power, as compared to other heterodyne detectors and arrays.

HEB mixers operate at much wider radiation frequency range though, as a rule, their quantum noise temperature limit is $T_n \sim 10 h\nu/k_B$, K. The state-of-the-art performance ($h\nu/k_B \sim 5$ K) of astronomical receivers in the sub-mm and low-frequency THz regimes ($\nu \sim 100...800$ GHz) was demonstrated by the ALMA receivers with SIS mixers [62] and, at $\nu = 2.5$ THz, with NbN HEB with bandwidth up to 7 GHz [63].

The primary benefit of the heterodyne-detection systems is that the frequency and phase information at the signal frequency $\nu_s$ is converted to the frequency $\nu_{IF}$, which is in much lower frequency band ($\nu_{IF} \ll \nu_s$) appropriate to the time response of electronics. For heterodyne detectors, the LO power $W_0$ required for the operation of a heterodyne detector with the ultimate performance is of importance. The dominant noise source (apart from the noise depending on $W_0$) is due to the IF amplifier in many practical systems. Therefore, to obtain the limiting performance, it is necessary to increase $W_0$.

In Ref. [64], the dependences of $NEP_{het}$ on the local oscillator power $W_0$ for an IR heterodyne HgCdTe detector were investigated. It was shown that, at 77 K, the $NEP_{het}$ is asymptotically approaching the quantum limit estimations $NEP_{het} = W_{s,het}^{min} = \dfrac{h\nu \cdot \Delta f_{IF}}{\eta} \times \dfrac{N}{S}$ with increasing the LO power (signal-to-noise ratio S/N = 1 and $\Delta f_{IF}$ = 1). At this temperature, the LO power is dictated by the amplifier noise [64]. At T = 195 K, the $NEP_{het}$ no longer has the asymptotic LO power dependence, but reaches a minimum at $W_{LO} \approx 5$ mW, which is a result of the detector heating. At this high LO power, the level of the quantum efficiency has dropped considerably below the lower power value of over 70 %.

To reach the heterodyne operation, the LO power is required to overcome the g-r noise that rises rapidly with the temperature in this narrow-gap HgCdTe semiconductor (g-r noise is proportional to the volume of a photoconductor) reaching a value of 30 mW at 300 K. This power level creates severe heating problems and a loss in sensitivity, as compared to an ideal photo-mixer [64].

In the quantum noise limit when ($\langle I_S^2 \rangle \gg \langle I_T^2 \rangle$), where $\langle I_S^2 \rangle$ is the mean square shot noise current, the current $\langle I_T^2 \rangle$ is the coherently detected thermal noise from any thermal source with equivalent black-body temperature lying inside the antenna pattern of the receiver or the unwanted background one. In this approximation, the required lower limit of the LO power $W_0$ for photoemission (barrier) mixers is given by [60]

$$\langle W_0 \rangle > \frac{2k_B T_A h\nu}{e^2 \eta R_A M},$$

$$(4.28)$$

where $T_A$ is the amplifier noise temperature, $R_A$ is the amplifier input resistance, M is the impedance mismatch (M ≤ 1) between the mixer and the amplifier, $\eta$ is the quantum efficiency (coupling efficiency). Assuming typical values M = 1, $\eta$ = 0.1, $R_A$ = 50 Ω, and $T_A$ = 500 K, $W_{PL}$ must be greater than about 2 mW at $\lambda \approx 10$ μm or 0.2 mW at $\lambda \approx 100$ μm. In photoconductors because of the generation-recombination noise in the current $\langle I_s^2 \rangle$, the requirement to $W_0$ will be 2 times larger. For photoconducting mixers, the shot-noise current includes the gain factor G. As a rule, G < 1, and this increases the required amount of the LO power.

In Fig. 4.10, following the estimations in [60] in order to illustrate the spectral ranges of preferable detectors, is shown the ultimate NEP of a heterodyne receiver with photoemissive (barrier) mixer for $\eta$ = 1 and with a thermal source temperature $T_s$ filling the field of view plotted as a function of the frequency and the wavelength. At infrared wavelengths, the quantum noise of coherent receivers is far larger than the typical background thermal noise. Therefore, in the IR and THz spectral range above ~0.5 THz, the direct detection is preferable, as compared to the heterodyne one.

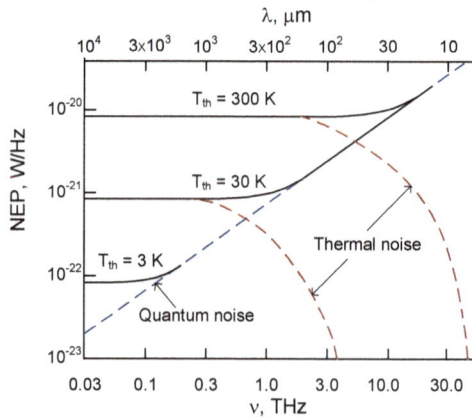

*Figure 4.10. The ultimate NEP of a heterodyne receiver using a photoemissive mixer, for the case $\eta$ = 1 and with a thermal source of temperature $T_{th}$ filling the field of view plotted as a function of the frequency and the wavelength. Quantum efficiency $\eta$ = 1, $\beta$ = 1 is the coefficient characterizing the background radiation filling field of view of the receiver ( ordinary $\beta \le 1$).*

Materials Research Forum LLC
https://doi.org/10.21741/9781644900758

Detector technologies used for IR and THz applications strongly depend on the wavelength region. For IR regions ($\lambda \leq 30$ μm), as a rule, semiconducting photon detectors with response time $\tau \geq 10^{-7}$ s that are not deeply cooled ($T \geq 20$ K) are used. Thermal uncooled detectors have the response time of about several or tenths of milliseconds.

As a rule, semiconducting photon detectors are used as direct detectors. But, if they are used as coherent (heterodyne) ones, their central frequency is about $\nu \sim 3 \cdot 10^{13}$ Hz, and the double sideband is only several or tenth MHz which is controlled by a response time $\tau \sim 10^{-6} \ldots 10^{-8}$ s that makes them practically unusable. Moreover, in this radiation frequency range, there are only a few narrow-band LO sources.

The primary distinction between coherent and incoherent detection is the presence or absence of the quantum noise. Coherent (heterodyne) receivers preserve information about both the amplitude and phase of the electromagnetic field, by ensuring a large photon number gain. As a result, coherent receivers are subject to the quantum noise. This was revealed in 1957 [65] in connection with the development of a maser. A minimum noise temperature $T_n = h\nu/k_B$ or 48 K/THz (see Fig. 3.15).

The quantum noise is equivalent to the shot noise produced by the background radiation flux. At radio and mm wavelengths, the background noise is significantly larger than the quantum noise (see Fig. 4.10). Thus, the use of the coherent receivers at radio and up to mm wavelengths is preferable for high spectral resolution systems and does not lead to a loss of sensitivity [39, 61, 66]. In Ref. [67], it was shown that the quantum noise is a general limitation of all phase insensitive linear amplifiers. In contrast to mm and radio wavelengths, the quantum noise of coherent receivers is intolerably large at visible or infrared wavelengths, far larger than the typical backgrounds. Therefore, the direct detection is preferred.

Whether or not, the quantum noise limit for the coherent detection is actually a factor that depends largely on the level of the background radiation. A lower limit, to the thermal background is provided by the CMB radiation. This means that the quantum noise is actually irrelevant at radio or mm wave frequencies $h\nu/k_B < 2.7$ K or $\nu < 56$ GHz. In the THz range, the quantum noise limit is important, *e.g.*, for cooled telescopes.

THz mixer receivers can operate in different modes, depending on the configuration of a receiver and the nature of the measurement. The signal and image frequencies may be separated in a correlator, or the image may be removed by the appropriate phase switching of pairs of local oscillators. The function of separating or dumping the image in the receiver is to remove some uncorrelated noise to improve system's sensitivity.

In single-sideband (SSB) operation, the receiver is configured so that, at the image sideband, the mixer is connected to a termination within the receiver. There is no external connection to the image frequency, and the complete receiver is functionally equivalent to an amplifier followed by a frequency converter.

In double-sideband (DSB) operation, on the other hand, the mixer is connected to the same input port at both upper and lower sidebands. The DSB receivers can be operated in two modes:

- In SSB operation to measure narrow-band signals contained entirely within one sideband – for detection of such narrow-band signals, the power collected in the image band of a DSB receiver degrades the measurement sensitivity;

- In DSB operation to measure broadband (or continuum) sources whose spectrum covers both sidebands – for continuum radiometry, the additional signal power collected in the image band of a DSB receiver improves the measurement sensitivity.

The technology that is traditionally available for terahertz receivers utilizes Schottky-barrier diode (SBD) mixers. The noise temperature of such receivers has essentially reached a limit of about 50 $h\nu/k_B$ in the frequency range below 3 THz (see Fig. 3.16). Above 3 THz, there occurs a steep increase, mainly due to the increasing losses of the antenna and the reduced performance of the diode itself.

In the last two decades, the improvements in receiver sensitivities have been achieved using superconducting mixers, with both SIS and HEB mixers. In Fig. 3.16, the selected receiver noise temperatures are plotted. Nb-based SIS mixers yield the almost quantum limited performance up to a gap frequency of 1.3 THz.

Unlike SBD and SIS mixers, the HEB mixer is a thermal detector. Up to 2.5 THz, the noise temperature follows closely the 10 $h\nu/k$ line. In comparison with the Schottky-barrier technology, HEB mixers require the LO power less by three to four orders of magnitude.

Heterodyne detectors are mostly used in investigations of the CMB radiation at mm and sub-mm wavelengths or in the cases where a high spectral resolution to measure vibration and rotation atomic and molecular fine line transitions is necessary.

## 4.5 Coherent vs. direct detection

In spite of the advantages of the coherent detection with high resolution ($\nu/\Delta\nu \sim 10^6$–$10^7$) over the direct detection ($\nu/\Delta\nu < 10^3$), there are some drawbacks, as compare to the direct detection. Theoretically, the upper limit performance of heterodyne detectors is defined by the lower limit quantum noise $h\nu/k_B$. Therefore, the quantum noise depends on the radiation frequency operation and the detector type.

Moreover, the IF detection bandwidth of coherent receivers is relatively small. It depends on the response time $\tau$ of detectors. The lowest $\tau$ is observed in SIS detectors ($\tau > 3 \cdot 10^{-11}$ s) and hot-electron low-temperature superconductive bolometers. That is why the IF band can be as wide as ~11 GHz [68, 69]. For other detector types, the response time $\tau \geq 3 \cdot 10^{-10}$ s, and the IF bandwidth is $\leq 5$ GHz [29].

To compare the preferable spectral range operation for direct-detection systems and the coherent ones, there is sometimes the need to express NEP important for direct detectors by temperature units that are inherent for coherent-detection systems. The caution is warranted, however, when comparing direct-detection systems with heterodyne receivers, since the former are primarily designed to measure the broadband continuum and the latter – the vibration and rotational atomic and molecular fine line transitions [70].

Consider a background limited direct detector (*e.g.*, low-temperature bolometer or TES detector) with an antenna and a heterodyne receiver (with an antenna) operating at the quantum noise limit ($T_n = h\nu/k_B$) in the spectral band, where the Gaussian statistics (Rayleigh–Jeans regime) is valid ($h\nu \ll k_B T$). For example at $\nu \leq 1$ THz, the photon energy $h\nu \leq 4.1$ meV $\ll k_B T \approx 26$ meV at the background temperature $T \approx 300$ K. In this case, the antenna temperature is linearly related to the input flux density $P_\nu$ at the detector [25, 71]:

$$T_A = \frac{A_{eff} \cdot P_\nu}{2k_B \cdot \Delta\nu_{dir}}, \qquad (4.29)$$

where $A_{eff}$ is the antenna effective area. It can be taken as $A_{eff} = (\lambda^2/4\pi) \cdot G_r$, where $\lambda$ is the wavelength, $G_r$ is the antenna gain that, in the lossless case, is equal to the directivity $D_0$ [71]. Here, $P_\nu$ is the flux density from the source (*e.g.*, background), the factor 1/2 arises due to a single mode illumination (single polarization), and $\Delta\nu$ is the frequency bandwidth. For a coherent receiver, the single-to-noise ratio $(S/N)_{coh}$ may be given in terms of the antenna and system noise temperatures ($T_A$ and $T_n$, respectively) by the Dicke relation for radiometers [25, 71]:

$$(S/N)_{coh} = \frac{T_A}{T_n} \cdot (\Delta\nu_{IF} \cdot \Delta t)^{1/2}, \qquad (4.30)$$

where $\Delta v_{IF}$ is the intermediate frequency bandwidth, $\Delta t$ is the integration time of the observation, and $T_n$ is the system noise temperature.

The signal-to-noise ratio for an incoherent (direct) detector (*e.g.*, low-temperature bolometer) is

$$(S/N)_{dir} = \frac{2k_B T_A}{NEP} \times \Delta v_{dir} \times (\Delta t)^{1/2}.$$

(4.31)

Assume the case of the direct detector ultimate performance in the spectral range, where the Gaussian statistics is valid.

Then

$$\frac{(S/N)_{coh}}{(S/N)_{dir}} = \frac{NEP \times (\Delta v_{IF})^{1/2}}{2hv \times \Delta v_{dir}}.$$

(4.32)

Here, it was taken that the system noise temperature $T_n = hv/k_B$. If the spectral resolution is kept constant, then $\frac{(S/N)_{coh}}{(S/N)_{dir}}$ behaves itself as $1/v^2$.

For the diffraction-limited approximation, when the throughput for a plane wave passing a circular aperture with area A, and a solid angle $\Omega$, a single spatial mode is governed by the antenna theorem $A \cdot \Omega = \lambda^2$ (Gaussian optics). For this case, NEP can be taken as photon's $NEP_{ph}$ governed only by fluctuations in the photon flux [72] and can be estimated, *e.g.*, for the background environment. It will be the ultimate performance figure for $(S/N)_{dir}$.

In this case at the background temperature $T_b = 300$ K and the spectral bandwidth $\Delta v_{dir} \approx 0.1 \cdot v$, $v = 0.3$ THz ($\lambda = 1$ mm) for a direct detector, and $NEP_{ph} \approx 3 \cdot 10^{-16}$ W/Hz$^{1/2}$. Then, at the typical intermediate frequency for a coherent detector, $\Delta v_{IF} = 3 \cdot 10^9$ Hz, $\frac{(S/N)_{coh}}{(S/N)_{dir}} = 1$, it follows from Eq. (4.32) that, at the radiation frequency $v$ exceeding the "critical" radiation frequency $v > v_{cr} \approx 0.35$ THz, the direct detectors will be preferable. The $NEP_{ph}$ decreases (improves), as $\Delta v_{dir}$ decreases, and vice versa. At $T_B = 300$K, $\Delta v_{dir}/v = 0.3$, $v = 0.3$ THz ($\lambda = 1$ mm), $NEP_{ph} \approx 5.5 \cdot 10^{-16}$ W/Hz$^{1.2}$, and $v_{cr} \approx 0.27$ THz.

Since $NEP_{ph}$ of direct detectors in the Rayleigh–Jeans regime is proportional to $(T_B, \Delta v_{dir})^{1/2}$,

$$NEP_{ph} = h^{1/2} \cdot v \cdot [4 \cdot k_B \cdot T_B \cdot (\Delta v_{dir}/v)]^{1/2}, \qquad\qquad 4.33)$$

the radiation frequency for the preferable direct detection will shift at lowering the temperature $T_B$ as $\dfrac{(S/N)_{coh}}{(S/N)_{dir}} \sim \dfrac{NEP}{\Delta v_{dir}}$ .

Actually, these calculations are only estimations, since a number of different factors should be also considered. They depend on the detector quantum (coupling) efficiency (in estimations, it was taken $\eta = 1$), the spectral bandwidth of an incoherent detector, as it depends on an antenna, the diffraction-limited approximation cannot be valid, *etc.* The $\Delta f_{IF}$ value is dependent on the detector type. Moreover, the noise limit of a mixer detector depends on its type (Fig. 3.16). Thus, at wavelengths longer than about 2 to 3 millimeters ($v < 0.15$ to $0.1$ THz), the coherent detectors are preferable.

In any case, the coherent detectors are preferred for high-resolution spectroscopy and interferometry. For high spectral resolving powers and for relatively small astronomical sources, the heterodyne systems are likely to offer a distinct advantage in the signal-to-noise ratio, especially at mm wavelengths [60].

For IR systems, one of the main tasks is to obtain the information of heated objects, and these imaging systems are, as a rule, passive in nature. The spectral width in which the surveillance takes part, as a rule, is about $\lambda/\Delta\lambda \sim 2...10$ (in $1.8...2.2$, $3...5$ or $8...12$ μm bands). For these purposes, semiconducting photon detectors cooled down to $T \sim 80$ K are generally used. In the last two decades, the uncooled microbolometer arrays are widely applicable, mainly for the spectral region $\lambda \sim 7.5...14$ μm.

In the THz spectral range, the measurements with low spectral resolution $\lambda/\Delta\lambda \sim 3...10$ are called photometry. They are used to characterize broad-spectrum sources such as, *e.g.*, electron synchrotron and thermal bremsstrahlung emission, and thermal emission from the interstellar dust grains. Photometry at mm and sub-mm wavelengths (sub-THz range) plays a key role in astrophysics.

The THz and sub-THz bands are unique in that galaxies may be seen out to very large distances, because the dust emission spectrum has a very steep long-wavelength slope, so that the cosmological frequency redshift essentially compensates for the dimming due to the increasing distance. Measurements of this dust emission by the sub-mm photometry give the total luminosity of the galaxy, which is related to the rate, at which stars are forming, and provides information about the behavior of the dust as well.

Depending on the wavelength for the photometry purposes, deeply cooled detectors are used, as a rule. The diode detectors (*e.g.*, SBDs) operate in a wide temperature band (T ~ 20…300 K) and were often used at $\nu$ ~ 0.1…5 THz [39, 73]. The thermal direct detectors – semiconducting bolometers [72, 74] are in current use for ~ 0.1 to 1.5 THz. Semiconducting photon detectors are generally used for the radiation frequency range $\nu \geq$ 1.5 THz ($\lambda \leq$ 200 µm). The superconducting bolometers (TES) and the semiconducting cooled ones are applied at radiation frequencies $\nu$ ~ 0.3…1.5 THz. HEBs can operate up to the visible spectral range (see Ch. 5). The thermal uncooled detectors (pyroelectric, thermocuples, pneumatic, *etc.*) can operate in a wide spectral range, from IR to sub-THz theoretically not dependent on the radiation frequency [29]. The very qualitative dependences of the operational spectral and temperature ranges of different kinds of detectors are presented in Fig. 4.11.

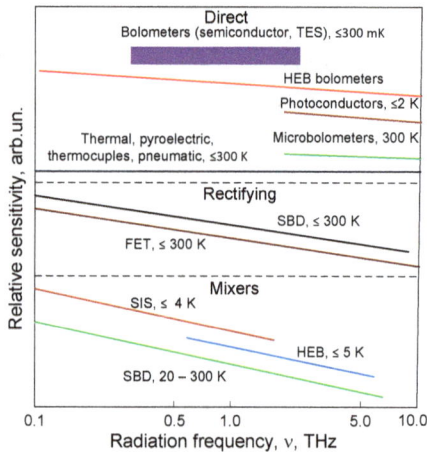

*Fig. 4.11. Approximate representation of different type of THz detectors with their rough spectral dependences. Their relative sensitivity may differ noticeably in dependence on many factors.*

Measurements with higher spectral resolution ($\lambda/\Delta\lambda$ ~ $10^6$ or higher) are generally different from photometry. They are typically used to characterize the molecular and atomic spectral lines [75] and are executed with heterodyne receivers. The molecular rotation lines dominate at the sub-THz range, and the atomic fine structure lines can be observed at the THz spectral range [76]. Resolutions as large as $v/\Delta v > 10^6$ are often in need to measure the Doppler shifts and spectral line profiles.

A key advantage of THz sensors over the infrared and visible ones is their ultimate noise limits. Both direct and heterodyne sensorsors operate against the fundamental noise limits that depend on the background radiation and the photon energy. In the coherent case, the limit is simply the photon shot noise that has NEP equal to $h\nu/\eta$. This limit is plotted in Fig. 3.16 for $\eta = 1$ in terms of the equivalent temperature $T = h\nu/k_B$. The same advantage is shared by RF receivers, which operate at lower frequencies. That is one of the causes for RF communications. Be wired or wireless, it is generally superior to photonic communications in terms of the sensitivity [77].

The sensitivity of heterodyne detectors is frequently given in terms of the mixer noise temperature $T_{mix}$, which correlates with the mixer noise equivalent power

$$NEP_{mix} = k_B T_{mix}. \tag{4.34}$$

For the wavelength band $\lambda \approx 3$ mm ($\nu \approx 100$ GHz), where it is atmospheric transparency window, the value $T_s^{min} = E_{s,het}^{min}/k_B = h\nu/k_B \approx 4.8$ K is the fundamental limit to the noise temperature imposed by the uncertainty principle on any simultaneous measurement of the amplitude and phase of an electromagnetic wave. In the case of heterodyne detection with SIS tunnel junctions as the mixing elements, "the true quantum-noise-limited mixer temperature is $T_{mix}^{min} = h\nu/2k_B$ " [78].

One of the important questions particularly regarding on-board systems for space spectroscopic investigations in the THz radiation spectral range is the choice between coherent (heterodyne) and direct-detection systems. Heterodyne systems are characterized by high spectral resolution ($v/\Delta v \sim (10^5...10^7)$ [75, 79], which is possible due to $v_{IF} \ll v$. However, for heterodyne systems, especially for those based on SBDs, the critical component is the LO power.

At the same time, the systems with low-temperature direct detectors operating, as a rule, in a wider spectral range, where the level of the background radiation is small, can provide a sufficient spectral resolution and BLIP regime. They are more preferable for the medium spectral resolution $v/\Delta v \sim (10^2...10^4)$ or lower [80, 81]. They are also preferable for technical vision systems. Direct-detection systems also can be used in applications in which the sensitivity is more important, as compared to the spectral resolution.

To have the background limited detector array is important from the viewpoint of removing, *e.g.*, the sky background noise, taking into consideration that any spatially correlated component of this noise detected in all detectors in the array can be substantially suppressed.

Among the direct detectors, the low temperature [T ~ (100...300) mK] transition edge sensors (TES) have near ultimate performance characteristics in a broad spectral range. Their NEP reaches the values ~ $(0.4...3) \times 10^{-19}$ W/Hz$^{1/2}$ [79, 82, 83].

Direct-detection bolometers have been used for decades in measurements of the CMB spectrum and anisotropy, including space flight missions, *e.g.*, COBE-FIRAS instrument [84]. In CMB experiments, the coherent detector systems and incoherent bolometric systems are used. For cosmic ground-based experiments, both detector types are viable [85].

The advantage of direct detectors is their practical simplicity. For effective operation of heterodyne detectors, some requirements both to the signal and LO radiation beams should be satisfied [40]:

- Both beams should have the same transverse spatial mode structure. Ideally, both should be in the fundamental (TEM$_{00}$) mode. If the beams have different structures, there exists a possibility of the generation of IF current components with different signs that will reduce the resulting IF current.

- Both beams should be coincident and have equal diameters, since only the detector surface, where both beams overlap, can yield the IF current. Areas covered by one beam alone yield direct-detection currents, rather than IF signals.

- Poynting vectors of both beams should be coincident for the best ability. Otherwise, wave fronts do not coincide, and the phase difference between two waves varies as a function of the position across the detector surface. The greater the angle between two beams, the smaller is the area over which the phase difference between two beams is constant.

- Both wave fronts should have the same radius of curvature. Otherwise, the wave fronts cannot coincide, and the phase difference varies as a function of the position across the detector surface.

- Both beams should be identically polarized. The IF current is derived from the dot product of two electric fields. Therefore, it is maximum, only if both fields are aligned, falling off with the cosine of the angle between them.

The disadvantage of the coherent detection is the difficulty to satisfy exactly all the conditions listed above. Thus, in many cases, it is generally desirable not to undertake the coherent detection except for the detection of very weak signals or for the detection of the frequency or phase shifts or modulation, and high resolution of spectral lines, which cannot be done with direct detectors.

All the remarks pointed out above are important for the process of photomixing to obtain the THz radiation from beating the beams of two visible or near IR lasers or one laser with a little detuned two-wavelength radiation also in visible or near IR ranges.

The requirement for the Poynting vectors of both beams to coincide for the best ability can be better realized with increasing the wavelengths of the signal and the LO radiation. It is one of the causes why such detectors were developed, first of all, in the sub-THz and cm spectral ranges. If the heterodyne detector is illuminated by the LO plane wave, the effective mixing of the signal and LO beams will be only for beams within the solid angle $A_d \cdot \Omega \approx \lambda^2$ [86, 87]. For the detector area $A_d \approx 1$ mm$^2$ and for the wavelength $\lambda \approx 10$ µm, the directions of the signal and LO wave vectors should differ for solid angles $\leq 0.6^0$, whereas, for $\lambda \approx 100$ µm, they should coincide within the $6^0$ cone [60].

In connection with that, the heterodyne detection refers to the phenomena with distributed feedback. The detector sizes should be relatively large, as compared to the radiation wavelength, and they ordinarily exceed the diffraction limited sizes. The detector different parts can generate signals with intermediate frequency, but each with the own phase. It is important to provide the identical phase difference at the detector surface from the signal and LO sources in order to ensure a sufficient gain. This means that the signals should be spatially coherent. This condition can be executed relatively simply for the mm, sub-THz and THz spectral regions. It is desirable to provide the time coherence as well. The coherence length of a signal radiation source is desirable to be larger, as compared to the distance from it to the detector.

In connection with the above-listed requirements, it is relatively difficult to provide the development of large-format matrix arrays from heterodyne detectors. Coherent systems (for example, with SIS or SBD mixers) are limited by the radiation frequency ranges approximately $\nu < 1$ THz and $\nu < 3$ THz, respectively. Heterodyne HEB mixers and TES direct-detection detectors practically do not have limitations in the short THz spectral range.

For the estimations of effective applications of systems with direct or heterodyne detection detectors, one should generally compare NEPs or the signal-to-noise (S/N) ratios of these detectors. Comparing NEPs of heterodyne or direct detector systems, it should be accounted for many parameters of the systems and their assignment, *e.g.*, the required spectral resolution, bandwidth, responsivity, working conditions, *etc.* Provided of the usage of sufficiently high LO signals, the heterodyne detector NEP can be reduced (the lower the NEP, the more sensitive is the detector) to values which are determined by quantum and thermal fluctuations. Such limit can be reached in direct detectors in which the influence of thermal fluctuations in the background radiation is unnoticeable due to their practical absence.

Though such ultimate NEP values in systems with direct detectors in the THz and mm wave spectra practically are difficult to be achieved because of sufficiently larger noise levels at the detector itself and the noise connected with fluctuations in radiation fluxes from the background and from unit elements, optics, *etc.* In this case, for getting the ultimate required NEP values, the various cold filters should be used.

One of the critical questions, especially for space-borne observatories at sub-THz ranges, is whether to use heterodyne or direct detector instruments, *e.g.*, for spectroscopic studies. In general, the heterodyne detection offers a higher spectral resolution R = $\nu/\Delta\nu$ ~ $10^5...10^6$ (see, *e.g.*, [75, 79]). Very high spectral resolution is possible, since $\nu_{IF} \ll \nu$. However, for heterodyne systems, especially for SBD receivers in the THz region, a critical component is the LO source power.

The sensitive elements in these two detection systems can be similar in nature. However, for providing the possible wider IF spectral band near high radiation frequency, the coherent systems need fast detectors ($\tau$ ~ $10^{-10}...10^{-11}$ s). They can be Schottky barrier diodes, since $\tau \leq 10^{-11}$s can be obtained in them [88]. Field effect transistors can also be used as, *e.g.*, in GaAs-based uncooled FETs, the response time $\tau \leq 10^{-11}$ s [89]. Frequently are used superconductor-insulator-superconductor niobium-based ($\Delta f \leq 8$ GHz [75, 90]) or superconductive NbN hot electron bolometer (effective IF $\Delta\nu \leq 5$ GHz [91]) detectors.

Semiconductor InSb HEB [92, 93] or HgCdTe one [94] detectors are not used, as a rule, in coherent systems because of the long time ($\tau$ ~ $10^{-6}...10^{-7}$ s) of their response resulting in a narrow band-width $\Delta\nu$ ~ 1...10 MHz. It is also applicable to uncooled thermal detectors, which are slow ($\tau$ ~ $10^{-2}$ s). The latter can be successfully used as matrix arrays in the direct-detection wide spectral range systems.

At the lower part of the THz/sub-THz radiation frequency range ($\nu \leq 300$ GHz), the heterodyne detection is extensively used, *e.g.*, in astronomy, for the identification of atomic and molecular lines, radar applications, and satellite communications. The wider the intermediate frequency bandwidth, the more the lines and the larger the line-widths can be observed. At IF $\Delta\nu$ ~ 0.5 GHz, the observed line-width in velocity terms can be V = c·$\Delta\nu/\nu$ ≈ 430 km/s at the rest frequency $\nu$ ≈ 350 GHz (here, c is the speed of light in free space). At higher radiation frequencies $\nu$ the value of V will be less. For IF bandwidths of, *e.g.*, 8 GHz and higher (SIS mixers), this allows one to observe distant galaxies with V ~ 7000 km/s at the same $\nu \approx 350$ GHz. In the case of high resolution, R = $\nu/\Delta\nu \sim 10^6$ at $\nu \approx 300$ GHz in the velocity term V = c/R $\approx 0.3$ km/s.

In a heterodyne detection system, the possible minimum detectable signal is twice lower in comparison with a direct-detection one [see Eqs. (4.14) and (4.27)]. Note also that, for the heterodyne detection, units of NEP are W/Hz instead of W/Hz$^{1/2}$ as for the direct detection. However, frequently, NEP for the heterodyne detection is cited still in units of W/Hz$^{1/2}$.

Today, there exist many different types of THz/sub-THz detectors (see Ch. 5]). The most sensitive of these detectors are those deeply cooled to cryogenic or sub-K temperatures. Their noise equivalent power in some spectral THz/sub-THz regions under the low background conditions attains NEP ~ $10^{-19}...10^{-20}$ W/Hz$^{1/2}$. However, under the Earth environmental conditions (T ~ 300 K), their NEP cannot be less (better) than NEP ~> $10^{-17}...10^{-18}$ W/Hz$^{1/2}$ (strongly dependent on the radiation spectral range and spectral bandwidth), because of the background noise (photon noise) restrictions.

In THz spectrometers with modest resolution, cooled and uncooled detectors are used as well. The advantages of uncooled detectors, in spite of a relatively low sensitivity (high NEP), are relatively simple conditions of operation, the lack of refrigerators or cryogenic liquids, and the feasibility of operation at large background photon fluxes.

To be widely used in the IR and THz/sub-THz applications, it is desirable to use uncooled detectors. However, many of uncooled THz/sub-THz direct detectors are not very sensitive and have a relatively long response time $\tau$ (in dependence of the wavelength, NEP ~ $10^{-8}...10^{-10}$ W/Hz$^{1/2}$, $\tau > 10$ ms, *e.g.*, pyroelectrics, Golay cells, thermopiles).

Being sensitive (*e.g.*, for SBDs and FETs, NEP is equal up to ~$10^{-11}$ W/Hz$^{1/2}$ in the low-frequency THz range), thin metal bolometers, (NEP ~ $10^{-10}...10^{-11}$ W/Hz$^{1/2}$), contrary to IR detectors, are difficult to be assembled into large focal plane arrays (FPAs) with readouts in the focal plane. Additionally, NEP for all THz detectors is substantially dependent on the radiation frequency – NEP ~ $1/\lambda^n$ (n ~ 2...4) because of the antenna presence. If microbolometer arrays and other uncooled thermal detectors can be used in IR for the passive imaging, they should be used in the THz range with additional THz sources to realize the real-time active imaging performance.

The IR detectors have been used for the passive imaging first in the scanning mode with single or low-number detectors in arrays, but they had a slow response, poor temperature and spatial resolutions, and slow frame rate. Nowadays, the uncooled matrix arrays of IR microbolometers achieve capabilities to be used in systems with the real-time operation and high definition for portable thermography facilities in biomedical diagnostics and are widespread among IR medical users [95]). The ongoing research has improved the performance of room-temperature devices in terms of attainable parameters (see, *e.g.*, [96]).

IR cooled and uncooled detectors and arrays in vision systems mostly operate in the direct-detection mode with sensitive elements having dimensions about the wavelength or smaller. In the THz arrays on the base of superconductive hot-electron bolometers, the direct-detection mode is also used. In these both cases, different kinds of conventional read-out electronics are used.

For the THz vision systems with CMOS-process direct detectors, there exist several solutions. One of them is based on power detectors with low-noise amplifiers (LNAs) not operating at radiation frequencies larger than the cut-off frequency, which can be up to about a hundred GHz [97, 98]. The small detector number arrays and systems on their base operating at several tenths GHz radiation frequency range are relatively bulky [99]). NEPs of THz detectors with LNAs can be as low as tenths of fW [100], but the cost and dimensions of even a small number of detectors in an array and a vision system on their base are not favorable. The arrays and systems with such detectors can be used for the passive imaging.

Another solution is the implementation of antenna-coupled SBD III-V arrays or MOSFET and hetero-junction bipolar transistor (HBT) arrays designed according to the standard CMOS technologies [101–103], which allows one to manufacture them at the foundry level.

## 4.6  Summary

The general questions concerning the information capacity in the IR and THz range and the operation of direct and coherent THz and IR detectors have been briefly discussed. Determining the appropriate sensing detector is important for the instrumentations of THz and IR detectors and is dependent on their applicability. The comparison of the effective, in general, applicabilities of direct (incoherent) and coherent THz and IR detectors in their spectral ranges of sensitivity is executed and may be applied to the estimations of the scientific IR or THz probable instrument application.

## References to Ch. 4

[1] I.I. Taubkin, M.A. Trishenkov, Information capacity of electronic vision systems, Infrared Phys. Technol. 37 (1996) 675–693. https://doi.org/10.1016/S1350-

4495(96)00002-3

[2] A.M. Filachev, I.I. Taubkin, M.A. Trishenkov, Solid State Photo-Electronics. Physical Basis, Fizmatgiz, Moscow, 2005 (in Russian).

[3] F. Sizov, Photoelectronics for Vision Systems in the Invisible Spectral Ranges, Akademperiodyka, Kiev, 2008 (in Russian).

[4] I.I. Taubkin, M.A. Trishenkov, Ultimate sensitivity of the imagers and informativeness of electronic vision systems, Proc. SPIE. 4369 (2001) 94–105. https://doi.org/10.1117/12.496302

[5] [5] G.D. Skidmore, Uncooled 10 µm FPA development at DRS. Proc. SPIE. 9819 (2016) 98191O. https://doi.org/10.1117/12.2229079

[6] G.C. Holst, R.G. Driggers, Small detectors in infrared system design, Opt. Eng. 51 (2012) 096401. https://doi.org/10.1117/1.OE.51.9.096401

[7] J. Robinson, M. Kinch, M. Marquis, D. Littlejohn, K. Jeppson, Case for small pixels: system perspective and FPA challenge, Proc. SPIE. 9100 (2014) 91000I. https://doi.org/10.1117/12.2054452

[8] A. Rogalski, Next decade in infrared detectors, Proc. SPIE. 10433 (2017) 104330L, in: "Electro-Optical and Infrared Systems: Technology and Applications XIV", Eds. David A. Huckridge, Reinhard Ebert, Helge Bürsing. https://doi.org/10.1117/12.2300779

[9] T. Sprafke, J.W. Beletic, High performance IR focal plane arrays, Optics Photonics News. 19 (2008) 24–27 (Open Access). https://doi.org/10.1364/OPN.19.6.000022

[10] B. Starr, L. Mears, C. Fulk, J. Getty, *et al.*, RVS large format arrays for astronomy, Proc. SPIE, 9915 (2016) 9915, in: High Energy, Optical, and Infrared Detectors for Astronomy VII, Eds. A.D. Holland, J. Beletic. https://doi.org/10.1117/12.2233033

[11] L. Kozlowski, HgCdTe focal plane arrays for high performance infrared cameras, Proc. SPIE. 3179 (1997) 200–211. https://doi.org/10.1117/12.276226

[12] A. Bergeron, L. Marchese, É. Savard, L. LeNoc, *et al.*, Resolution capability comparison of infrared and terahertz imagers, Proc. SPIE. 8188 (2011) 81880I. https://doi.org/10.1117/12.898937

[13] H. Moseley, Detectors for future missions: Increasing scale, extending performance, and integrated functions, https://asd.gsfc.nasa.gov/conferences/FIR/talks/11-HMoseley.

[14] J.M. Lamarre, Photon noise in photometric instruments at far-infrared and submillimeter wavelengths, Appl. Opt. 25 (1986) 870–876. https://doi.org/10.1364/AO.25.000870

[15] A.K. Panwar, A. Singh, A. Kumar, H. Kim, Terahertz imaging system for biomedical applications: Current status, Int. J. Eng. Technol. 13 (2013) 33–39.

[16] J.P. Guillet, B. Recur, L. Frederique, B. Bousquet, *et al.*, Review of terahertz tomography techniques, J. Infrared, Millimeter, and Terahertz Waves. 35 (2014) 382–411. https://doi.org/10.1007/s10762-014-0057-0

[17] A. Kreisler, M. Pyee, M. Redon, Parameters influencing for infrared video detection with submicron size Schottky diodes, Intern. J. Infrared and Millimeter Waves. 5 (1984) 559–584. https://doi.org/10.1007/BF01010152

[18] M. Sakhno, A. Golenkov and F. Sizov, Uncooled detector challenges: Millimeter-wave and terahertz long channel field effect transistor and Schottky barrier diode detectors, J. Appl. Phys. 114 (2013) 164503. https://doi.org/10.1063/1.4826364

[19] D.B. But, C. Drexler, M.V. Sakhno, N. Dyakonova, et al., Nonlinear photoresponse of field effect transistors terahertz detectors at high irradiation intensities, J. Appl. Phys. 115 (2014) 164514. https://doi.org/10.1063/1.4872031

[20] S. Regensburger, A. Mukherjee, S. Schonhuber, M.A. Kainz, et al., Broadband terahertz detection with zero-bias field-effect transistors between 100 GHz and 11.8 THz with a noise equivalent power of 250 pW/√Hz at 0.6 THz, IEEE Trans. Terahertz Sci. Technol. 8 (2018) 465–471. https://doi.org/10.1109/TTHZ.2018.2843535

[21] A.J.M. Kreisler, Submillimeter wave applications of submicron Schottky diodes, Proc. SPIE. 666 (1986) 51–63, in: Far-Infrared Science and Technology, Ed. J.A.R. Izatt, 1986. https://doi.org/10.1117/12.938820

[22] K.J. Button, Infrared and Millimeter Waves, Vol. 6: Systems and Components, New York–London, Academic Press, 1982. https://doi.org/10.1007/978-1-4615-7766-9

[23] R.J. Keyes, Optical and Infrared Detectors, Springer, Berlin, 1983.

[24] E.L. Dereniak, G.D. Boreman, Infrared Detectors and Systems, Wiley Interscience, New York, 1996.

[25] G.H. Rieke, Detection of Light. From the Ultraviolet to the Submillimeter. Cambridge University Press, Cambridge, 2003. https://doi.org/10.1017/CBO9780511606496

[26] Lee Y.-S., Principles of Terahertz Science and Technology, Springer, New York, 2009.

[27] A. Rostami, H. Rasooli, H. Baghban, Terahertz Technology: Fundamentals and Applications, Spinger, Berlin, 2011. https://doi.org/10.1007/978-3-642-15793-6

[28] C. O'Sullivan, J.A. Murphy, Field Guide to Terahertz Sources, Detectors, and Optics, SPIE Press, Bellingham, 2012. https://doi.org/10.1117/3.952851

[29] F. Sizov, Terahertz radiation detectors: the state of the art, Semicond. Sci. Technol. 33 (2018) 123001. https://doi.org/10.1088/1361-6641/aae473

[30] M.A. Kinch, Fundamentals of Infrared Detector Materials, SPIE Press, Bellingham, 2007. https://doi.org/10.1117/3.741688

[31] F. Sizov, V. Reva, A. Golenkov, V. Zabudsky, Uncooled detector challenges for THz/sub-THz arrays imaging. J. Infrared, Millimeter, and Terahertz Waves. 32 (2011) 1192–1206. https://doi.org/10.1007/s10762-011-9789-2

[32] X.G. Guo, J.C. Cao, R. Zhang, Z.Y. Tan, H.C. Liu, Recent progress in terahertz quantum-well photodetectors, IEEE J. Selected Topics Quant. Electron. 19 (2013) 8500508. https://doi.org/10.1109/JSTQE.2012.2201136

[33] F. Castellano, Quantum well photodetectors, in: M. Perenzoni, D.J. Paul (eds.), Physics and Applications of Terahertz Radiation, Springer, Dordrecht, 2014, pp. 3–34. https://doi.org/10.1007/978-94-007-3837-9_1

[34] A. Betz, R. Boreiko, Y. Zhou, J. Zhao, et al., HgCdTe photoconductive mixers for 3–15 terahertz, 14th International Symposium on Space Terahertz Technology, April 22–24, 2003, Loews Ventana Canyon Resort Tucson, Arizona, pp. 102–111.

[35] P. Bhattacharya, X. Su, G. Ariyawansa, and A.G.U. Perera, High-temperature tunneling quantum-dot intersublebel detectors for mid-infrared to terahertz frequencies, Proc. IEEE. 95 (2007) 1828–1837. https://doi.org/10.1109/JPROC.2007.900968

[36] B.A. Volkov, L.I. Ryabova, D.R. Khokhlov, Mixed-valence impurities in lead telluride-based solid solutions, Physics-Uspekhi. 45 (2002) 819–846. https://doi.org/10.1070/PU2002v045n08ABEH001146

[37] P.L. Richards, L.T. Greenberg, Infrared detectors for low-background astronomy. Incoherent and coherent devices from one micrometer to one millimeter, Ch. 3, in: K.J. Button (Ed.), Infrared and Millimeter Waves, Vol. 6: Systems and Components, New York–London, Academic Press, 1982, pp. 149–210.

[38] J. Zmuidzinas, Thermal noise and correlations in photon detection, Appl. Optics. 42 (2003) 4989–5008. https://doi.org/10.1364/AO.42.004989

[39] G. Chattopadhyay, Submillimeter-wave coherent and incoherent sensors for space applications, in: S.C. Mukhopadhyay, R.Y.M. Huang (Eds.), Sensors, Advancements in Modeling, Design Issues, Fabrication and Practical Applications, Springer, Berlin–Heidelberg, 2008, pp. 387–414. https://doi.org/10.1007/978-3-540-69033-7_19

[40] N.S. Kopeika, A System Engineering: Approach to Imaging, SPIE Optical Eng. Press, Bellingham, Washington, 1998. https://doi.org/10.1117/3.2265069

[41] Z. Chen, C.-C. Wang, H.-C.Yao, P. Heydari, A Bi-CMOS W-band 2×2 FPA with

on-chip antenna, IEEE J. Sol.-St. Circuits. 47 (2012) 2355–2371.
https://doi.org/10.1109/JSSC.2012.2209775

[42] T.W. Crowe, R.J. Mattauch, H.-P. Roser, W.L. Bishop, et al., GaAs Schottky diodes for THz mixing applications, Proc. IEEE. 80 (1992) 1827–1841.
https://doi.org/10.1109/5.175258

[43] A. Van Der Ziel, Noise in Measurements. Wiley Interscience, New York, 1976.

[44] N.B. Lukyanchikova, Noise Research in Semiconductor Physics. Gordon and Breach Science Publishers, Amsterdam, 1996.

[45] B.S. Karasik, A.V. Sergeev, D.E. Prober, Nanobolometers for THz photon detection, IEEE Trans. Terahertz Sci. Technol. 1 (2011) 97–111.
https://doi.org/10.1109/TTHZ.2011.2159560

[46] M. Harwit, G. Helou, L. Armus, C.M. Bradford, et al., Far-infrared/submillimeter astronomy from space tracking an evolving Universe and the emergence of life, https://asd.gsfc.nasa.gov/cosmology/spirit/FIR-SIM_Crosscutting_White_Paper.pdf. (White paper).

[47] M.C.E. Huber, A.J. Pauluhn, L.J. Culhane, G. Timothy, K. Wilhelm, A. Zehnder, Observing Photons in Space. A Guide to Experimental Space Astronomy, Springer, New York–Heidelberg–Dordrecht–London, 2013. https://doi.org/10.1007/978-1-4614-7804-1

[48] D. Farrah, K.E. Smith, D. Ardila, Ch.M. Bradford, *et al.*, Review: far-infrared instrumentation and technological development for the next decade, J. Astronom. Telescopes, Instrum. Systems. 5 (2019) 020901. (Open Access).
https://doi.org/10.1117/1.JATIS.5.2.020901

[49] D. Leisawitz, NASA far-IR/submillimeter roadmap missions: SAFIR and SPECS, Adv. Space Research. 34 (2004) 631–636. https://doi.org/10.1016/j.asr.2003.06.023

[50] A. Van Der Ziel, Noise in Solid State Devices and Circuits, Wiley Interscience, New York, 1986.

[51] M.J. Buckingham, Noise in Electronic Devices and Systems, E. Horwood, New York, 1983.

[52] G.M. Voellmer, C.A. Allen, M.J. Amato, S.R. Babu, *et al.*, Design and fabrication of two-dimensional semiconducting bolometer arrays for HAWC and SHARC-II, Proc. SPIE. 4855 (2003) 63–72. https://doi.org/10.1117/12.459315

[53] J.G. Staguhn, D.J. Benford, F. Pajot, T.J. Ames, *et al.*, Astronomical demonstration of superconducting bolometer arrays, Proc. SPIE. 4855 (2003) 100–107.
https://doi.org/10.1117/12.459377

[54] D.M. Glowacka, M. Crane, D.J. Goldie, S. Withington, A fabrication route for arrays of ultra-low-noise MoAu transition edge sensors on thin silicon nitride for space applications, J. Low Temp. Phys. 167 (2012) 516–521. https://doi.org/10.1007/s10909-012-0580-0

[55] M.D. Audley, G. de Langeb, J.-R. Gao, P. Khosropanah, *et al.*, Optical performance of prototype horn-coupled TES bolometer arrays for SAFARI, Rev. Sci. Instrum. 87 (2016) 043103. https://doi.org/10.1117/12.2231088

[56] C.M. Posada, P.A.R. Ade, A.J. Anderson, J. Avva, *et al.*, Large arrays of dual-polarized multichroic TES detectors for CMB measurements with the SPT-3G receiver, Proc. SPIE. 9914 (2016) 9914E.

[57] Information on https://en.wikipedia.org/wiki/Heterodyne.

[58] A.T. Forrester, R.A. Gudmundsen, P.O. Johnson, Photoelectric mixing of incoherent light, Phys. Rev. 99 (1955) 1691–1700. https://doi.org/10.1103/PhysRev.99.1691

[59] J. La Tourrette, S. Jacobs, Technical Note on Heterodyne Detection in Optical Communications. Syosset, New York, 1962.

[60] T.G. Blaney, Signal-to-noise ratio and other characteristics of heterodyne radiation receivers, Space Sci. Rev. 17 (1975) 691–702. https://doi.org/10.1007/BF00727583

[61] G. Chattopadhyay, Sensor technology at sub-millimeter wavelength for Space applications, Int. J. Smart Sensing Intell. Systems. 1 (2008) 1–20. https://doi.org/10.21307/ijssis-2017-275

[62] U.U. Graf, C.E. Honingh, K. Jacobs, J. Stutzki, Terahertz heterodyne array receivers for astronomy, J. Infrared, Millimeter, and Terahertz Waves. 36 (2015) 896–921. https://doi.org/10.1007/s10762-015-0171-7

[63] I. Tretyakov, S. Ryabchun, M. Finkel, A. Maslennikova, *et al.*, Low noise and wide bandwidth of NbN hot-electron bolometer mixers, Appl. Phys. Lett. 98 (2011) 033507. https://doi.org/10.1063/1.3544050

[64] D.L. Spears, IR detectors: heterodyne and direct, in: D.K. Killinger, A. Mooradian (Eds.), Optical and Laser Remote Sensing, Springer, Berlin, 1983, pp. 278–286. https://doi.org/10.1007/978-3-540-39552-2_36

[65] K. Shimoda, H. Takahasi, C.H. Townes, Fluctuations in amplification of quanta with application to maser amplifier, J. Phys. Soc. Japan. 12 (1957) 686–700. https://doi.org/10.1143/JPSJ.12.686

[66] G. Chattopadhyay, Terahertz Technology and Applications, California Institute of Technology (2014), https://www.e-fermat.org/files/multimedias/153374fc8ebe1c.pdf.

[67] H.A. Haus, J.A. Muller, Quantum noise in linear amplifier, Phys. Rev. 128 (1962)

2407–2413. https://doi.org/10.1103/PhysRev.128.2407

[68] A.M. Baryshev, R. Hesper, F. P. Mena, T.M. Klapwijk, *et al.*, The ALMA band 9 receiver, Astronomy and Astrophys. 577 (2015) A129.

[69] E. Novoselov, S. Cherednichenko, Low noise terahertz $MgB_2$ hot-electron bolometer mixers with an 11 GHz bandwidth, Appl. Phys. Lett. 110 (2017) 032601. https://doi.org/10.1063/1.4974312

[70] J.W. Kooi, Advanced receivers for submillimeter and far infrared astronomy, Thesis, Groningen, 2008.

[71] A.C. Balanis, Antenna Theory: Analysis and Design, third ed., Wiley, New York, 2005.

[72] P.L. Richards, Bolometers for infrared and millimeter waves, J. Appl. Phys. 76 (1994) 1–24. https://doi.org/10.1063/1.357128

[73] P.H. Siegel, Terahertz Technology, IEEE Trans. Microwave Theory Tech. 50 (2002) 910–928. https://doi.org/10.1109/22.989974

[74] A. Rogalski, F. Sizov, Terahertz detectors and focal plane arrays, Opto-Electr. Rev. 19 (2011) 346–404. https://doi.org/10.2478/s11772-011-0033-3

[75] J. Zmuidzinas, P.L. Richards, Superconducting detectors and mixers for millimeter and sub-millimeter astrophysics, Proc. IEEE. 92 (2004) 1597–1616. https://doi.org/10.1109/JPROC.2004.833670

[76] T.G. Phillips, J. Keene, Submillimeter astronomy, Proc. IEEE. 80 (1992) 1662–1678. https://doi.org/10.1109/5.175248

[77] E.R. Brown, Fundamentals of terrestrial millimetre-wave and THz remote sensing, Inter. J. High Speed Electronics and Systems. 13 (2003) 995–1097. https://doi.org/10.1142/S0129156403002125

[78] J.R. Tucker, M.J. Feldman, Quantum detection at millimeter wavelength, Rev. Modern Phys. 57 (1985) 1055–1113. https://doi.org/10.1103/RevModPhys.57.1055

[79] J. Wei, D. Olaya, B.S. Karasik, S.V. Pereverzev, *et al.*, Ultrasensitive hot-electron nanobolometers for terahertz astrophysics, Nat. Nanotechnol. 3 (2008) 496–500. https://doi.org/10.1038/nnano.2008.173

[80] G. Chattopadhyay, Future of heterodyne receivers at submillimeter wavelengths, Digest IRMMW-THz-2005 Conf., Williamsburg, 2005, pp. 461–462.

[81] C.M. Bradford, B.J. Naylor, J. Zmuidzinas, J.J. Bock, et al., WaFIRS: A waveguide far-IR spectrometer: Enabling spectroscopy of high-z galaxies in the far-IR and sub-millimeter, Proc. SPIE. 4850 (2003) 1137–1148. https://doi.org/10.1117/12.461572

[82] M. Kenyon, P.K. Day, C.M. Bradford, J.J. Bock, H.G. Leduc, Progress on

background-limited membrane-isolated TES bolometers for far-IR/submillimeter spectroscopy, Proc. SPIE. 6275 (2006) 627508. https://doi.org/10.1117/12.672036

[83] B.S. Karasik, D. Olaya, J. Wei, S. Pereverzev, *et al.*, Record-low NEP in hot-electron titanium nanobolometers, IEEE Trans. Appl. Supercond. 17 (2007) 293–297. https://doi.org/10.1109/TASC.2007.897167

[84] J.C. Mather, E.S. Cheng, D.A. Cottingham, R.E. Eplee, et al., Measurement of the cosmic microwave background spectrum by the COBE FIRAS instrument, Astrophysical J. 420 (1994) 439–444. https://doi.org/10.1086/173574

[85] J. Dunkley, A. Amblard, C. Baccigalupi, M. Betoule, *et al.*, A program of technology development and of sub-orbital observations of the cosmic microwave background polarization leading to and including a satellite mission, A Report for the Astro-2010 Decadal Committee on Astrophysics, April, 2009, http://cmbpol.uchicago.edu/depot/pdf/white-paper-3.pdf.

[86] A.E. Siegman, The antenna properties of optical heterodyne receivers, Proc. IEEE. 54 (1966) 1350–1356. https://doi.org/10.1109/PROC.1966.5122

[87] E.R. Brown, Fundamentals of terrestrial millimeter-wave and THz remote sensing, in: D.L. Woolard, W.R. Loerop, M.S. Shur (Eds.), Terahertz Sensing Technology, Vol. 2, World Scientific, New York, 2003, pp. 1–106.

[88] L. Liu, J.L. Hesler, H. Xu, A.W. Lichtenberger, R.M. Weikle, A broadband quasi-optical terahertz detector utilizing a zero bias Schottky diode, IEEE Microw. Wireless Components Lett. 20 (2010) 504–506. https://doi.org/10.1109/LMWC.2010.2055553

[89] S. Preu, M. Mittendorf, S. Winnerl, H. Lu, *et al.*, Ultra-fast transistor-based detectors for precise timing of near infrared and THz signals, Opt. Exp. 21 (2013) 17941–17950. https://doi.org/10.1364/OE.21.017941

[90] H.-W. Hübers, Terahertz heterodyne receivers, IEEE J. Sel. Top. Quantum. Electron. 14 (2008) 378–391. https://doi.org/10.1109/JSTQE.2007.913964

[91] G.N. Gol'tsman, Yu.B. Vachtomin, S.V. Antipov, M.I. Finkel, et al., NbN phonon-cooled hot-electron bolometer mixer for terahertz heterodyne receivers, Proc. SPIE. 5727 (2005) 95–106. https://doi.org/10.1117/12.590490

[92] T.G. Phillips, K.B. Jeffers, A low temperature bolometer heterodyne receiver for millimeter wave astronomy, Rev. Sci. Instrum. 44 (1973) 1009–1014. https://doi.org/10.1063/1.1686288

[93] T.G. Phillips, InSb heterodyne receivers for submillimeter astronomy, Proc. SPIE. 280 (1981) 101–107. https://doi.org/10.1117/12.931954

[94] B.A. Weber, S.M. Kulpa, A high-sensitivity near-millimeter-wave photoconductive detector using HgCdTe, Int. J. Infrared Millim. Waves. 3 (1982) 235–240.

https://doi.org/10.1007/BF01007098

[95] F. Khodayar, S. Sojasi, X. Maldague, Infrared Thermography and NDT: 2050 Horizon, Quantitative Infra-Red Thermography. 13 (2016) 210–231. https://doi.org/10.1080/17686733.2016.1200265

[96] A. Karim, J.Y. Andersson, Infrared detectors: Advances, challenges and new technologies, IOP Conf. Series: Mater. Sci. Eng. 51 (2013) 012001. https://doi.org/10.1088/1757-899X/51/1/012001

[97] L. Zhou, C.-C. Wang, Z. Chen, P. Heydari, A W-band CMOS receiver chipset for millimeter-wave radiometer systems, IEEE J. Sol.-St. Circ. 46 (2011) 378–391. https://doi.org/10.1109/JSSC.2010.2092995

[98] B. Yook, K. Park, S. Park, H. Lee, *et al.*, A CMOS W-band amplifier with tunable neutralization using a cross-coupled MOS–varactor pair, Electronics. 8 (2019) 537. https://doi.org/10.3390/electronics8050537

[99] J.-H. Qiu, J. Qi, N.-nan Wang, A. Denisov, Passive Millimeter-Wave imaging technology for concealed contraband detection, in: A. Sidorenko (Ed.), Functional Nanostructures and Metamaterials for Superconducting Spintronics, Cham, Switzerland, 2018, pp. 129–159. https://doi.org/10.1007/978-3-319-90481-8_7

[100] J. Grzyb, U. Pfeiffer, THz direct detector and heterodyne receiver arrays in silicon nanoscale technologies, J Infr. Milli Terahz Waves. 36 (2015) 998–1032. https://doi.org/10.1007/s10762-015-0172-6

[101] S.-P. Han, H. Ko, J.-W. Park, N. Kim, *et al.*, InGaAs Schottky barrier diode array detector for a real-time compact terahertz line scanner, Opt. Express. 21 (2013) 25874–25882. https://doi.org/10.1364/OE.21.025874

[102] P. Hillger, J. Grzyb, R. Jain, U.R. Pfeiffer, Terahertz imaging and sensing applications with silicon-based technologies, IEEE Trans. Terahertz Sci. Technol. 9 (2019) DOI:10.1109/TTHZ.2018.2884852. https://doi.org/10.1109/TTHZ.2018.2884852

[103] Z. Ahmad, A. Lisauskas, H.G. Roskos, K.K. O, Design and demonstration of antenna coupled Schottky diodes in a foundry complementary metal-oxide semiconductor technology for electronic detection of far-infrared radiation, J. Appl. Phys. 125 (2019) 194501. https://doi.org/10.1063/1.5083689

# Chapter 5. Detectors

## 5.1 Remarks

To get more information in IR and THz spectral ranges, an increase in array size is required, since large arrays substantially multiply the information capacity and data output. *E.g.*, the development of large THz (far-IR) format sensors for ground-based astronomy is desirable for many astronomic observatories.

For uncooled thermal IR (microbolometer) arrays, which are currently produced in larger volume as compared to other cooled down devices, the number of sensitive pixels in separate arrays that reaches the value $\geq 10^6$ with the pixel pitch down to 12 μm [1] and less. These arrays and cameras based on these are widely used for reconnaissance, surveillance, medical diagnostics, airborne applications, firefighting, automotive industry, and other applications. The trades between system MTF and smaller pixels allowing smaller die suggests that the pixel sizes may be reduced to between 5 and 10 μm for uncooled LWIR applications. The benefits of doing so are smaller optics, smaller die and systems, and less weight and power, assuming that performance metrics are maintained [2].

Uncooled thermal detector (*e.g.*, semiconducting and metal bolometers, pyroelectric detectors) arrays for IR and THz ranges mainly have disadvantage as compared to their cooled ones especially for astronomy or military applications. They are less sensitive in IR military or special applications and much less sensitive as arrays and detectors, *e.g.*, for astronomical applications (several orders less sensitive) and in the most cases they are intrinsically slow and can not be used in coherent devices. These uncooled thermal arrays are relatively cost-effective in many applications and can provide an acceptable image quality. However, in applications needed of high image quality, multispectral operation or frame rate they are less acceptable or inacceptable at all.

## 5.2 Incoherent (direct) detectors

Detectors with direct signal detection are basically used in spectroscopic and vision systems ($\nu/\Delta\nu < 10^3$), in which high spectral resolution is not needed. In general, one of the advantages of the direct detection systems is their relative simplicity and the possibility to design large format arrays. These arises mainly because unlike the coherent systems, in the direct detection systems, there does not exist the problem of multielement array formation conditioned by local oscillator (LO) power and fast detector response ($\tau \sim 10^{-10}...10^{-11}$ s) defining the broad IF spectral bandwidth in coherent receivers.

The incoherent detectors and arrays operate mainly in the IR ranges and are preferable in the THz range $v > 0.3$ THz (see part 4.5), as there upper limit performance is background noise limited in contrast to coherent receivers, which performance is quantum noise limited, and the latter is increased with radiation frequency increase. Moreover, in the coherent receivers the noise would emerge from the amplifier output even when no radiation signal is present at the input.

For astronomical applications, the most suitable as incoherent detectors are the superconducting bolometers and extrinsic detectors. They are among the most sensitive direct detectors with NEP $\sim 10^{-19} \ldots 10^{-14}$ W/Hz$^{1/2}$. In the direct detection systems, contrary to the coherent detection receivers, principally there is no fundamental limit to the sensitivity, when the background fluctuation noise predominates. With non-photoconductive detectors, the minimum detectable power in the BLIP conditions is defined by Eq. 4.13 and no signal is produced (at $W_B = 0$) unless photons are being absorbed. Coherent receivers are well applicable in spectroscopy and astronomy in resolving and determining ("finger print") the atomic and molecular lines.

### 5.3 Photon (quantum) detectors

Photon (quantum) detectors are widely used to detect radiation in various spectral ranges – from UV to high frequency THz. Among the photon detectors, the semiconductor-based ones are used to detect radiation in the visible and THz ($v \geq 1.5$ THz) ranges as intrinsic (interband), extrinsic (optical transitions between the states in the gap and conduction or valence band) and intersubband (QW, QD) detectors. In a low-frequency part of the THz spectral range, they can be used as pair breaking detectors in which the photon energy overcomes the binding gap.

Radiation in the waveband $\sim (0.2 \ldots 10)$ THz in the Earth's atmosphere is strongly absorbed by the water vapor, which makes this wavelength band not very attractive for Earth THz remote sensing applications. Operation of semiconductor photon detectors in this region is strongly limited by the thermo-generation rate. Advances in research and fabrication of detectors and arrays in this radiation frequency range are mainly concentrated on cooled systems, which are used for space and ground based astronomy applications [3, 4]. Extrinsic cold THz detector arrays also have found their niche in the long-wavelength strategic systems, *e.g.*, for defence space vehicles. QWs detectors have found mainly some kind of practical field THz applications at $v > 3$ THz. For the IR range applications the fundamental and technological issues associated with the development of IR detector technologies are implemented, *e.g.*, in Refs [5, 6]. In the IR spectral range, attention today is focused on intrinsic HgCdTe ternary alloy detectors and type II SLs (T2SLs) detectors, barrier and quantum well detectors and the extrinsic ones.

The intrinsic HgCdTe arrays have been grown to megapixel dimensions and are ubiquitous in ground-based and space-based astronomical observatories.

### 5.3.1   Intrinsic (interband) detectors

The intrinsic detectors based on interband transitions are not practically used as THz detectors at $v \leq 10$ THz, because of too high thermogeneration rate (noise) of carriers and also for the reason that their narrow-gap is not well controllable in this frequency region (*e.g.*, in HgCdTe, PbSnTe semiconductors). Theoretically, one could also make zero-band-gap III-V alloys (*e.g.*, InTlP, InTlAs, or InTlSb), but these materials have been difficult to fabricate epitaxially [7].

The intrinsic IR semiconductor detectors are mostly used in the spectral range $\lambda \sim 1...20$ μm and operate almost at BLIP conditions (Fig. 3.14). The development of IR detector technology was in fact dominated by photon detectors up to the end of the 20th century. The essential drawback of photon detectors is the need of cryogenic cooling. This is necessary to prevent the charge carriers thermal generation and is a reason of cost ineffective and bulky systems.

Beginning from the late 1970s, the progress in the number of detectors in the IR detector arrays, which revolutionized IR technologies and made them much more cost effective. It was primarily connected with application of silicon readout circuits (ROICs). Assembling ROICs with different types of detectors allowed to build up the IR focal plane arrays (FPAs), which now can contain about $10^8$ IR detectors (mosaic arrays, see, *e.g.*, Fig. 4.2). Applications of these technologies allowed a discretization of the process of image creation as well as its processing by the instrumentality of linear and matrix detector arrays from discrete elements.

Parameters of IR detectors and FPAs (*e.g.*, their sensitivity, which can be characterized by noise equivalent power (NEP), detectivity (D*) or noise equivalent temperature difference (NETD), dimensions of sensitive elements, and some others) are critical in the final analysis of lateral resolution, objects detection, recognition and identification ranges.

From the beginning of the development of IR technologies (see Ch. 2), many materials have been realized for IR detectors. Here, the main attention is attained to uncooled detectors, which can be used for thermovision applications in medicine. Therefore, the discussion for quantum detectors is narrowed actually to IR MWIR and LWIR detectors and FPAs (spectral range $\lambda \approx 3...5$ and $8...12$ μm, respectively) based on HgCdTe ternary alloys as the most currently advanced and used. The typical spectral dependences of sensitivity for intrinsic IR photo-detectors are shown in Figs. 3.5 and 3.7.

At present, the use of MCT-based FPA detectors for IR imaging applications in (1.3…2.5), (3…5) and (8…12) μm wavelength regions, is well established. These ternary solutions have tailorable energy band gap over the entire spectral range λ ≈ 1…30 μm and the lattice constant changes are small over these spectral ranges. The short, medium and long-wave infrared HgCdTe photodiode arrays have the most suitable parameters for ultimate performance IR technology applications. With CMOS ROICs they are commercially available from several manufactures. They can be produced as single crystal arrays with format 4096 × 4096 [8] and pixel pitch up to 10 μm for the spectral range 1.0…5.4 μm operating at T < 40 K. For multielement two-color or multicolor photodetectors, the planar method of photodiode formation is the most widespread and developed successfully. For the spectral range 3…5 μm the 1920 × 1536 large format arrays based on of InSb binary semiconductor with the pitch 10 μm now are also attainable [6].

The short wavelength infrared (SWIR) (λ ≈ 1.0…2.5 μm) and mead wavelength infrared (MWIR) (λ ≈ 3…5 μm) ranges are now typically observed using HgCdTe detectors instead of InSb, owing to the higher dark currents in the latter one for comparable wavelength cutoffs and operation temperatures. Operation of InSb detectors have been mainly limited by MWIR applications in the atmospheric window range 3…5 μm. InSb is a binary compound and is more stable as compared to ternary HgCdTe alloy. The InSb detectors and arrays operate at temperatures not much in excess of T ~ 80 K [5]. To a great degree, the reason of it is that $dE_g/dT < 0$ (the band gap $E_g$ of InSb decreases with the temperature increase and, therefore, the thermal generation rate increases).

The various x in compositions of the ternary alloy $Hg_{1-x}Cd_xTe$ cover the entire IR range (λ ~ 1…20 μm) with a capability of operating at temperatures for SWIR and MWIR ranges, essentially higher as compared to T ~ 80 K. At present time, the only material that meets this requirement is HgCdTe, as in it $dE_g/dT > 0$ (the band gap enlarges with the temperature increase). Therefore, smaller dark currents, especially in p-i-n structures can be realized [5, 9]. Moreover, because of the requirements of the astronomy community for space observation, it is a need of large number of infrared pixels in arrays. It is affordable with HgCdTe arrays grown on large silicon substrates [10] up to the very large formats (> 8K × 8K, pitch ≈15 μm) for SWIR and MWIR spectral ranges. Today attainable are single crystal HgCdTe IR arrays with 2048 × 2048 pixels and 20-μm pitch.

Nowadays, there occurs the tendency to reduce the detector pitch down to d ≈ 4 μm [5] for detectors in arrays with $λ_{co} ≤ 5$ μm. For LWIR arrays, the possible pitch can be shortened to d ~ 5 μm for detectors with the center wavelength λ ≈ 10 μm [2]. A similar trend is also representative for THz arrays. This tendency is based on the absolute resolution limit according to the parameter λF/# / (d), where F/# is the optics f-number.

Materials Research Forum LLC  
https://doi.org/10.21741/9781644900758

To describe the parameters of IR or THz vision systems in different approximations, the MTF function is used (see Ch. 3).

### 5.3.2 Extrinsic detectors

The large extrinsic semiconductor infrared detector arrays are mainly manufactured for astronomy applications. They are used in instruments for investigation of birth and evolution of planets, stars and galaxies. Spectral dependences of sensitivity of some extrinsic detectors are shown in Fig. 3.7. Up to date the most progress is achieved with Si:As block impurity band (BIB) detector arrays of up to 2048 × 2048 pixels [11].

Extrinsic THz direct photon detectors (photoconductors) are fabricated by shallow impurities doping the semiconductors Ge, Si, GaAs. In these detectors, the electrons bounded by shallow-impurity centres (e.g., B, In, Ga, Be acceptors in Ge, Si and Sb donors in GaAs and Ge, respectively), at low temperatures can be excited to conductive band by THz radiation. The position of energy states of impurities can be tuned by strain [12, 13], doping concentration [14, 15] or magnetic field. Now extrinsic detector arrays are the principal photon semiconducting detectors operating in THz frequency range $v >$ 1.5 THz at low level background fluxes.

Extrinsic photoconductors have been developed several decades ago and can detect energies down to 6 meV ($v \sim$ 1.5 THz, $\lambda \sim$ 200 μm). It corresponds to the ionization energies of shallow donors (for example, Si or Te donors in GaAs) and acceptors (Ga acceptors in stressed Ge). BIB GaAs detectors would show an onset beyond 30 cm$^{-1}$ ($\sim$ 330 μm) [16, 17]. The donor ground state to conduction band transition in n-type unstressed gallium arsenide (e.g., GaAs:Te) has the ionization energy 5.7 meV, as compared to the ionization energy 11.3 meV of gallium in unstressed germanium, but the purity of semiconductor is poor for these aims. Therefore, at THz radiation frequencies $v <$ 1.5 THz, where now there are no suitable photoconductors, the thermal sensors (bolometers, pyroelectrics or thermocouples) must be used.

The majority carrier time in these detectors is determined by the density of ionized empty donor or acceptor states and lies within several ns [18] to tenths and hundreds ns range. The frame rate of large Si doped arrays in dependence of application aims is within f $\sim$ 0.1...500 Hz [11]. The responsivity increases with bias up to a certain voltage value [19].

The most used extrinsic detectors are Ge and Si. Between these two materials, Si has several advantages over Ge:

- Three orders of magnitude higher impurity solubilities (hence higher absorption coefficient and thinner detectors, which leads to better spatial resolution),

- Si has a lower dielectric constant than Ge,

- Si detectors are characterized by superior hardness in nuclear radiation environments,

- Si-MOS technologies and related device technology of Si detectors are more developed and the surface passivation is much more robust and stable,

- For Ge, the diffusion lengths are larger, which requires larger pixel pitch to suppress crosstalk. However, for wavelengths longer than 40 μm ($v \leq 7.5$ THz) there are no appropriate shallow dopants for Si (the longest cut-off wavelength BIB detector has been implemented in Si:Sb, lowering the detection onset to ~41 μm) [20], as compared to doped Ge photoconductors. Therefore, Ge extrinsic devices are the only photoconductive detectors for the long-wave region $\lambda \sim 40...220$ μm (see Fig. 5.1 and Fig. 3.7).

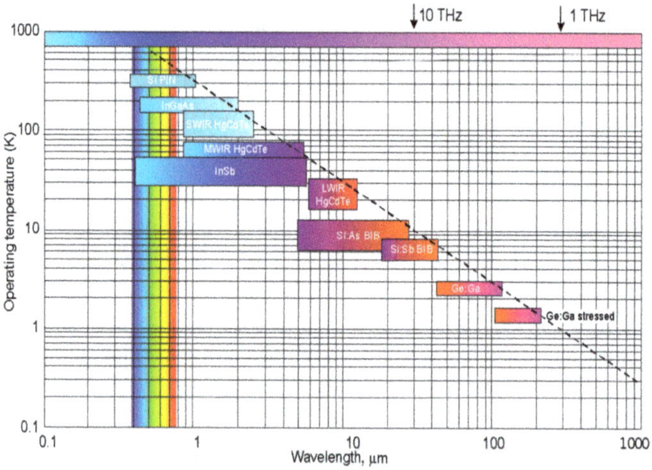

*Figure 5.1. Operation temperature for low-background photon detectors after [21]. The dashed line indicates the trend for longer wavelength detection. THz detectors ($v < 10$ THz, $\lambda > 30$ μm) are mainly composed as BIB detectors. Here, PIN is the p-type-intrinsic-n-type semiconductor diode. (By permission of Elsevier).*

Ga acceptors in Ge have the ionization energy close to 11 meV, corresponding to a cutoff wavelength of 113 μm [22]. The response of Ge:Ga photoconductors can be extended to longer wavelengths (~220 μm) by applying the uniaxial compressive stress [13]. This effects the splitting of four-fold degeneracy of the $\Gamma_4$ valence band edge and reduces the energy difference between the bound acceptor states and top of the valence band. The drawback of stressed Ge detectors is the difficulties in fabricating large arrays.

To increase the absorption coefficient in doped photoconductors, one should increase the impurity concentrations. However, high impurity concentration is a reason of overlapping the wave functions of impurities and arising an impurity band. It increases the dark current and noise, limiting the detector resistance and photoconductive gain. Therefore, relatively low level of impurity concentration is desirable.

This contradiction was resolved using separate layers for photon absorption and carrier conductivity. In the BIB detector, the thin intrinsic blocking layer is adjacent to the heavily doped absorbing layer. Because of it, optimization of optical and electrical properties of the detector independently in BIB detectors was achieved [23–27].

Beside optimization of optical and electrical properties, the presence of thin intrinsic blocking layer adjacent to a heavily doped absorbing layer (impurity band) leads to other advantages. E.g., a higher doping concentration of the absorbing layer can be a reason for broadening the absorption band, shifting a cut-off edge to longer wavelength response, reducing the detectors area and making the device to be radiation harder. The BIB photoconductor schematic is shown in Fig. 5.2. As any extrinsic THz detector, the BIB one to get ultimate performance has to be used at low temperatures $T \leq 4$ K.

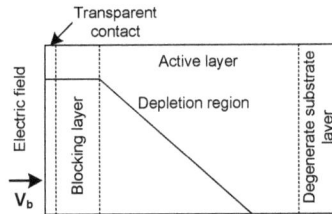

*Figure 5.2. Schematic cross-section of typical blocked impurity band detector.*

The IR or THz active region of BIB detector, e.g., using n-type material layer, is sandwiched between a degenerate substrate layer and an undoped (high resistance) blocking layer. Excitation of electrons by radiation flux in active layer is related with optical transitions from the donor impurity band to the conduction band.

To transfer electrons from the IR or THz active layer to the electrodes, the electric field with the bias voltage $V_b$ should be applied. The IR active layer has low resistance, and blocking layer has high resistance. Therefore, the electric field mostly exists in the blocking layer, and excited electrons move from n-type IR active region to the transparent contact. The applied voltage defines the width of the depletion layer (and hence the quantum efficiency of the detector).

The heavier doped the active layer is in the BIB detector, the longer cutoff wavelength can be attained. Thus, operation over a broader spectral range is possible, and lower impedances lead to faster response time. In BIB detectors, recombination occurs in the layer of high conductivity. Therefore, just the noise from the generation process is present, and thus, the generation-recombination noise is suppressed.

The THz BIB detectors were mainly designed with Ge:Ga [13]. Still, the BIB Ge-based IR detector ($\lambda \approx 2...9.5$ μm) can be obtained by sulphur ion implantation [28]. For long IR and THz spectral ranges, Si:As BIB arrays [29] that are sensitive in the spectral range $\lambda \sim 2...30$ μm were designed too.

It seems important to pay attention to GaAs:S [30] that can be sensitive up to $\lambda \sim 300$ μm. BIB test structures were fabricated by liquid phase epitaxy (LPE) growth of high purity blocking layer onto the vapor phase epitaxy (VPE) layer [31]. The blocking layer was polished to the thickness of about 10 micrometers. Then, a transparent blocking layer top contact was formed by ion-implantation of sulfur. The doping and thickness of this contact was limited to provide sufficient transparency for far infrared (FIR) photons. There realized were active layer majority doping at $5 \cdot 10^{15}$ cm$^{-3}$, blocking layer majority doping at $5 \cdot 10^{13}$ cm$^{-3}$, and blocking layer minority doping at $5 \cdot 10^{12}$ cm$^{-3}$. The active layer thickness was 40 μm, with a blocking layer thickness of 10 μm. However, the expected sensitivity and wavelength response were not achieved (T = 1.3 K), probably because of high concentration of acceptors in the active layer and at the interface in the blocking layer. The crystalline quality, precise control of the doping concentration profile and device processing technique are still not satisfactory to get GaAs-based BIB for THz detector applications.

GaAs BIB detector with the absorption region formed by ion implantation have been designed and investigated in [32]. By four-time-implantation procedure with different implantation energies and doses, a 790 nm deep region has been obtained. The results show that this planar GaAs BIB detector can response within the spectral range 750 GHz and 1.8 THz.

Among BIB detectors, the Si-based BIB arrays and also other extrinsic detectors are most mature, and thus have been actively used. This is because quality of Si material is far superior to that of either Ge or GaAs. The response wavelength of Si:Sb BIB detectors can cover range 2...40 μm [33], and their performances are excellent in terms of quantum efficiency, dark current, linearity, and operability. Other Si-based BIB detectors were, *e.g.*, phosphorous ion implanted Si BIB devices with the cut-off wavelength $\lambda \approx 31$ μm [34]. Si:As and Si:Sb BIB arrays were widely used in different space missions.

When concerning the THz spectral region of unstressed Ge:Ga photoconductors (Fig. 5.1) they can be used as low background photon detectors for the wavelength range from 40 up to 120 μm ($v \approx 7.5...2.5$ THz) though low-temperature cooling is necessary because of large free carriers thermogeneration rate and, therefore, necessity of decreasing large dark currents. For doped Ge, the diffusion lengths are large (typically 250...300 μm) requiring larger pixel dimensions of 500...700 μm to reduce the crosstalk, compared to doped Si BIB detectors operating at shorter wavelengths [35].

Ge:Ga arrays commonly are manufactured by hand. Individual Ge:Ga pixels or linear strips of pixels are typically mounted next to one another to form the far-infrared photoconductor arrays [36]. The resulting mechanical assembly procedures are relatively complex and thus limit the available array size, so Ge:Ga arrays manufactured are the 16 × 25, 32 × 32 pixel arrays used in the 55 μm to >120 μm waveband. Recent developments have enabled fabrication of monolithic 2D arrays, with pixels formed using the lithographic and wet-etching processes. Monolithic Ge:Ga arrays can be contacted directly to Si read-out circuits using In-bump bonds to form direct hybrid arrays [26].

The quantum efficiency (QE) of extrinsic photoconductors is lower (Fig. 3.7) as compared to intrinsic IR detectors (*e.g.*, intrinsic InSb and HgCdTe detectors for IR region, QE $\approx 95...98$ %) because of limits in radiation absorption related with ultimate solubility of impurities that can be incorporated into semiconductor. Moreover, unwanted conductivity can arise when impurity atoms are too close to each other, therefore, their wavefunctions can overlap.

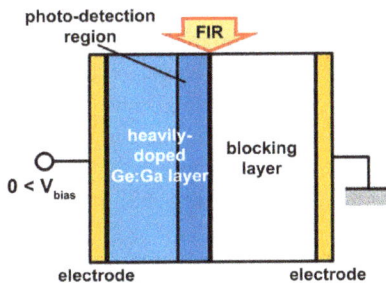

*Figure 5.3. Schematic image of the structure of the surface activated wafer bonding Ge junction device, clarifying relation of the electrodes, heavily-doped Ge:Ga layer and blocking layer, bonded interface, and direction of the incident FIR light [37]. (By permission of Springer Nature)*

Fabricated recently [37, 38] are Ge junction devices with Ge:Ga BIB structures (Fig. 5.3) by surface activated wafer bonding, covering expanded spectral range up to ~250 μm similar to the spectral region of stressed Ge:Ga BIB detector. These detectors operate at T ~ 2 K with NEP < $2 \cdot 10^{-17}$ W/Hz$^{1/2}$ and capability to achieve NEP ~ $2 \cdot 10^{-19}$ W/Hz$^{1/2}$.

Typical Ge:Ga arrays were used [39] for Hershel space telescope in the spectral regions 60...85, 85...130, 130...210 μm. The 25 × 16 pixels Ge:Ga photoconductor arrays are a modular design (Fig. 5.4). 25 linear modules of 16 pixels each are stacked together to form a contiguous, 2-dimensional array. Each linear module is read-out by a cryogenic CMOS amplifier/multiplexer circuit. Median NEP values are $8.9 \times 10^{-18}$ W/Hz$^{1/2}$ for the stressed and $2.1 \times 10^{-17}$ W/Hz$^{1/2}$ for the unstressed detectors, respectively.

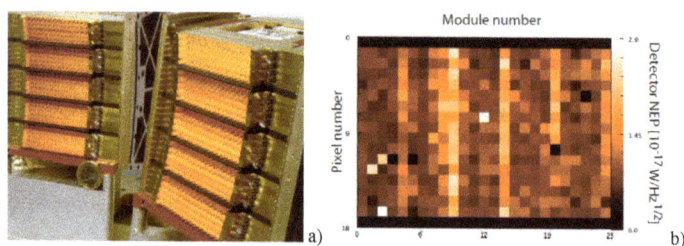

*Figure 5.4. (a) 25 × 16 stressed (back) and unstressed (front) Ge:Ga photoconductor arrays with integrated cryogenic readout electronics. (b) NEP map of the stressed arrays [39]. (By permission of SPIE).*

Much larger THz detector arrays were designed on the base of doped Si. *E.g.*, Si:As BIB megapixel arrays (1024 × 1024 and 2048 × 2048), which can operate in IR at $\lambda \approx 10...25$ μm spectral range (out of THz band), were fabricated [11, 15]. There is renewed interest in the potential for Si:Ga BIB FPAs as an alternative for imaging systems in the LWIR ($\lambda \approx 8$ to 13 μm) atmospheric window which uses either Si:As or mercury cadmium telluride FPAs. Both present technologies have advantages and disadvantages. The latter, with the cut-off wavelength tuned to the LWIR window, can operate above 40 K with compact cryocoolers, but pixel operability, uniformity and stability are poorer than those of Si:As [15].

The main advantage of germanium and silicon for extrinsic THz and IR detectors are well developed growth technologies that allows obtaining materials with perfect crystalline quality and low uncontrolled impurity concentration. Yet, probably as THz detector there can be used p-type $Hg_{1-x}Cd_xTe$ layers with shallow acceptor levels in the gap. In them, Hg vacancy is the double acceptor. In HgCdTe layers with $x \approx 0.19$ to $x \approx 0.29$ at T = 4.2

K, there was observed the photoconductivity with peaks of response at $v \approx 2.4$, 4.6 and 6.5 THz, which corresponds to double acceptor (Hg vacancy) levels [40]. As well, the THz devices based on PbSnTe:In semiconductors with the persistent response time (hundreds of seconds) at the temperatures $T < 20$ K were realized [41]. This effect is related with the specific position of the impurity level in these semiconductors, which changes with the chemical composition "x" and temperature. Because of it and the position of the Fermi level, there appears a semi-insulating region, in which a barrier exists between the conducting and localized states. Some characteristics of typical extrinsic detectors are shown in Table 5.1.

*Table 5.1. Some characteristics of typical extrinsic detectors.*

| Detectors | Operation radiation frequency, THz or wavelength range, | NEP, $W/Hz^{1/2}$ | Refs. | Remarks |
|---|---|---|---|---|
| Ge:Ga, unstressed, Ge:Ga stressed | ~80...130 μm, ~130...230 μm | $2.1 \times 10^{-17}$, $8.9 \times 10^{-18}$ | [42] | T = 65 K, Array 25 × 16 |
| Ge:Ga stressed | $\lambda_{co} \approx 220$ μm, | $5.7 \times 10^{-17}$ | [13] | T = 2.1 K, |
| Ge:Ga, unstressed | $\Lambda \sim 50...110$ μm | $7.5 \times 10^{-17}$ | [26] | T = 2.2 K, 20 × 3, array |
| Ge:B, BIB | $\lambda_{co} \approx 220$ μm | $5.23 \times 10^{-15}$ | [43] | T ≤ 1.2 K, |
| Si:Sb, BIB | $\Lambda \sim 14...40$ μm | | [26] | 128 × 128 array, T = 3.8...5 K, |
| Si:As BIB | ≈2...28 μm | - | [15, 29] | 1024 × 1024 array, T ≈ 8...11 K, 10 ms integration time |
| Si:Sb BIB | ≈5...40 μm | - | [15] | 1024 × 1024 array, T ≈ 4.5 K, |

### 5.3.3    Low dimensional (QW, SL, QD) IR and THz detectors

Among the intrinsic photon interband detectors currently used for detection of IR radiation up to $\lambda \sim 20$ μm, there exist only HgCdTe alloys, which, however, are less stable as compared to much more stable III-V alloys that enable to obtain uniform pixel characteristics in arrays. However, for detectors based on these alloys the longest III-V system (InAsSb) has the cut-off wavelength only 9 μm. Moreover, they can't be used for detection of THz radiation. Still, the developments in IR optoelectronics related with large format FPAs have led to demands for new device concept with account of uniformity, controllability and yield. One of propositions to meet these demands are based on the researches of [44], which observed direct optical transitions in low dimensional structures – quantum wells based on GaAs.

In these researches, they observed for the first time the intersubband optical transitions in the so-called "type I quantum wells". By 1987, the basic operation principles for QW IR photodetectors (now frequently cold QWIPs) demonstrating sensitive infrared detection were formulated. Many aspects of physics and applications of low dimensional structures have been considered in Ref. [45]. Schematics of optical intersubband transitions in QW and SL IR and THz structures are shown in Figs. 3.4.c,d,e. The processes of radiation absorption in quantum dots (QDs) are similar to those shown in Fig. 3.4.c for type I QW.

The nature of different types of SLs and multiple QWs (MQWs) is associated with discontinuity at the heterointerfaces between neighboring semiconductors. This discontinuity in thin layers (several or tens atomic layers) heterojunctions involves creation of specific mini-band structures taking place in different types of SLs or MQWs (see Figs. 3.4.c,d,e).

In type I SLs or MQWs (*e.g.*, GaAs/AlAs, GaSb/AlSb), electrons and holes are confined in one of the semiconductors. In type II SLs and MQWs (*e.g.*, InAs/AlSb) the top of the valence band is located above the bottom of the conduction band of another semiconductor. This type II structures shown schematically in Fig. 3.4.d are called "staggered" SLs or MQWs. In another kind of type II heterostructures, there is realized the discontinuity, at which the bottom of conduction band of one of semiconductors is located above the top of the valence band of another one ("broken" SLs or MQWs, *e.g.*, InAs/GaSb heterojunctions). Both these type II SLs are applicable as IR detectors at normal incidence of radiation. It is possible that type II SLs can be considered as alternative to intrinsic HgCdTe IR detectors.

To get the same level of information capacity for arrays with comparable format as in the case of intrinsic interband detectors, *e.g.*, on the base of HgCdTe, from the Exp. (4.1) it follows that the accumulation time in low dimensional detectors should be appreciably longer. This is because the quantum efficiency $\eta$ of these detectors is $\eta \sim 5...10$ %, which is noticeably lower in comparison with $\eta$ in quantum interband detectors ($\eta$ reaches 98 %). That is why the devices based on low dimensional structures can't follow properly fast processes like, *e.g.*, camera based on of HgCdTe arrays that can provide the frame close $\sim 2$ kHz. Moreover, the spectral bandwidth of these detectors are narrower but can be tuned to a certain wavelength. The main advantage of QW and SL detectors, especially type I ones, is the mature technology on the base of III-V alloys, which enable to obtain high uniformity of physical parameters of pixels in arrays.

The QWIPs and QDIPs arrays are mainly applied in the IR range. In spite of the fact that there are quite a lot of researches devoted to these devices, their application in THz range as detectors is relatively limited, as compared to IR spectral range. In these detectors, low-temperature operation is needed, and, as a rule, their operation is limited by spectral range of $v \geq 2$ THz because of several hurdles:

- To reach the THz low frequency ranges (shallow wells) the control of components concentration in wells should be precise because, as a rule, their concentration is small;

- Presence of large dark current density by thermally assisted tunneling of localized electrons through the barriers;

- The absorption coefficient in the well is lowered with the barrier height, thus, it is decreased with radiation frequency (for Refs. of early research see, *e.g.*, [46]). Different kinds of design inherent to QW, SL and QD detectors are presented, *e.g.*, in [47].

Because of these reasons, the sensitivity (detectivity) of such detectors is quickly decreased with decreasing the radiation frequency range (see Fig. 5.5).

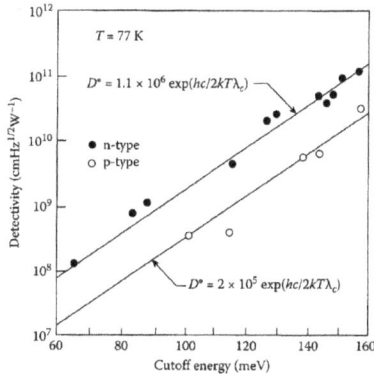

*Figure 5.5. Detectivity of n-typ and p-type GaAs/AlGaAs QWIPs. The straight lines are fits to the experimental data [48]. (By permission of AIP Publishing).*

To realize the response in these type I QWIP detectors there should be used the facet light-coupling geometry (Fig. 3.6.d) or some kind of gratings. The detectors based on III-V semiconductors can't operate within the Restrahlen band ($\lambda \sim 30...50$ μm, $v \sim 6...10$

THz, in dependence of chemical composition [49–51]), in which the absorption coefficient is very high.

Due to the nature of intersubband transitions, MQWs and SLs have a rather narrow band of response (see, *e.g.*, Fig. 3.9). Therefore, filters are not required in some spectroscopic or imaging applications. Due to the short intrinsic lifetime of photon-excited electrons, terahertz QW detectors can operate with the high response speed $\tau \leq 1$ ns [52, 53] or less ($\tau \approx 70$ ps, GaAs/AlAs QW) [54] and, therefore, can be used in high-frequency applications (heterodyne detection, THz communication, *etc.*) [55, 56] or imaging [57]. In SLs with QDs, the response time can be less than 1 ps [58].

THz QW photodetectors based on intersubband transitions have been considered relatively recently [59–61]. The detectivity was rather poor, $D^* \approx 5 \cdot 10^7$ cm $Hz^{1/2}/W^{1/2}$ with the peak absorption wavelength $\lambda = 84$ µm ($\approx 3.6$ THz) at 10 K [60].

It is typical for the THz spectral range that due to the low doping levels in these low-dimensional structures, the latter are characterized by low intersubband absorption coefficients, which limits their THz performance. These detectors based on III-V semiconductors (GaAs/$Al_xGa_{1-x}As$, GaN/$Al_xGa_{1-x}N$) can operate only in the high frequency part of THz spectra ($\nu > 2$ THz) (see, *e.g.*, [62–64].

For HgTe/HgCdTe QWs bolometric and cyclotron resonant contributions to the THz response were observed in [65] at $1.5K \leq T \leq 4.2$ K and $\nu \approx 1.7...2.5$ THz. Quantum oscillation of photocurrents in HgCdTe quantum wells at $\nu = 2.54$ THz were considered in [66]. A fast THz detector ($\tau \sim 50$ ns) with spectral tunability ($\sim 1$ THz) at the photon radiation frequency $\nu \approx 1.9$ THz) was realized on the base of HgTe quantum wells, operating at the magnetic fields (B < 2 T) [67].

There was quite a lot of researches on THz QD detectors, especially in 1990s and early 2000s, because of the need to have fast and sensitive detectors in the systems, in which QCLs are used as sources, *e.g.*, for TDS, THz communication and heterodyne systems, *etc.* [58, 68, 69].

It was supposed that in these detectors one could get improvements in operation temperature, since the thermal generation rate of electrons can be reduced due to the energy of quantization in three dimensions, which may lead to high-performance imagers operating at or close to room temperature. Important is the fact that these detectors operate at normal-incidence radiation. For THz spectral range, these detectors operate at low temperatures. *E.g.*, extremely low NEP $\sim 10^{-22}$ W/$Hz^{1/2}$ was achieved with lateral QD detector and single-electron transistor (SET) used as a sensitive charge detector [70]. This performance was reached under laboratory conditions, high magnetic fields, dilution refrigerator temperatures (T ≈ 300 mK) and precise control of the bias. Within the IR range, QD and QW detectors as a rule operate at T ~ 50...60 K.

The spectral response of QD IR detectors is tuned by varying the sizes of low dimensional structures (QWs or QDs). Presented in Fig. 5.6 are IR spectral sensitivities of the hybrid dot-in-a-well (DWELL) QDIP device heterostructure (GaAs/InGaAs QW with InAs QD). The internal absorption peak quantum efficiency η with no grating or cavity for a 30–stack DWELL QDIP is η < 3 %. If using the grating or tilted (45⁰) irradiation, η increases by several times. These spectra are similar to those observed in QWs (see Fig. 3.9)

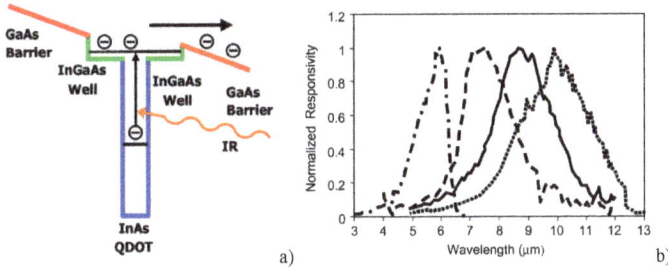

*Figure 5.6. (a) Schematic illustration of a DWELL infrared detector operation. (b) Experimentally measured spectral responsivity of some DWELL QDIPs demonstratesg spectral tunability by varying the well width from 55 to 100 Å [71]. (By permission of Elsevier).*

The device exhibit responsivity with the peak of detectivity at λ ~ 8 μm reaching D* ~ $10^{11}$ cm·Hz$^{1/2}$/W at T ~ 50 K. It was fabricated as 640 × 512 pixel QD infrared imaging FPA. Typical parameters of the QW and QD THz and IR detectors are presented in Table 5.2.

### 5.3.4    Low-temperature superconducting photon detectors

Low temperature superconductive detectors have gained applications mainly in astronomy. In these detectors, various low-temperature phenomena can take place to enable highly sensitive photon detection. Superconductive detectors can be separated into two main categories: thermal and athermal (photon) detectors.

In thermal semiconductor detectors, the superconducting material absorbs photons, and its temperature slightly increases above its transition temperature $T_c$, and a sharp change in resistance occurs. Then, the superconductor cools down by dissipating heat to the heat sink and the thermal detector returns to the superconducting state.

*Table 5.2. Typical parameters of the QW and QD THz and IR detectors.*

| Detectors | Operation radiation frequency, THz or wavelength range, | NEP, W/Hz$^{1/2}$ or D*, cm·W$^{1/2}$/W | Refs. | Remarks |
|---|---|---|---|---|
| GaAs/AlGaAs QW | ~1…6 | Responsivity R = 13 mA/W | [75] | At peak $v_p \approx$ 3 THz, T = 3 K. |
| GaAs/AlGaAs QW | $\lambda_p$ = 84 μm (≈3.6 THz) | D* = 5 × 10$^7$ | [60] | QCL like band structure, T = 10 K, |
| GaAs/AlGaAs QWIP | $\lambda_{co} \approx$ 14.1 μm (≈21.3 THz) | D* ≈ 10$^9$ | [48] | T = 77 K |
| In$_{0.4}$Ga$_{0.6}$As/GaAs multiple QD | ≈6.5…15 | D* = 1.3 × 10$^8$ (80 K), 2·10$^7$ (150 K) | [76] | At $v_p \approx$ 7.5 THz |
| InGaAs/GaAs QD | ~3…30 | D* = 2.3 × 10$^{11}$ | [74] | At $v_p \approx$ 14 THz, T = 10 K. |
| In$_{0.6}$Al$_{0.4}$As/GaAs QD | 4…15 | D* ~ 10$^8$ (T = 4.6 K) | [72] | Sensitive up to T = 150 K, peak D* at λ ≈ 50 μm, 4.6 K. |
| QD+SET | ~0.5 THz | ~10$^{-19}$ (T = 0.3 K) | [69] | Count rate illumination N~0.1 s$^{-1}$, NEP = (2N)$^{1/2}$×hv/η, η~10$^{-3}$. |
| QD+SET | ~0.6 THz, BBR emitter, T ~ 70 K | ~10$^{-19}$ | [73] | Count rate illumination N~0.1 s$^{-1}$, NEP = (2N)$^{1/2}$×hv/η, η ~ 10$^{-3}$. |

In the photon superconductive detectors, photons with hv > 2Δ (Δ is the superconductive energy gap), excite Cooper pairs to break them into quasi-particles (QPs, two electrons) which then can tunnel to the other side of the junction before recombination, resulting in a sharp increase in the current. The superconductive binding energy gap 2Δ is two to three orders of magnitude lower than the band gap in semiconductors. Many of these detectors exploit the Cooper pair breaking in a thin-film (of about several or tens nanometers) low transition temperature superconductors operating at low reduced temperatures T/T$_c$ ≃ 0.1, where T$_c$ is the superconducting transition temperature.

Shown in Fig. 5.7 are T$_c$ data for different superconductive materials. Summarized in Table 5.3 are the critical temperatures T$_c$ and barriers 2Δ of some bulk superconductive metallic compounds and alloys that can be used for design of thin-film superconductive detectors.

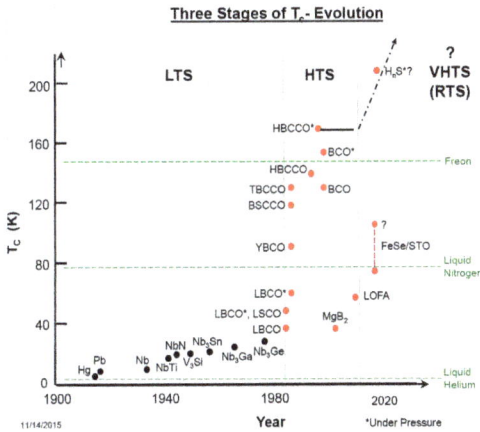

*Figure 5.7. Transition temperature $T_c$ of low-temperature superconducting materials discovered and three stages of $T_c$ evolution [77] (By permission of CCAS, updated).*

*Table 5.3. Low-temperature superconducting bulk materials and their respective critical temperatures and barriers (after [78]).*

| Material | $T_c$, K | $2\Delta(0)$, meV |
|---|---|---|
| Nb | 9.1 | 2.77 |
| Ti | 0.4 | 0.12 |
| Pb | 7.2 | 2.19 |
| Al | 1.2 | 0.36 |
| MoC | 14.3 | 4.35 |
| NbN | 16 | 4.9 |
| TiN | 5 | 1.52 |
| *NbTiN (film) | 15.0…15.3 K | ≈4.55 |
| **$MgB_2$ (bulk) | 39 K | - |

*[79]; **[80].

Critical temperature $T_c$ of thin films of superconductors is strongly influenced by their microstructure, crystallographic orientation, thickness, and deposition method. The lattice constant should match well with that of the substrate. Therefore, the influence of technology, the substrate and its crystallographic orientation are important. DC magnetron and reactive sputtering are the most widely used techniques for deposition of high-quality superconducting layers. Frequently, as substrates, high resistive Si, sapphire and SiC are used. Shown in Fig. 5.8 are the dependences of $T_c$ on deposition methods and thicknesses (see also 5.26.c below, the $T_c$ dependences of NbN films deposited on different substrate surfaces).

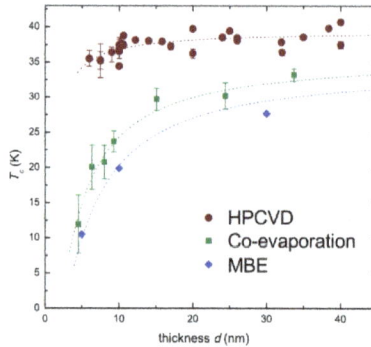

*Figure 5.8. Dependences of critical temperature $T_c$ on deposition methods and thicknesses [81]. HPCVD – Hybrid physical–chemical vapour deposition, MBE – molecular beam epitaxy. (By permission of AIP Publishing).*

### 5.3.5   Superconducting photon pair breaking detectors

The photon (athermal) detectors count QPs generated by photons incident onto a superconductor. Photon breaks a number of Cooper pairs in the superconductor (pair breaking detectors), generating QPs leading to their tunneling through the insulating barrier to generate current signal. When using a superconducting detector to measure photons with energies far above the binding energy gap $2\Delta$, a number of QPs proportional to the photon energy is generated. The binding energy of the Cooper pairs is the spectroscopic gap $E_g = 2\Delta$, where $\Delta$ is the energy gap for a single excitation. QP generation refers to the creation of two quasiparticle excitations when a Cooper pair is broken by a photon or thermal phonon. Quasiparticle recombination relates to the annihilation of two quasiparticles as they form a Cooper pair and emit a phonon.

### 5.3.5.1 STJ and SIS detectors

STJ (superconducting tunnel junctions) and SIS (superconductor-insulator-superconductor) are the types of Josephson junction. STJ and SIS are from the family of cryogenic pair-breaking detectors. Their operation rely on generation of charge carriers by breaking Cooper pairs in a superconductor under absorbed photon energy leading to appearance of QPs. Incoming radiation produces charge carriers that are collected. The produced charge (current) is proportional to energy (power) of photons. STJ consists of two superconducting films separated by a thin (~20 Å) insulating barrier. The charge carriers can be detected through the tunnel-current pulse they produce, if the STJ is under the voltage bias. SIS device is the structure of symmetric STJ detector that consists of two identical superconducting layers separated by an insulating barrier. The current flow in STJ passes through the insulating barrier due to quantum tunneling process (Figs. 5.9 and 5.10). The STJs can be used as photon-counting detectors with intrinsic energy resolution over a wide energy band from the near infrared to well into the X-ray band. The operating temperature is typically at 10 % of the critical temperature $T_c$ of the superconducting material and may range from 0.1 K to 1 K [82].

There are two principal components of the tunneling current. The first is tunneling of Cooper pairs [83]. The second is the QPs current, which arises when the energy from the bias voltage exceeds $2\Delta$. QPs states exist for energies higher than the superconducting gap $\Delta$ counted from the Fermi energy $E_F$. QPs tunnel back and forth through the barrier until they are out of the process. At $T \neq 0$ K, a small tunneling current is present even for voltages less than $2\Delta$, due to QPs thermal excitation above the gap.

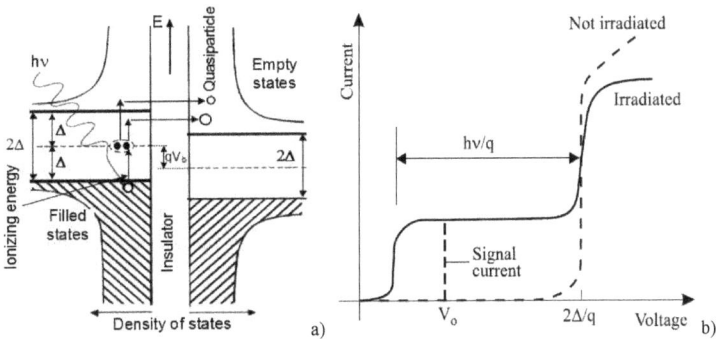

*Figure 5.9. SIS junction. (a) Energy diagram with applied bias voltage and illustration of photon-assisted tunnelling. Cooper pairs exist at the Fermi level $E_F$, indicated by the dashed lines. Different energy excitations of Cooper paires are shown for one of it because of the lack of space. (b) Current-voltage characteristics of a non-irradiated and irradiated barrier. The intensity of the incident radiation is measured as an excess of the current at a certain bias voltage $V_0$. At no incident radiation but with a bias voltage $V_0 = 2\Delta/q$ applied ther quasi-particle can tunnel through the insulator barrier [84]. (By permission of Elsevier).*

One of the earliest pair-breaking detectors was the STJ detector [85]). The STJ is a type of Josephson junction [83]. It creates the tunneling current in the junction barrier. This process resembles interband absorption in semiconductors, when absorbed photons create electron-hole pairs.

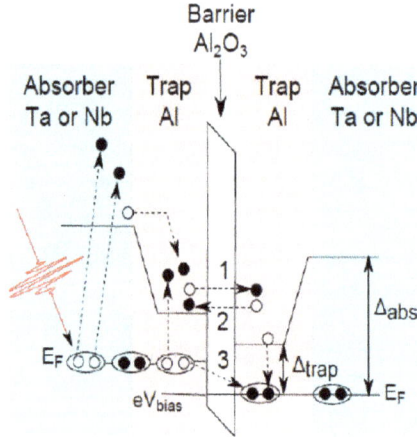

*Figure 5.10. Schematic energy diagram of STJ detector with trapping layers. Incident photons generate quasi-particles (hot electrons) in the superconducting absorbing layer, which tunnel through the insulating $Al_2O_3$ barrier to generate current. Photon absorption creates excess charges that scatter inelastically into the low-gap trap and produce a current pulse by direct tunneling (process 1), back-tunneling (process 2) or reverse tunneling (process 3) (after [86]). (By permission of Springer Nature).*

The simplified STJ schematic diagram is shown in Fig. 5.9.a. The horizontal axis shows the density of states. Cooper pairs exist at the Fermi level. The states below the energy gap $2\Delta$ are occupied and those above the gap are empty. The curves indicate the electron density of states. When a bias voltage, $V_b$, is applied across the junction, there is an energy shift of $qV_b$ between the Fermi levels of these two superconductors. If $qV_b < 2\Delta$, no current flows (only tunnelling process into unoccupied states at the same energy that is weakly dependent on bias). However, if the junction is irradiated, photons of energy hv may assist the tunnelling, which now may occur for $qV_b > 2\Delta - hv$.

The simplified I-V characteristic of a device is shown in Fig. 5.9.b. When the bias voltage reaches the gap voltage $2\Delta/q$, a steep increase of the current occurs. At this particular voltage, the divergent densities of states of both superconducting layers cross, and Cooper pairs on one side of the insulating layer break up into quasi-particles. These quasi-particles tunnel from one side of the insulator to the other, where they recombine. A sharp onset of normal tunnelling current arises beyond a DC threshold voltage equal to the binding energy gap $2\Delta$.

The tunnelling processes in STJ in reality are a little bit more complicated. The nature of QPs leads to the effect of QPs multitunneling. QPs start to tunnel back and forth through the insulating barrier each time transferring the difference in its energy between two electrodes into heat via phonons emission [87]. The schematic energy diagram with trapping layers and possible tunnel processes in STJ is shown in Fig. 5.10.

The number of charge carriers generated in STJ structure is proportional to the energy of the absorbed photon, and ranges from several hundreds to a few thousand per eV [88] in dependence of the material used. As the QP relaxation is a fast process ($\tau \sim 10^{-11}$ s), it makes STJs be used with high photon-counting rates.

For single photon detection, the STJ is DC biased. When a photon with energy $h\nu$ larger than the binding energy $2\Delta$ strikes the STJ, it breaks up Cooper pairs and generates a number $N_{QP} = \eta h\nu/\Delta$ QPs, which then can tunnel through the barrier along the direction of the bias voltage. It allows STJs to be used as photon-counting detectors with intrinsic energy resolution over a wide energy band from the near infrared to the X-ray band. The operation temperature T is typically $\leq 0.1 \cdot T_c$ to minimize bias currents due to thermal QPs generation and may range from 0.1 K to 1 K. Imaging arrays of >100 pixels of close-packed STJs have been made and operate in ground-based astronomical applications [88].

The magnitude of the tunnelling current depends on tunnelling, recombination and scattering of charge carriers. STJ consists of a thin insulating barrier layer (*e.g.*, $Al_2O_3$) sandwiched between two thin superconducting absorbers (*e.g.*, Nb, Ta, NbN or AlN) [89–91]. Up to the pair-breaking frequency of niobium of approximately 700 GHz, the embedding circuit is made of the same material like to the Nb junction electrodes.

One of the advantages of these detectors is that the fundamental noise due to the random generation and recombination of thermal QPs decreases exponentially with temperature as $\exp(-\Delta/k_BT)$ [92]. The best single-side band (SSB) noise temperature that can be gained is $k_BT_n \geq h\nu/\eta$. With $\eta = 1$, the quantum limit can be achieved, but it never can be overcome [93]. In pair-breaking detectors, it is possible to get $\eta \to 1$ and thus one can approach nearly the quantum operation limit (Fig. 3.16). In order to achieve the fundamental sensitivity limit of a SIS mixer [94], it is crucial that the tunnel junction is embedded in a high-Q tuning circuit that enables a near 100 % coupling to the signal. For this reason, optimum embedding circuits are made of superconducting transmission lines.

Several structures, which use different ways to separate QPs from Cooper pairs, have been proposed: SIS, superconductor-insulator-normal (SIN), kinetic inductance detector (KID), superconducting quantum interference device (SQUID), *etc*. Physics of these devices operation is considered, *e.g.*, in [95, 96].

Various STJ detectors were designed (see, *e.g.*, [89, 97, 98]). With the antenna coupled STJ detectors and single-electron transistor readout, one can get NEP $\approx 10^{-20}$ W/Hz$^{1/2}$ in the absence of background [99]. The effect of direct conversion of sub-THz photons into electrical current through the photon-assisted tunnelling process is used both for direct as well as for heterodyne detection.

Due to high energy resolution, the STJs are used, *e.g.*, in IR astronomy. They are used for detection and imaging applications [100]. However, STJs need low operation temperature and require a magnetic field to suppress the Josephson current.

The SIS device is the structure of the symmetric STJ detector that consists of two identical superconducting layers separated by an insulating barrier. SIS detectors (the tunnel junction is in the ~1 $\mu m^2$ range) are ordinary used in mixers due to the sharp nonlinear voltage dependence of the tunnel current and fast QPs relaxation processes. That makes the SIS an almost ideal mixer [101] with the IF (intermediate frequency) band up to 12 GHz [102]. In this case, they are used as rectification detectors.

The DSB noise temperature of the SIS mixers are close to the minimum feasible mixer noise temperature: $T_{mix}^{min} = h\nu/2k_B$ [94]. The highest detectable frequency in the SIS mixer is set by the superconducting pair breaking energy $2\Delta$. For higher frequencies, for which $h\nu > 2\Delta$, QP excitations in the embedding circuit are created due to breaking the Cooper pairs, and large ohmic losses are observed, which would significantly decrease the sensitivity of the mixer device [103]. The near quantum-limited sensitivity for these devices has been demonstrated over a broad frequency range (~100...1200 GHz) [96, 104].

The NbTiN mixers are less sensitive at frequencies beyond the superconductor band-gap ($2\Delta \approx 4.5$ meV) where the reverse tunnelling becomes an essential factor, and mixer performance is dominated by circuit losses. The lossless signal transport is provided only up to the superconducting pair-breaking frequency $2\Delta/h$. The heterodyne mixing above the pair-breaking frequency is intrinsically not limited by the superconducting electrode material itself. It is despite that the quasi-particle excitations are created too absorbing a part of the detection signal [103].

The needs of high sensitivity detectors for astronomy applications and low available local oscillator power in the THz range require using the superconducting devices as mixers. For them, a much lower LO power (several $\mu$W) is needed, as compared to SBD mixers

($W_{LO}$ ~ 1 mW, scalable with v) used in earlier astronomical devices. Below approximately 1 THz, the SIS junctions (see Fig. 3.16) are the detectors of choice, and at higher frequencies, the hot electron bolometer (HEB) mixers should be used [105]. Today, heterodyne instruments are mainly based on single-pixel SIS receivers.

Large heterodyne THz FPAs, not yet realized, would be preferable for astronomical surveys to study the galaxies, stars, *etc.*, allowing, *e.g.*, for sub THz ground-based astronomy, to shorten the dwell time image obtaining. FPAs with hundreds or thousands of pixels would provide a large reduction of observing time [106, 107].

The SIS tunnel junction mixers need a magnetic field of several hundred Gauss to suppress the unwanted Josephson current, which is caused by Cooper pair tunneling. Conventionally, this is made with superconducting electromagnets. For arrays, this is cumbersome both in terms of volume and wire count. A feasible alternative is the use of tiny permanent magnets close to the tunnel junctions where again the arbitrarily shapable Si substrates are advantageous, if several SIS junctions need separate magnets. The close proximity of the magnets to the junction helps reduce magnetic cross-talk between adjacent mixers [105, 108].

*5.3.5.2 KID detectors*

The kinetic inductance detector (KID) is a type of non-equilibrium superconducting incoherent pair-breaking detector, in which the Cooper pairs are broken up into QPs due to the radiation absorption. Microwave KIDs (MKIDs) are non-equilibrium superconducting detectors made out of high quality factor superconducting microwave resonant circuits. MKIDs are investigated now for frequencies from tens of GHz to the X-ray region. Their primary advantage over other low-temperature superconducting detector technologies is their built-in frequency domain multiplexing at GHz frequencies, allowing thousands of detectors to be read out through a single transmission line [109–112]. The Al-NbTiN based MKIDs have shown almost background limited performance at 350 GHz [113] and 850 GHz [114]. The MKIDs can count individual photons with essentially no false counts and determine the energy and arrival time of every photon with good quantum efficiency.

In MKIDs, the QPs produced by photons that break the Cooper pairs may also be detected by measuring the complex AC surface impedance of the superconductor. The number of QPs ($N_{QP}$) created by photon absorption (h$\nu$ > 2$\Delta$), is $N_{QP}$ ~ $\eta h\nu/\Delta$. At finite frequencies, the surface impedance is non-zero and is largely inductive (kinetic inductance effect). The kinetic inductance effect occurs because the energy can be stored in the super-current of a superconductor. Reversing the direction of super-current requires extracting this stored energy, which yields an extra inductance. This effect can be used to make detectors in which the resonance frequency of a superconducting resonator changes when photon is detected, and can be monitored with microwave readout circuits. The sub-gap photons (non-pair-breaking photons) are also absorbed, and they generate an excess population of QPs, too, which influences the operating characteristics of the device [112, 115]. Therefore, the detector response requires the knowledge of non-equilibrium distributions [116]. The primary attraction of MKIDs is that they are to multiplex into large arrays (*e.g.*, $31 \times 31 = 961$ antenna coupled KIDs) [117].

The MKIDs operate on the principle that incident photons change the surface impedance of a superconductor through the kinetic inductance effect. In Fig. 5.11 photon with the energy $h\nu > 2\Delta$ is absorbed in a superconducting film cooled to $T \ll T_c$, breaking Cooper pairs and creating a number of QP excitations $N_{QP} \sim \eta h\nu/\Delta$. The efficiency of creating QPs $\eta$ will be less than unit, since some part of the photon energy will end up as phonons. In this diagram, CPs are shown at the Fermi level, and the density of states for QPs, $N_s(E)$, is plotted as the shaded area being a function of QP energy E.

Large arrays of MKIDs are significantly easier fabricated and read out than in any competing technology, *e.g.*, frequency-domain multiplexing [79, 112, 117]. Several designs of MKID devices are possible, *e.g.*, a lumped element MKIDs (LEKID). It uses a separate inductor and capacitor to form a resonator. The device uses both lumped element and transmission line resonators [119, 120].

A superconducting microresonator may be produced by depositing a superconducting thin film on an insulating substrate and applying standard lithographic patterning techniques to produce a resonator structure. As illustrated in Fig. 5. 11, the resonator may either be a lumped-element circuit, *e.g.*, a meandered inductor and an interdigitated capacitor, or a transmission-line resonator. These simple single-layer structures permit use of high-quality crystalline substrates as well as a wide variety of superconducting films, and, therefore, provide an opportunity to achieve extremely low dissipation [112].

Al, Nb, TiN, NbTiN are used for MKIDs manufacturing. MKIDs can have average detector NEP $\sim 2 \times 10^{-17} \ldots 3 \times 10^{-19}$ W/Hz$^{1/2}$ at sub-K temperatures in the radiation frequency range $\nu \sim 80\ldots200$ GHz in dependence of radiation power [117, 120, 121]. The ultimate sensitivity achievable with MKIDs is not yet known. Measurements of $T_c = 1.1$ K for TiN gave values of NEP $\approx 4 \times 10^{-19}$ W/Hz$^{1/2}$ by using ther relatively low resonator Q value of Q $\sim 5 \times 10^4$. NEP $\sim 10^{-20}$ W/Hz$^{1/2}$ range should, therefore, be possible using using higher-Q resonators [112]. At $\nu = 850$ GHz, for 961 KIDs array the optical NEP$_{opt}$ $= 5 \times 10^{-19}$ W/Hz$^{1/2}$ of the central pixels was achieved. The electrical NEP$_{el} = (3.3\pm1.3) \times 10^{-19}$ W/Hz$^{1/2}$ is provided for 85% of the pixels [122]. In Fig. 5.12, it is shown the measured NEP as a function of the absorbed power together with the calculated NEP, assuming the background limited detector performance.

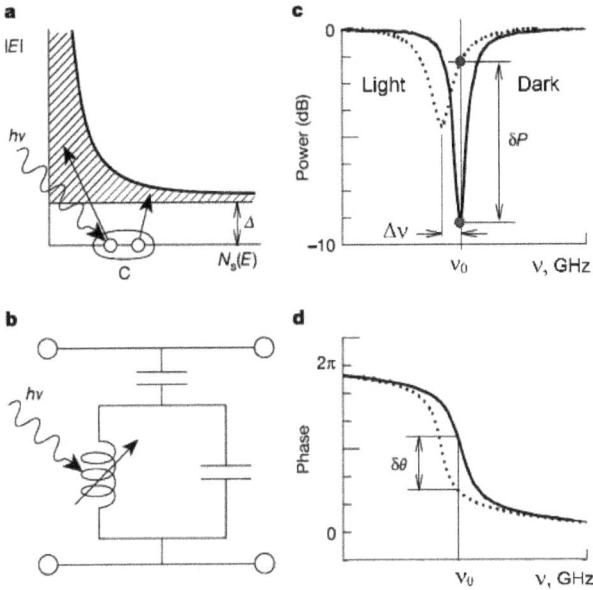

Figure 5.11. The basic operation of MKID. The change of resonance of a superconductor resonator takes place, when photon is absorbed. (a) Photon with the energy $h\nu > 2\Delta$ strikes a superconducting film breaking up a number of CPs and generating QPs. (b) To sensitively measure these QPs, the film is placed in a high frequency planar resonant circuit. Due to generating QPs, the number of CPs within the superconductor is decreased, and the impedance is increased. (c) An increase in the inductance leads to a decrease in resonant frequency $\nu_0$ and the signal amplitude. (d) The shift of the frequency can be viewed as a phase signal $\delta\theta$, which can be read out. If the detector (resonator) is excited with a constant on-resonance microwave signal, the energy of absorbed photon can be determined by measuring the degree of phase and amplitude shift (after [118]).

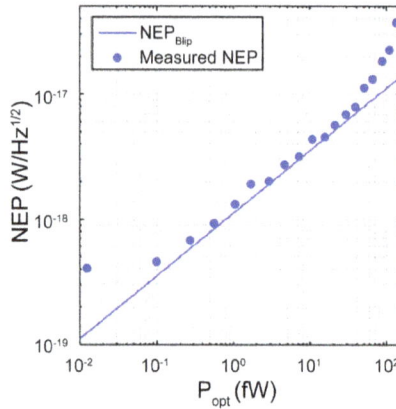

*Figure 5.12. NEP as a function of absorbed power at a post-detection frequency of 60...80 Hz. The measured NEP approaches the NEP$_{Blip}$ for powers exceeding 100 aW up to 40 fW [117].*

NbTiN is known as a wide $2\Delta$ gap (see Table 5.3) superconductor that exhibits small phase noise [123] and microwave losses [124] making microwave-frequency NbTiN resonators on silicon good for MKID detectors [79, 113].

The QP recombination time is high in this material and T-dependent (from ~10 µs to ~1 ms in low-temperature range). The QP recombination time evolution in Ta ($T_c$ = 4.48 K) and Al superconducting films, in dependence on implanted dose levels for nonmagnetic and magnetic impurities, was studied in [125] in response to an optical pulse of 2 µs duration. It was concluded that the enhancement of recombination of QPs into Cooper pairs is not caused by the atom magnetic moment, but arises from an enhancement of film disorder, and within the range of $T/T_c$ ~ 0.03...0.1 is $\tau$ ~ 100...2000 µs for Al film and $\tau$ ~ 10...30 µs for Ta film.

## 5.4 Thermal (power) uncooled and cooled detectors

This class of widely used devices is composed of the thermal (power) detectors that measure the temperature rise of the detector element. They can be subdivided into several types. Among these types of detectors, there are the pneumatic, thermistor and metal bolometers, pyroelectrics, thermocouples, composite bolometers, semiconductor and superconducting low temperature bolometers, mechanical cantilevers. The latter require complex readout systems and are relatively rarely used as IR and THz detector arrays.

Materials Research Forum LLC
https://doi.org/10.21741/9781644900758

Many types of absorbing materials can be chosen for broad-band and narrow-band THz spectra operation. These detectors preferably operate at room temperature or under extremely low (T ~ 50…300 mK) temperature conditions. Uncooled operations of these detectors are mostly used for Sun/Earth/planet observation [126]. Ultralow temperature imaging bolometers are typically used in THz and sub-THz (mm-wave) regions [127]. Superconductive thermal detectors (bolometers) are the most sensitive incoherent (direct) detectors with NEP ~ $10^{-19}…10^{-20}$ W/Hz$^{1/2}$ [128].

Thermal detectors were the first detectors for registration radiation in the IR spectral range. Today, their various modifications designed from a large number of materials for different temperatures from T ~ 0.1 K up to ~300 K are among the basic detectors and arrays operating in IR and THz spectral ranges. To get high sensitivity, the thermal detectors should have a low thermal mass (low heat capacity $C_{th}$, J/K) and a good thermal isolation (low thermal conductance $G_{th}$, W/K). Plenty of materials can be used for manufacturing thermal detectors. Schematic temperature dependences of basic group of solid-state materials used for thermal detectors are shown in Fig. 5.13.

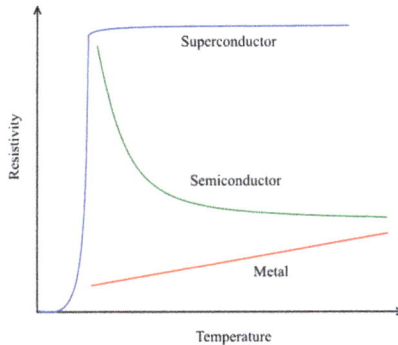

*Figure 5.13. Relative temperature dependences of typical solid-state materials used for bolometers manufacturing.*

Now thermal detectors (*e.g.*, semiconductor oxide and metal microbolometers, microthermocouples, *etc.*) manufactured by micromachining technologies (for THz range been integrated with planar antennas) are devices suitable for use throughout the IR to THz bands.

Thermal detectors are generally supposed to be wavelength independent (*e.g.*, for pyroelectric detectors with absorbing layers and without antennas, Fig. 5.14) and Golay cells.

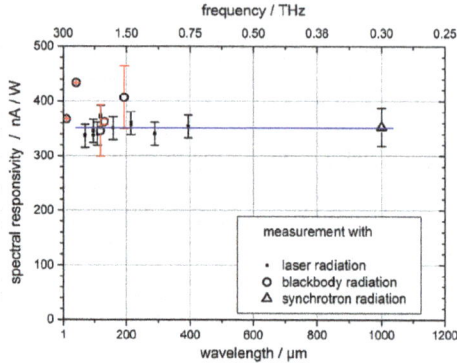

*Figure 5.14. The spectral responsivity of the large-area (**diameter d** = 31 mm) thin-film pyroelectric detector (PVDF - polyvinylidene fluoride foil coated on both sides with a thin layer of a metal oxide) [129].*

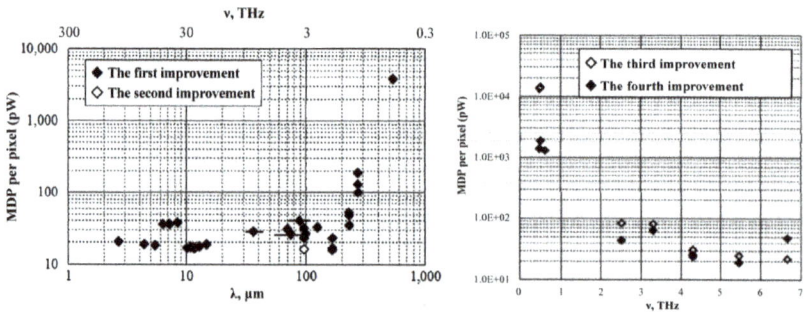

*Figure 5.15. Spectral dependencies of NEP for some uncooled microbolometer arrays at different steps of improvement of technology. See: [130]. Here, MDP =NEP/(2$\tau_{int}$)$^{1/2}$ where MDP is the minimum detectable power and $\tau_{int}$ is the integration time for read-out electronics, which is close to the inverse frame rate. (By permission of SPIE).*

The signal in an ideal case depends only upon the radiant power (or its rate change in pyroelectric detectors, see Ch.3), but for different reasons (depending on design, physical properties, antennas, coatings, *etc.*) in different types of thermal detectors they can be spectral dependent as, *e.g.*, for uncooled semiconductor microbolometers (Fig. 5.15) with antennas and resonant cavities.

Nonetheless, the microbolometer arrays and cameras can operate in wide THz spectral range of 0.094…4.25 THz [131] though the NEP or MDP (minimal detectable power) data for the minimal radiation frequency range were not displayed. In the range of radiation frequencies $\nu \approx 0.3…4.25$ THz the MDP values MDP $\sim 90$ pW and lower in the dependence of matrix array [130–132]. These NEP values correspond to NEP per one pixel NEP $\approx 2.5 \times 10^{-11}$ W/Hz$^{1/2}$, when taking into account the real imaging time (frame rate f = 25 Hz, f $\approx 1/2\tau_{int}$).

Here,

$$MDP = NEP/(2\tau_{int})^{1/2}, \qquad\qquad\qquad (5.1)$$

and $\tau_{int}$ is the integration time of a read-our circuit and the noise equivalent bandwidth $\Delta f_e = (2 \cdot \tau_{int})^{-1}$ in the case when the white noise is dominant.

At lower part of the THz spectral range, the NEP (or MDP) values are worser and can't be comparable with the lowest values of NEP $\approx 4 \times 10^{-13}$ W/Hz$^{1/2}$ for Schottky barrier diodes (see, *e.g.*, [133]. However, they are much better compared to typical values of NEP $\approx 1.3 \times 10^{-8}$ W/Hz$^{1/2}$ per pixel in the THz spectral range $\nu \sim 0.05…10$ THz for pyroelectric matrix arrays [134].

To get low NEP (large sensitivity), quick response time, the thermal detectors should be small and thin in sizes – typically several μm square and several or tenths to hundreds nm thick. These small volume devices (the air-bridge) will have large thermal impedance, and therefore, by using an antenna to couple power into it, the large temperature rises (and, thus, high responsivity) can be achieved. In addition, due to small thermal mass, the micromachined thermal detectors can be relatively fast. However, because uncooled semiconductor thermal detectors still are relatively slow sensors ($\tau > 10^{-3}$ s) they can be used only as incoherent devices providing only signal amplitude detection.

Temperature-dependent resistive sensors (bolometers) have the advantage of monolithic array design and for low temperature semiconducting and superconducting have high sensitivity (low NEP). Pyroelectric detectors sense changes in temperature, therefore, requiring a signal modulation. Pyroelectrics also require an assembly technique that raises the $G_{th}$ and $C_{th}$ values. That is one of the reasons that they have rather moderate sensitivity. Earlier designed thermopile detectors suffered from low sensitivity, but contemporary MEM technologies enable to obtain the high linearity, small 1/f noise and sensitivity admissible for many applications.

A critical part of all the bolometers is their absorber that converts incident radiation into heat. Composite bolometers use a uniform metallic film deposited on a dielectric substrate as a broadband absorber. To obtain frequency independent absorption, the film must be resistive and its impedance in parallel with the admittance of the dielectric detector film should match the impedance of free space [135].

*5.4.1 General*

There mainly exist two types of detectors: quantum (photon) detectors and thermal ones. Quantum photon detectors (see Fig. 3.4), as a rule, are semiconductor detectors. Photon mechanism of radiation detection is realized when the output electrical signal of a detector appears at direct conversion of photon energy into initial reaction of the photodetector. Absorbed photon directly interacts with electrons, transfers them into the states of excited energy (nonequilibrium state). Other mechanisms of radiation energy absorption are practicaly insignificant. This process is not accompanied by noticeable temperature changes of a detector sensitive element. The time of increase and decrease of the number of nonequilibrium carriers (electrons or holes) depends on their relaxation time $\tau$ (see Fig. 5.16).

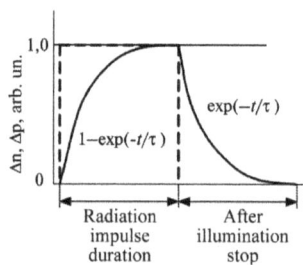

*Figure 5.16. The curves of rising and falling down of photoresponse signal in detectors.*

The nature of fundamental limitations of thermal and quantum detectors operation is different. The reason of it is that the thermal detectors are sensitive to absorbed radiation power, while the quantum detectors are sensitive to the number of radiation quanta absorbed, which move electrons from one state to another, almost not changing the detector temperature.

The ultimate performance of IR and THz quantum detectors is mainly defined by fluctuations of the photon number in radiation fluxes from the background and corresponding fluctuations in the number of photo-excited carriers (see Figs. 3.2 and 3.14). The ultimate performance of thermal detectors is governed by the fluctuations in radiation power absorbed by the detector (in fact, by fluctuations of temperature in a thermal detector, when the heat exchange with the heat sink is vanishingly small) at conditions of only radiation heat removal.

To evaluate the thermal detectors ultimate performance, there can be used two approaches. One of these approaches is related with the analysis of power fluctuations in the radiation flux coming to the detector. Another one is related with the temperature fluctuations in the detector. Both of these approaches provide the same results.

The average value of temperature fluctuation $<(\Delta T)^2>$ that defines the temperature noise (phonon noise) $\Delta T_n = <(\Delta T)^2>^{1/2}$ due to thermal losses, is [136]

$$< (\Delta T)^2 >= \frac{k_B T^2}{C_V} = (\Delta T_n)^2 \quad , \tag{5.2}$$

where in solids $C_V \approx C_P = C_{th}$, J/K ($C_V$ is the heat capacity at a constant volume and $C_P$ is the heat capacity at a constant pressure) and $k_B = 1.380648 \times 10^{-23}$ J/K is the Boltzmann constant.

The heat flow from the detector having the temperature T, which is higher than the temperature of heat sink by $\Delta T$, is defined by the heat conductivity equation

$$\frac{d(\Delta E)}{dt} = G_{th} \times \Delta T \quad , \tag{5.3}$$

where $d(\Delta E)/dt$ is the rate of the heat flow, $G_{th}$ = W/K is the thermal conductivity. The energy changes of the detector at the temperature alteration is

$$\Delta E = C_{th} \times \Delta T. \tag{5.4}$$

Then, the heat flow from the thermal detector to the environment

$$-C_{th} \times \frac{d(\Delta T)}{dt} = G_{th} \times \Delta T \quad , \tag{5.5}$$

the solution of which is

$$\Delta T = \Delta T_0 \cdot \exp(-t/\tau_{th}) \quad , \tag{5.6}$$

where $\Delta T_0$ is the $\Delta T$ at $t = 0$ and $\tau_{th} = C_{th}/G_{th}$. Here, $R_{th} = 1/G_{th}$ is the thermal resistance.

To decrease $\tau_{th}$, one should decrease $C_{th}$. One can choose the material with lower $C_{th}$ or lower the temperature T of detector, since $C_{th}$ quickly decreases with T.

When the radiation is coming to a thermal detector from the environment, the heat balance equation is

$$C_{th} \times \frac{d(\Delta T)}{dt} + G_{th} \times \Delta T = W'(t) \quad , \tag{5.7}$$

where W'(t) is the radiation power density of fluctuations from the environment.

One can find the solution of (5.7) assuming that W'(t) is dependent on time as W'(t) = $W_0 \times \exp(i\omega t)$, where $i = (-1)^{1/2}$ and $W_0$ is the frequency independent noise power fluctuation ($W_0$ = const). Then, the solution of (5.7) is

$$\Delta T_\omega = \frac{W_0 \cdot \exp(i\omega t)}{G_{th} + i\,\omega \times C_{th}} \quad , \tag{5.8}$$

where $\omega = 2\pi f$. For the equation for average square temperature fluctuation, it follows

$$<(\Delta T_\omega)^2> = \frac{<(W_0)^2>}{G_{th}^2 + \omega^2 \times C_{th}^2} \quad . \tag{5.9}$$

To find the temperature dispersion for these fluctuations, one should integrate over the whole spectral range

$$< (\Delta T)^2 >= \int_0^\infty \frac{<(W_0)^2> \cdot df}{G_{th}^2 + \omega^2 \times C_{th}^2} = \frac{W_0^2}{2\pi G_{th} C_{th}} \times [\tan^{-1}\infty - \tan^{-1}0] = \frac{W_0^2}{2\pi G_{th} C_{th}} \times \frac{\pi}{2} = \frac{W_0^2}{4 G_{th} C_{th}} \qquad (5.10)$$

Comparing (5.2) and (5.10) for average square power fluctuation of the radiation coming to the thermal detector in the frequency band $\Delta f$, it follows

$$< (W_0)^2 >= 4 k_B \times G_{th} \times T^2 \times \Delta f , \qquad (5.11)$$

From this expression, follows an important result: in the case of only temperature fluctuations (because of the radiation power density fluctuations) the ultimate performance of thermal detector does not depend on the material nature for the detector as it does not depend on $C_{th}$.

The temperature noise is important for estimations of low-temperature thermal detectors, when taking account of the noise in radiation flow from the environment prevail over other mechanisms (*e.g.*, thermal conductivity of the parts of the detectors).

Eq. (5.11) resembles the expression for Johnson-Nyquist noise (spectral density of thermal fluctuations) in a photoconductor [136]

$$G_{v,T} = 4 \times R \times k_B T \qquad (5.12)$$

For thermal detector, the average square of temperature fluctuations using (5.9) and (5.11) can be written as

$$< (\Delta T)^2 >= \frac{4 k_B G_{th} T^2}{G_{th}^2 + \omega^2 C_{th}^2} \qquad (5.13)$$

From the obtained expressions, one can estimate the ultimate parameters of thermal detectors (ideal detector, temperature fluctuations in the detector is due only for radiation exchange between the sensitive element and the environment).

The estimated NEP value can be obtained from the condition that the average power of radiation W is equal to the averaged power of the thermal noise. For thermal detector with its sensitive area $A_d$ from the Stephan-Boltzmann law $W = A_d \varepsilon \sigma T^4$, where $\sigma = 5.670367 \cdot 10^{-8}$ $W \cdot m^{-2} \cdot K^{-4}$ is the Stephan-Boltzmann constant. When the temperature of the detector is changed by dT, the radiation from it is changed by $dW = 4A_d\varepsilon\sigma T^3 dT$. Then for radiation heat removal

$$dW = 4A_d\varepsilon\sigma T^4 dT, \tag{5.14}$$

and for "radiation" thermal conductivity it follows

$$G_r = \frac{d}{dT}(A_d\varepsilon\sigma T^4) = 4A_d\varepsilon\sigma T^3, \tag{5.15}$$

where the emissivity $\varepsilon$ does not depend on the temperature and wavelength.

For the detector area $A_d = (50 \times 50)$ $\mu m^2$ the value $G_r = 1.53\cdot10^{-8}$ W/K and the thermal resistance $R_r = 1/G_r = 6.53\cdot10^7$ K/W ($\varepsilon = 1$).

From (5.11) for power fluctuation, when the detector is in equilibrium with the environment, it follows

$$W_r = (4k_B G_r T^2 \Delta f)^{1/2}. \tag{5.16}$$

Then, when taking into account the emissivity from the front surface of a detector $\varepsilon NEP_r = W_r = (16A_d k_B \varepsilon\sigma T^5 \Delta f)^{1/2}$ for the noise equivalent power

$$NEP = \left(\frac{16\sigma \times k_B T^5}{\varepsilon}\right)^{1/2} \sqrt{A_d}\sqrt{\Delta f}, \tag{5.17}$$

and for the detectivity D*

$$D^* = \frac{\sqrt{A_d \times \Delta f}}{NEP} = \left(\frac{\varepsilon}{16\sigma \times k_B T^5}\right)^{1/2}. \tag{5.18}$$

One can see that for mentioned conditions D* does not depend on the detector area and wavelength. For such an ideal detector $NEP = 5.55 \times (A_d\Delta f)^{1/2} \times 10^{-11}$ W/Hz$^{1/2}$, which is dependent on the detector area, and D* = $1.813 \times 10^{10}$ cm·Hz$^{1/2}$/W ($\Delta f = 1$) at $T_d = T_B =$ 300 K. *E.g.*, at $A_d = 20 \times 20$ µm the value NEP $\approx 10^{-12}$ W/Hz$^{1/2}$.

These are the ultimate parameters, and in the real thermal uncooled detectors operating at room temperature they are notably worse. In typical uncooled thermal semiconductor or metal bolometers they are within NEP $\sim (1...100) \times 10^{-11}$ W/Hz$^{1/2}$ and D* $\sim (1...100) \times 10^8$ cmHz$^{1/2}$/W (see Table 5.4 below).

Taking into account only temperature noise of the detector at temperature $T_d$, which is in the media with the background temperature $T_b$ one can obtain the expression (only radiation heat exchange) [137, 138]

$$D_b^* = \left[ \frac{\varepsilon}{8k_B\sigma \times (T_d^5 + T_b^5)} \right]^{1/2} . \tag{5.19}$$

The NEP value that is proportional to $(A_d\Delta f)^{1/2}/D^*$ can be improved by many orders by lowering both the temperature of detector and background.

From this expression it can be seen that, if even the temperature of the detector or background is zero the detectivity is improved only by $2^{1/2}$ times. This is in contrast to quantum detectors, in which even at $T_B \sim 300$ K the detectivity can be improved by many orders by cooling the detector that is due both to the limited spectral sensitivity of quantum detectors and mechanisms of radiation absorption.

Decreasing the detector and background temperature and narrowing the spectral range of sensitivity of thermal detectors as well as their field of view by using cold diaphragms can improve the detectivity D* by many orders, thus improving the NEP value up to NEP $\sim 10^{-19}$ W/Hz$^{1/2}$ (Figs. 3.2 and 3.3). For the low temperature detectors, the parameter D* is not a good figure of merit, since for them frequently the rule of proportionality between the average square noise value and average square of detector area is not valid [139].

When a radiation input power, P, is received by the bolometer detector, the absorbing element of bolometer temperature $T_B$ increases with time with the rate $dT_B/dt = P/C_{th}$ and approaches the limiting value $T_B = T_S + P/G_{th}$ with the thermal time constant $\tau_{th} = C_{th}/G_{th}$. Here $T_S$ is the heat sink temperature. The thermal conductance is $G_{th} = G_{eph} + G_{diff} + G_{photon}$ [140], $G_{th} = dP/dT$. Here, $G_{eph}$ is the thermal conductance due to emission of phonons by the heated electrons, $G_{diff}$ is the conductance defined by cooling the electrons by diffusion into the colder contacts at a bath temperature $T_0$ [141, 142]. $G_{photon}$ is the thermal conductance related with emission of microwave photons (Johnson-Nyquist noise), which removes energy from the detector until it returns to the quiescent temperature.

The absorbed photon energy is thermalized and heats the detector that is connected to a heat sink through a poorly conducting thermal insulation links (*e.g.*, $Si_3N_4$ microbridges). The resulting change of detector temperature is sensed to produce a signal. When radiation is turned off, it relaxes back to heat sink temperature $T_S$ with the time constant $\tau_{th} = C_{th}/G_{th}$.

In uncooled microbolometers based on $VO_x$ or $\alpha$-Si, in dependence of the detector dimensions ($\sim 40 \times 40 \times 0.5$ μm) the typical values $C_{th} \sim 10^{-9}$ J/K, $G_{th} \sim 10^{-7}$ W/K and thus $\tau_{th} = C_{th}/G_{th} \sim 10$ ms.

Within some temperature range, the resistance changes are proportional to the absorbed radiation power, changing the thermal detector temperature by $\Delta T$. To provide large $\Delta T$ the heat capacity $C_{th}$ and the thermal conductance $G_{th}$ should be as low, as possible.

The intrinsic temperature fluctuations of the thermal detector with a heat capacity $C_{th}$ are related to the power fluctuations in a thermal conductance $G_{th}$ in legs. The legs connect the detector and the heat sink (cooled bath) that defines the estimations of the upper phonon noise-limited detector $NEP_{th}$ [136, 143]. From Eq. (5.16):

$$NEP_{th} = (4k_B T^2 G_{th})^{1/2}. \tag{5.20}$$

The upper thermal $NEP_{th}$ of the bolometer, when taking into account the radiation losses from the front surface and emissivity $\varepsilon$ has the form [136]

$$NEP_{th} = (4k_B T^2 G_{th})^{1/2}/\varepsilon. \tag{5.21}$$

For lower $G_{th}$, the lower values of NEP can be achieved.

The key trade-off with respect to conventional uncooled thermal detectors is between the NEP and response time. The thermal detector $NEP \sim (G_{th})^{1/2}$, but the thermal response time of the detector $\tau_{th}$ is inversely proportional to $G_{th}$. Therefore, a change in the thermal conductance due to improvements in material processing technique improves the thermal detector NEP at the expense of time response [144, 145].

For lower $G_{th}$ (lowering the temperature), the lower values of NEP can be achieved. For $G_{th} \sim 60$ fW/K at $T = 220$ mK the electrical (not optical) $NEP_{th} \approx 4 \cdot 10^{-19}$ W/Hz$^{1/2}$ for superconductive TES with long thin legs to suppress thermal conductance was achieved [146].

### 5.4.2    Ultimate NEDT of thermal detectors

For thermal detectors in equilibrium with the environment (see, *e.g.*, [137, 147]) the noise equivalent power (NEDT) can be described by the expression

$$NEDT = \frac{4(F/\#)^2 \times (\Delta f)^{1/2}}{(A_d)^{1/2} D^* \times (\Delta W / \Delta T)},$$

(5.22)

Here, the thermal contrast

$$\Delta P / \Delta T = \int_{\lambda_1}^{\lambda_{CO}} (\partial P(\lambda(T)/\partial T) \times \varepsilon \times \tau_{opt} \tau_{atm} \tau_f d\lambda,$$

(5.23)

where the function $\frac{\partial P(\lambda(T)}{\partial T}$ is quickly decreasing with the wavelength decrease at $\lambda \leq 5$ μm (Fig. 3.15). Here, $A_d$ is the detector area, $\lambda_1$ and $\lambda_{CO}$ are the initial and the long cut-off wavelengths within the range of detector spectral sensitivity. The noise equivalent bandwidth $\Delta f$ in the case of white noise is defined by the dwell time $\tau_d$ (time of light-striking)

$$\Delta f = \frac{1}{2\tau_d}.$$

(5.24)

For the constant emissivity $\varepsilon = 1$, transmission coefficients of optics, atmosphere and filters $\tau_{opt}, \tau_{atm}, \tau_f = 1$ then it follows

$$NEDT = \frac{4(F/\#)^2}{(2A_d\tau_d)^{1/2} D^* \int_{\lambda_1}^{\lambda_{CO}} (\partial P(\lambda(T)/\partial T) d\lambda}.$$

(5.25)

The dependence of the NEDT on the cut-off wavelength $\lambda_{co}$ for ther spectral rang $\lambda = 3...5$ μm is shown in Fig. 5.17.

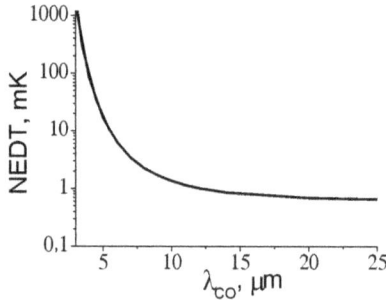

*Figure 5.17. NEDT of an ideal thermal detector on the cut-off wavelength $\lambda_{CO}$ within the spectral range $\lambda = 3...5$ μm ($D^* = 1.813 \times 10^{10}$ cm×Hz$^{1/2}$/W, FOV = $2\pi$ sr, ($\varepsilon$, $\tau_{opt}$ $\tau_{atm}$, $\tau_f = 1$, $f = 50$ Hz (dwell time $\tau_d \approx 20$ ms), F# = 1, $A_d = 30 \times 30$ μm$^2$, $T_d = T = 300$ K from $\lambda_l \approx 0$ μm.*

As it is seen from Fig. 5.17, at $\lambda > 25$ μm NEDT is weakly dependent but has no minimum like to the ideal quantum detector. The thermal detectors are not very effective devices at the cut-off wavelengths $\lambda_{co} \leq 5$ μm.

### 5.4.3    Thermal uncooled detectors

#### 5.4.3.1 Pneumatic detectors

There exist two kinds of pneumatic detectors: the Golay cell [148] and the capacitance microphone one [149]. These detectors are used at the ambient temperature. The Golay cell is one of the most efficient devices detecting THz radiation (see Table 5.4 below) with a broad (with the polyethylene window they operate at $\lambda \sim 20...1000$ μm) and well-characterized spectral response. It has good sensitivity at room temperature and flat optical response over a wide wavelength range. One of the schematics of a contemporary Golay cell is shown in Fig. 5.18.

In the Golay cell based on gas-filled closed cavity with a thin absorbing film and a membrane to which is attached light mirror. The gas is heating under radiation leading to its expanding. The expanding of gas results in changing of mirror bending or changing its angle position, which is measured by some kind of an optical system. The signal measured is proportional to the radiation intensity.

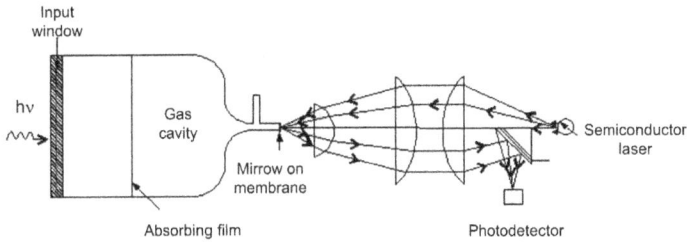

*Figure 5.18. Schematic of a Golay cell.*

The sensitivity of a Golay cell is mainly restricted by the noise, which is caused by the heat exchange between the absorbing film and the gas in enclosed chamber. That is why, the typical NEP is relatively low (better) as compared to other uncooled thermal detectors (NEP $\sim (2...4)\times 10^{-10}$ W/Hz$^{1/2}$ [150]) and the detectivity D* $\sim 3\times 10^9$ cm·Hz$^{1/2}$/W.

The Golay cells are difficult to assemble into large arrays. However, several decisions were proposed [151, 152]. The Golay cells were frequently used in astronomy and in the IR [139, 153]. However, they are fragile (thin membrane), have slow speed response ($\leq$ 15 Hz) and are sensitive to mechanical vibrations.

In capacitance microphone detectors, the gas volume in the enclosed chamber is changing by radiation heating that leads to a displacement of thin membrane towards fixed plate of capacitor, resulting in changes of a capacitance. The response time of these thermal detectors is $\tau \sim (20...30)$ ms [149].

*5.4.3.2 Uncooled semiconducting bolometers*

In a bolometer, the incident radiation power is absorbed and modifies the lattice or electron-gas temperature. As a result, the physical properties of the bolometer (*e.g.*, its conductance or electric polarization) change, and an electrical output signal is generated.

Bolometers can be classified into two types: composite bolometer (Fig. 3.11.a), in which a temperature sensor (thermometer) intended for measuring temperature variations is attached to the absorber, and monolithic bolometers, in which the temperature changes are measured in the absorber itself. Monolithic bolometers are usually applied as microbolometers in the IR and THz wavelengths ranges. In the THz spectral interval, antennas are used, as a rule, to introduce the power into a small thermally activated region.

Infrared and THz microbolometers are usually fabricated using various MEMS technologies. In particular, a thin $Si_3N_4$ membrane is produced by anisotropically etching silicon. The etched $Si_3N_4$ microbridges provide a weak coupling to a temperature reservoir. Among various kinds of IR and THz thermal detectors, there are sensitive detectors operating at room temperature and suitable for imaging and spectroscopic applications.

The larger the temperature resistance coefficient $\alpha = (1/R)\cdot dR/dT$ of a bolometer, where R is the bolometer resistance, the higher the bolometer sensitivity. In semiconductors, the specific resistances are larger as compared with those in metals, and their temperature derivatives are also larger. In metallic bolometers, $\alpha \sim T^{-1}$, and $\alpha \approx 0.0033$ $K^{-1}$ at room temperature. This value is close to $\alpha$ in ordinary metals, such as Pt, Ag, and Al. In semiconductors, the magnitudes of $\alpha$ are about $(0.02...0.05)$ $K^{-1}$, with $\alpha < 0$ in $VO_x$ bolometers and $\alpha > 0$ in $\alpha$-Si ones [154, 155].

As a rule, uncooled thermistor bolometers and microbolometers are composed of a sintered mixture of various semiconductor-oxides flakes, each about several μm's in thickness (*e.g.*, the mature $VO_x$ and $\alpha$-Si technologies). Semiconducting microbolometer arrays are widely applied for the real-time IR and THz imaging [130, 156–159].

Modern FPA microbolometers tailored the mature IR bolometer technology well to be applicable in the THz region $\nu > 1.5$ THz (Fig. 5.15) but can be sensitive up to $\nu = 96$ GHz [131]. Nowadays, FPA microbolometers created on the basis of $VO_x$ or $\alpha$-Si demonstrate the NEP values of $(1...3) \times 10^{-11}$ $W/Hz^{1/2}$ (the MDP values of 30...90 pW) per pixel and better in the spectral range $\nu > 1.5$ THz [130, 131, 160] (see Table 5.4).

Conventional IR microbolometers are based on a microbridge structure consisting of a membrane suspended above a substrate (a heat sink) by means of thermal insulation arms and metal studs (Fig. 5.19). Typically, they are designed for a broad absorption spectrum, 7.5 μm $< \lambda < 14$ μm. The membrane also supports a radiation absorber (*e.g.*, TiN or NiCr) and a thermometer. The latter is a thin film usually made from $\alpha$-Si or $VO_x$. The response spectrum is mainly determined with the help of an optical $\lambda/4$-cavity created between the microbolometer membrane and the reflector (Fig. 5.19.a). This design enhances radiation absorption in the thin membrane. The manufacture process is compatible with CMOS techniques.

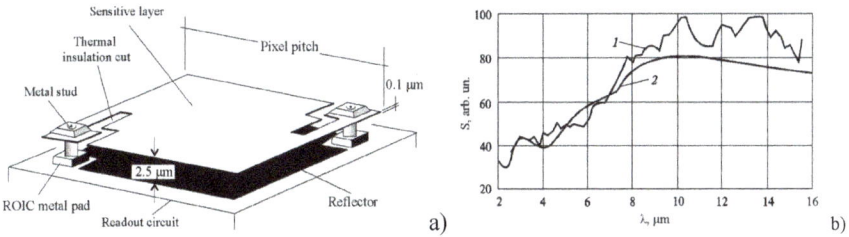

*Figure 5.19. (a) Schematic representation of a standard IR microbolometer pixel and (b) its relative spectral sensitivity [161]. ROIC is the read-out integration circuit, 1 is a spectrum obtained on a Fourier spectrometer, and 2 is a spectrum calculated from integrated sphere measurements. (By permission of SPIE.)*

To extend the spectral responsivity of microbolometers onto the THz spectral range, several designs have been proposed (see, *e.g.*, Refs. [130–132, 162]). In one of them, the spectral absorptivity of the bolometer in the THz spectral range was enhanced by introducing additional layers and eaves on the top of the conventional microbolometer (Fig. 5.20.a). In the pixel structure (on the basis of $VO_x$), an area, which is sensitive to an electromagnetic wave and a diaphragm on it, are suspended above the ROIC.

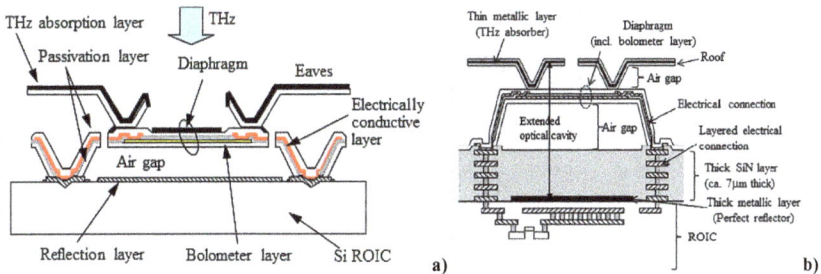

*Figure 5.20. Pixel structures in broad-band THz FPAs [130]. (By permission of SPIE.)*

In another structure, the diaphragm, which contains a bolometer layer (vanadium oxide) and a passivation layer (SiN), is covered with a roof to increase the fill-factor (Fig. 5.20.b) [132]. A thin metallic layer is formed in the roof, and a thick metallic layer (a perfect reflector) is formed on the ROIC surface. The structure composed of a thin metallic layer, a thick SiN layer, and a thick metallic layer acts as an optical cavity. In the previous design, the thick SiN layer was absent. The thick SiN layer and the air gap made the optical cavity much longer geometrically, by a factor of three, than in the previous case. Those designs allowed the maximum of the spectral sensitivity to be shifted towards higher radiation frequencies in the THz spectra in comparison with the IR bolometer design.

Other designs of THz microbolometer arrays with various kinds of antennas and cavities partially filled with solid-state fillers, were considered, *e.g.*, in Refs. [131, 162, 163] (see Fig. 5.21).

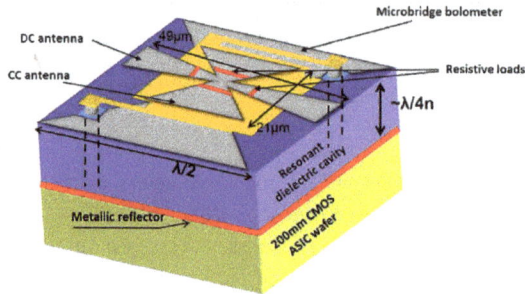

*Figure 5. 21. Structure of a 2-stage antenna-coupled bolometer [162]. To decrease the thickness of resonant cavity, the latter should be fabricated from a siutable dielectric material. (By permission of Springer Nature.)*

The operational parameters of modern uncooled microbolometers, such as their sensitivity and response time, are far from those inherent to cooled semiconductor- or superconductor-based devices; moreover, uncooled microbolometers suffer from thermal noise and self-heating when being under a constant bias [164–166]. Therefore, uncooled semiconducting bolometers operate at a pulsed voltage bias. Despite that uncooled $VO_x$ or $\alpha$-Si microbolometers have relatively low NEP values, their detectivity is as low as that of other thermal detectors; as a rule, it is worse than $D^* \sim 2 \cdot 10^9$ cm·Hz$^{1/2}$/W, which is about an order of magnitude lower than the relevant value for an ideal uncooled thermal detector. Those thermal arrays can be used for passive imaging only in the IR spectral range. In the THz spectral interval, their sensitivity is insufficient to detect THz radiation in passive mode. Therefore, those arrays can be used in the THz range only for active imaging. To be used in passive vision systems in the THz range, they should have NEP values less than $10^{-13}$ W/Hz$^{1/2}$ [167].

Typical parameters of uncooled IR and THz microbolometer arrays are quoted in Table 5.4.

***Table 5.4.*** *Typical parameters of some IR and THz uncooled thermal detectors.*

| Detectors | Modulation frequency f (Hz) or response time τ (s) | Operation radiation frequency, THz | NEP, W/Hz$^{1/2}$ | Refs | Remarks |
|---|---|---|---|---|---|
| Golay cell | τ ~ 0.1 s, f ≤ 20 Hz | – ≤20 | $2 \times 0^{-10}$ ~$10^{-8}$–$10^{-9}$ | [148] | Commercial |
| Micro-array Golay cell | f ≤ 30 Hz | 0.105 | $3 \times 10^{-7}$ | [151] | |
| Pyroelectric, LiTiO$_3$ Pyroelectric, TGS | τ > 0.1 s – – | 0.2–30 – – | ~$4 \times 10^{-10}$ ~$8 \times 10^{-9}$ ~$6 \times 10^{-9}$ | [168] [169] [169] | |
| LT pyroelectric | – | 2.52 | $1.5 \times 10^{-9}$ | [174] | 30-pixel linear array |
| VO$_x$ microbolometer | τ ~ 13 ms | 4.3 | $3.2 \times 10^{-10}$ | [157] | 320 × 240 |
| VO$_x$ microbolometer | 30 Hz | ~1.5…30 | MDP~20…30 pW (per pixel) | [130] | 320 × 240 |
| VO$_x$ microbolometer | | 2.52 | $4.57 \times 10^{-11}$ | [181] | 320 × 240 |
| α-Si microbolometer | 25 Hz | 2…4 | MDP~32 pW (per pixel, ν = 2.5 THz) | [176] | 320 × 240 |
| α-Si microbolometer | τ ~ 20…40 ms | 2.5 | $3 \times 10^{-11}$ | [176] | 320 × 240 |
| Ni microbometer | 30 Hz | 0.094 | $1.9 \times 10^{-11}$ | [170] | |
| Nb microbolometer | – | ≤30 | $5 \times 10^{-11}$ | [171] | |
| Nb microbolometer | τ ≈ 3 μs | | $1.5 \times 10^{-11}$ | [179] | |
| Ti microbolometer | τ < 1 μs | 0.2…2.1 (0.3; 06 resonant) | $1.4 \times 10^{-11}$ | [172] [173] | 2 × 16 array |
| Bi microbolometer | τ ~ 2 × 10$^{-7}$ s | 0.215 | D* = $4 \times 10^8$, cm·Hz$^{1/2}$/W | [182] | 400 pixels array |
| Nb$_5$N$_6$ microbolometer | 4 × 10$^3$ Hz | 0.22…0.3 3 | $1.3 \times 10^{-11}$; $1.7 \times 10^{-11}$ | [177] [184] | 5 × 6 array, 1 × 64 array |
| YBa$_2$Cu$_3$O$_{7-x}$ microbolometer | 10$^5$ Hz | 0.1…2 | $5 \times 10^{-11}$ | [175] | |
| Thermocouple | ~15 Hz, frame rate | 2.8 | $0.49 \times 10^{-9}$ | [178] | 32 × 32 (Commersial, pixel 0.7×0.7 mm) |

| BiSb/Sb thermocouple | $\tau \approx 22$ μs | 0.812 | $1.7 \times 10^{-10}$ | [180] | Sensor pixels consist of 8 single thermocouples |
| p-polySi/n-polySi thermocouple | 50 Hz | 2.52 | $5 \times 10^{-9}$ | [183] | $8 \times 8$ |

Notations: TGS is Triglycine sulphate $(NH_2CH_2COOH)_3 \cdot H_2SO_4$, LT is lithium tantalate, D* is the detectivity, and MDP is the minimum detectable power (MDP = $NEP/(2 \cdot \tau_{int})^{1/2}$, where $\tau_{int}$ = $1/f_r$ (for white noise), and $f_r$ is the frame rate).

*5.4.3.3 Metallic bolometers*

Other materials applied in uncooled THz detectors are metallic microbolometers. Metallic bolometers, despite their lower TCR in comparison with $VO_x$ or α-Si microbolometers, have similar NEP values (Table 5.4), At the same time, they are some fragile because of very thin metallic air-bridge layers used as sensitive elements. The characteristics of some modern metallic bolometers and arrays are considered, *e.g.*, in Refs. [172, 173, 185].

Thin layers made of Ni, Sb, Nb, Ti, Pt, and Pd metals ─ these are materials that keep stable parameter values, which do not strongly deteriorate with the operation time ─ are frequently used in metallic bolometers. The parameters of some materials that can be used for the fabrication of metallic microbolometers are summarized in Table 5.5. The values for the thermal conductivity $G_{th}$, the electrical resistivity ρ, and the TCR also are given.

Bismuth seems to be a promising material for THz microbolometers [186]. Teraherz bolometers with a thin Bi layer have a good sensitivity and a good noise performance. However, they also demonstrate unstable characteristics due to electro-migration.

The authors of Refs. [172, 185, 187, 188] considered uncooled Ti antenna-coupled microbolometers. They showed that those devices have a high sensitivity, can operate in the real-time mode, and can be applied as arrays in systems intended for the real-time operation in the THz interval.

Materials Research Forum LLC
https://doi.org/10.21741/9781644900758

*Table 5.5. Some parameters of materials that can be used for the fabrication of metallic microbolometers (taken from Ref. [173]). (By permission of SPIE).*

| Material | $G_{th}$, W/mK | $\rho$, $\Omega\cdot$m | TCR, $K^{-1}$ |
|----------|-------|-------|-------|
| Cu | 400 | $17.0 \times 10^{-9}$ | 0.04 % |
| Al | 235 | $26.5 \times 10^{-9}$ | 0.11 % |
| Sn | 67 | $110.0 \times 10^{-9}$ | 0.05 % |
| Pb | 35 | $210.0 \times 10^{-9}$ | - |
| Pt | 72 | $106.0 \times 10^{-9}$ | 0.39 % |
| Ti | 22 | $400.0 \times 10^{-9}$ | 0.13 % |
| Bi | 8 | $1.3 \times 10^{-6}$ | –0.30 % |

The parameters of some contemporary metallic uncooled bolometers are given in Table 5.4. A schematic view of an antenna-coupled Ti-microbolometer sensor together with the microscope images of a fabricated Ti-microbolometer and a central part of THz antenna is shown in Fig. 5.22. Here, important is the fact that Ti was proved to be compatible, to a great extent, with the silicon wafer processing technology.

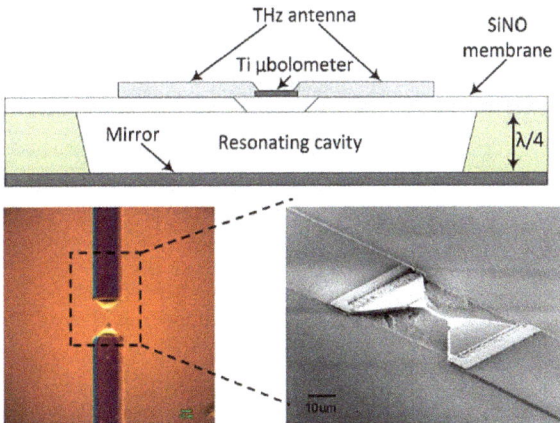

*Figure 5.22. Schematic view of an antenna-coupled Ti-microbolometer sensor (top); microscope images of a fabricated Ti-microbolometer and a central part of THz antenna: flat-top (bottom left) and side-3D (bottom right) views [172].*

### 5.4.3.4 Pyroelectric detectors

The wavelength-independent pyroelectric detectors are manufactured from ferroelectric (pyroelectric) crystals. These crystals are spontaneously polarized, i.e. every unit cell in the crystal has a permanent electric dipole moment aligned with a specific crystal axis. Actually, spontaneous polarization is a fundamental property of a good many crytalline systems. Ten of 32 crystal classes are pyroelectric; they do not have a centre of symmetry. If an external electric field can reverse the dipole, the material is said to be ferroelectric. All ferroelectric materials are pyroelectric, but not the other way around [189].

The spontaneous electric polarization P in a pyroelectric material is sensitive to temperature variations. At small temperature changes $\Delta T$, the polarization is determined by the pyroelectric coefficient $p = dP/dT$. For $BaTiO_3$ and TGS, $p \approx 20$ and $35$ $nCcm^{-2}K^{-1}$, respectively. The generated charge equals $Q = pA_d\Delta T$, where $A_d$ is the area of the detector's sensitive part.

The current response of pyroelectric detectors depends on $dT/dt$, where t is the time. Being irradiated, a pyroelectric crystal is heated up, and an electric charge is generated. When THz radiation is turned off, the crystal cools down, and a charge with the opposite sign is generated. This means that only modulated radiation creates a signal in pyroelectric detectors. That is why pyroelectric detector does not react on constant power radiation.

In the absence of electric field, pyroelectric crystals are spontaneously polarized, but at a constant temperature, this polarization cannot be revealed owing to the presence of surface charges, as these charges are neutralized due to the surface and bulk conductivity. Therefore, temperature changes in the sensitive pyroelectric structure should occur faster than the redistribution of compensating charges does. This condition restricts the operational capability of pyroelectric detectors to low radiation modulation frequencies.

As a rule, pyroelectric detectors sensitive to temperature changes have a high responsivity of about 5...300 kV/W, but they are noisy and, typically, highly resistive (R $\sim 10^{10}$ $\Omega$). Their typical response time equals $\tau \approx 10...50$ ms. However, if the detecting area is small, pyroelectric detectors can be rather fast ($\tau < 1$ ms). For examle, with the load R = 50 Ohm, they can be fast enough to measure short laser pulses, but possess a low responsivity level. Pyroelectric detectors can be both windowless and with an input window. In the latter case, they are wavelength-dependent because of a material used for the window. For flat spectra response, they can be coated with a black absorber. Their NEP values exceed $10^{-9}$ $W/Hz^{1/2}$.

The number of pyroelectric materials is rather large. In particular, these are triglycine sulfate (TGS), $LiTaO_3$, $LiNbO_3$, $Li_2SO_4xH_2O$, $BaTiO_3$, $NaNO_2$, $PVF_2$, $Sr_xBa_{1-x}Nb_2O_6$ (SBN), SbSi, $Ba_{1-x}Sr_xTiO_3$ (BST), and others. Many uncooled broadband (from IR to at least microwave) sensitive detectors on the basis of pyroelectric materials are commercially available, *e.g.*, lithium tantalate ($LiTaO_3$), $LiNbO_3$, and deuterated L-alanine doped triglycene sulfate (DLARGS) [168, 189]. Their thermal time constants decrease with the temperature growth and the detector thickness reduction [169].

As a rule, pyroelectric detectors operate at room temperatures, which are lower than the Curie temperature $T_C$ (*e.g.*, $T_C \approx 135^0$ C for $BaTiO_3$, and $T_C \approx 50^0$ C for TGS), at which a pyroelectric turns into a paraelectric with no polarization.

Pyroelectric detectors are among the most explored IR detectors. Their technology allows a large-scale-array integration. They are also applicable in commercial THz detectors. Typical responsivity and NEP values of a pyroelectric detector are 1 kV/W and $10^{-8}...10^{-9}$ $W/Hz^{1/2}$, respectively, at the modulation frequency f $\approx$ 10 Hz. The operational parameters of some pyroelectric THz detectors are listed in Table 5.4.

An important feature of pyroelectric detectors, as well as thermocouple THz detectors, where the output voltage is generated under the radiation action, is the absence of applied bias. The NEP of such detectors is worser as compared to that of microbolometer THz detectors (see Table 5.4). Besides that, pyroelectric detectors are sensitive to piezoelectric noise.

*5.4.3.5 Thermoelectric detectors (thermocouples, thermopiles)*

The operation of any thermocouple detector is based on the Joule heating by incident radiation and the thermoelectric effect (the Seebeck effect). In those devices, the temperature difference is directly converted into the thermoelectric voltage $\Delta V = \alpha_s \times \Delta T$, where $\alpha_s$ is the Seebeck coefficient. In particular, $\alpha_s \approx 500$ µV/K for Si and about an order of magnitude lower for metals (Bi, a series from Sb to Pb). The Seebeck coefficient for thin films is usually lower than the corresponding bulk coefficient, and it is technology-dependent. The electrical conductivity in such nanoscale devices strongly depends of the device thichness [190], which is important for designing fast thermoelectric detectors and arrays.

Unlike other thermal detectors - *e.g.*, bolometers, where the bias current is applied to the thermo-resistive sensor element-thermocouples (thermopiles) have no need in the external bias. One of the first thermocouple detectors fabricated using the micromachining technology was the bismuth-antimony micro-thermocouple [191].

For some thin-film pairs, the Seebeck coefficients are as follows: 7.4 µV/K (Ti/Al), 70 µV/K (Bi/Cr), and 190 µV/K (Ti/doped Si) [192]. The thermocouple detector parameters turn out better for materials with the lower thermal conductivity $G_{th}$. Therefore, the THz thermoelectric detectors can differ strongly in their NEP values, which depend of the manufacture technology and the applied materials. Thick coupling layers are characterized by slow response times and large NEP values. An important feature of thermocouple detectors is a capability of their arrangement into arrays.

When manufacturing very thin layers, air-bridge layers, typically a few nanometers in thickness, can be produced using the MEMS technique in order to provide a short response time and reasonable NEP values. Nano-thermocouples can be sufficiently fast thermal detectors with a response time of several picoseconds, and they can be potentially used in mixers with a high intermediate frequency [193].

To manufacture contemporary THz thermocouple air-bridge structures, surface micromachining technologies (MEMS) are applied (see, *e.g.*, Refs. [194, 195]). In order to provide the short thermal response time $\tau_{th} = C_{th}/G_{th}$, micro-bridge structures with small sensors (to ensure the low capacitance $C_{th}$) and high thermal insulation (to provide the low conductance $G_{th}$) are used.

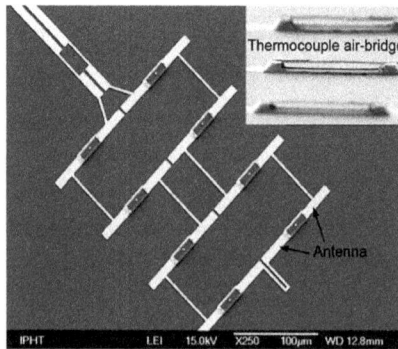

*Figure 5.23. Complete antenna structure with interconnects and Ni diffusion barrier upon polyimide pads. A zoom of a freestanding BiSb/Sb air-bridge structure is presented in the inset [180]. One pixel consists of eight thermocouples connected in series in order to increase the antenna gain and, therefore, increase the active sensor area. (By permission of AIP Publising).*

There can be different designs of thermocouple detectors. One of them for thermocouple THz sensors with thermocouple pixels connected in series is shown in Fig.5.23.

Microbridge thermocouple detector arrays with available parameters [196] can be used in systems for the real-time THz imaging. Potentially, graphene-based thermocouple arrays with a relatively good detectivity (D* $\approx$ 8×10$^8$ cmHz$^{1/2}$/W) can also be applied for this purpose [197].

### 5.4.4 Cooled thermal detectors

Cooled thermal detectors can be divided into several groups: extrinsic semiconducting detectors, semiconducting (non-superconducting) bolometers, and various kinds of superconducting detectors.

### 5.4.4.1 Cooled semiconducting bolometers

The parameters of cooled semiconducting bolometers can be substantially improved in comparison with those of room-temperature devices, because the resistance changes very much at low temperatures, and the heat capacity is substantially lower at temperatures below the Debye temperature $\Theta_D$ (e.g., $\Theta_D \approx 370$ K for Ge, and $\Theta_D \approx 640$ K for Si). The latter circumstance allows thicker devices to be appled in order to enhance absorption. Therefore, the sensitivity of such low-temperature devices can be several orders of magnitude higher (the NEP is much lower) as compared with that for bolometers operating at room temperatures.

Those conditions were used while designing IR bolometers intended for operation at liquid helium temperatures. They were developed on the basis of the core of a carbon composition resistor used as a sensitive element that operates in the ms time response interval [198]. The low-temperature semiconducting (Ge:Ga) bolometer seems to be first invented in 1961 for astronomical applications [199]. In that device, a NEP value of 5×10$^{-13}$ W/Hz$^{1/2}$ was obtained at T = 2 K, and the response time was $\tau \approx 400$ µs. Estimations showed that a NEP value of about 10$^{-15}$ W/Hz$^{1/2}$ could be obtained at T $\approx$ 0.5 K, with $\tau$ falling within the millisecond interval. The capabilities and performance of the compensated Si-impurity conduction bolometers as extremely sensitive detectors of far-infrared radiation (from 2 mm to 30 µm) were demonstrated [200]. Electrical and far-infrared measurements showed a NEP value of about 2.5×10$^{-14}$ W/Hz$^{1/2}$, and a response time of an order of 10$^{-2}$ s when operating at T = 1.5 K.

Semiconducting bolometers of two types are often used; these are neutron transmutation doped (NTD) Ge thermistors and ion-implanted silicon ones [127, 201, 202]. They usually operate at sensor temperatures ranging from ~100 mK to 300 mK and are typically biased with an AC signal in order to avoid the amplifier 1/f noise and offset.

Low-temperature semiconducting thermistor bolometers dissipate power and thus are charaterized by a significant self-heating. The sensor self-heating leads to an electro-thermal feedback, which reduces the sensor time constant and decreases the temperature change induced by incident radiation [143].

Low-temperature dependences of resistivity for some NTD Ge samples are shown in Fig. 5.24.

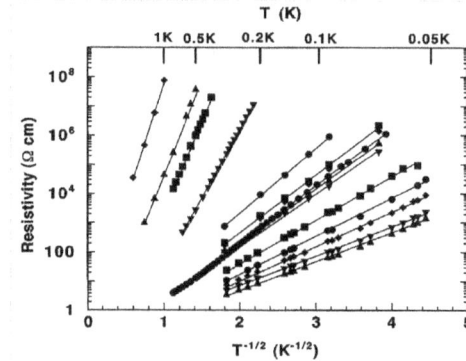

*Figure 5.24. Logarithm of resistivity plotted as a function of $T^{-2}$ for 14 insulating NTD $^{70}$Ge:Ga samples with gallium concentrations of (from top to bottom) 3.02, 8.00, 9.36, 14.50, 17.17, 17.52, 17.61, 17.68, 17.70, 17.79, 17.96, 18.05, 18.23, and 18.40 (in units of $10^{16}$ cm$^{-3}$) [203]. (By permission of American Physical Society).*

The heat capacity $C_{th}$ of NTD Ge temperature sensors at T = 100...300 mK were measured in Ref. [204]. The examined NTD Ge sensors were 30 × 100 × 250 μm$^3$ in dimensions, had a natural isotopic abundance, the doping level N = 5.6×10$^{16}$ cm$^{-3}$, and possessed ion-implanted and metallized contact pads. Every sensor was mounted on a freestanding silicon nitride (Si-N) pad supported by Si-N legs. For the Ge slab itself, the heat conductivity $G_{th}$ < 67 fW/K. The heat conductivity of contact metals, the bonding pad and legs (a cross-section less than 10 μm$^2$), and other elements was considerably higher. The heat capacity of leads in an NTD Ge bolometer with a Si$_3$N$_4$ micromesh absorber was $C_{th} \approx 1.8 \times 10^{-11}$ J/K at T = 400 mK [205].

The NTD $^{70}$Ge transforms into $^{71}$Ga (acceptor), and $^{74}$Ge into $^{75}$As (donor). The doping level depends on the neutron flux, and the compensation ratio can be changed by altering the isotope ratios [206].

The NTD Ge bolometers have the NEP value of about $10^{-16}$ W/Hz$^{1/2}$ and $\tau \approx 11$ ms at T = 300 mK [207]. The averaged, over ten bolometers, NEP was equal to 8.5×$10^{-17}$ W/Hz$^{1/2}$ at T = 400 mK, with the thermal conductance $G_{th}$ = 9.1·$10^{-10}$ W/K and $\tau \approx 15.5$ ms [205]. At T = 100 mK, the NEP of NTD Ge bolometers improved to 2×$10^{-17}$ W/Hz$^{1/2}$, with $\tau \approx$ 30 ms [208]. The spider web bolometers developed for the Herschel Space Observatory have a rated NEP of ~2×$10^{-18}$ W/Hz$^{1/2}$. The latter value is well below the thermal background provided by the space observatory. For ground observations under high-background conditions, a NEP of about several units of $10^{-15}$ W/Hz$^{1/2}$ or better is sufficient for revealing low radiation powers, since the atmospheric transmission and emissivity from the sky dominate [209].

To reach thermal isolation, semiconducting bolometers are typically fabricated using micromachining techniques on membranes of Si, SiN or SOI (silicon-on-insulator) substrates [168, 210]. To get a high sensitivity, low-temperature semiconducting bolometers usually operate at T = 100...300 mK. The sensitivity values S $\approx$ 0.1...0.2 A/W were obtained in the spectral interval from 0.1 to 10 THz at −196 °C under the black body radiation background. Typical NEP values for cold semiconducting bolometers operating in the temperature range T ~ 100...300 mK are about $10^{-15}$...$10^{-16}$ W/Hz$^{1/2}$, with a millisecond response time [39, 211]. They are 1...2 orders of magnitude worser in comparison with superconducting THz detectors.

To fabricate Ge or Si bolometers that would be not very thick in order to overcome the large heat capacity inherent to thick bolometer detectors, they are designed as composite bolometers consisting of a thin-film absorber layer, a substrate (which determines the active area of the bolometer), and a temperature sensor (a sensitive element with the same or a smaller size). Cooled-down (to T $\leq$ 4.2 K) bolometers and arrays are one of the best choices for detecting THz radiation, spectroscopy, and imaging at wavelengths from 200 μm to 3 mm. They historically played an important role in the sub-mm astronomy up to 1990s, when they were superseded (to a certain degree) by superconducting bolometers [212]. By reducing the temperature from 4.2 K to 0.3 K, the NEP of semiconducting bolometers can be improved by about three orders of magnitude.

Typical Si bolometer arrays were used for the Hershel space telescope [39, 42]. Two Si bolometer arrays (one of them consisting of 16 × 32 and the other of 32 × 64 pixels) were composed of 16 × 16-pixel sub-arrays and allowed imaging line spectroscopy and imaging photometry to be performed in the 60...210 μm wavelength band. The sub-arrays were mounted on a carrier, which was cooled down to 300 mK and thermally isolated from the surrounding structure with T = 2 K. The buffer/multiplexer electronics was moubted on a part of the indium-bump-bonded back plane of the FPAs operating at 300 mK, and on a heat sink running at 2 K. The multiplexing readout sampled each pixel

Materials Research Forum LLC
https://doi.org/10.21741/9781644900758

at a rate of 40 Hz or 20 Hz. The NEP dependences on the detector bias voltage are shown in Fig. 5.25 for a number of background fluxes [39]. Depending on the background level, the NEP values for Si bolometers can reach about $1.2 \cdot 10^{-16}$ W/Hz$^{1/2}$ at T $\approx$ 300 mK.

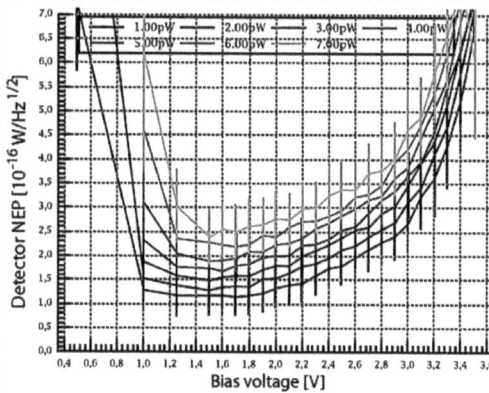

*Figure 5.25. NEP as a function of detector bias voltage for a number of background flux levels [39]. (By permission of SPIE).*

### 5.4.5  Superconducting thermal THz detectors

The changes of resistance in superconducting bolometers near the phase transition temperature are extremely large and, therefore, the sensitivity of those devices can be very high. The superconducting thermal detectors (bolometers) are the most sensitive among incoherent (direct) THz detectors, with NEP values of $10^{-19} \ldots 10^{-20}$ W/Hz$^{1/2}$ [128, 213]. They include hot-electron bolometers (HEBs) and transition edge sensors (TESs). The both are quite similar in the operational principle: small temperature changes caused by the incident radiation absorption significantly affect the resistance of the biased sensor operating near its superconducting transition temperature $T_c$. The main difference between them is their response rates.

In HEBs, radiation is directly absorbed by the superconducting layers, and if thin layers with intensive electron-phonon interaction (*e.g.*, NbN) are used, their thermal electron-temperature relaxation time $\tau$ can be as short as several ns [214] leading to the intermediate frequency $\nu_{IF} \approx 1/(2\pi\tau) \sim 10^{-1}$ GHz. Later [215], it was shown that $\tau$ can be as small as several ps, giving an intermediate frequency of several GHz. The phenomena occurring in NbN films were carefully investigated in a number of laboratories (see, *e.g.*,

Refs. [216–219]). A high response rate of HEBs in different materials led to their use as mixers in heterodyne receivers, which made it possible to extend the spectral resolution in radio astronomy beyond 1 THz [220] (see Fig. 3.16).

In TESs, a separate radiation absorber is used, which allows the energy to flow to the superconducting part of TES via phonons. This is the reason of a much larger response time as compared to those of HEBs.

The superconducting bolometers seem to be the most sensitive among available direct THz detectors but a requirement of their cooling to T ~ 100...300 mK in order to reach NEP values of about $10^{-19}$ W/Hz$^{1/2}$ make the systems on their basis bulky and expensive. However, the availability of those devices has opened a variety of possibilities to get new information from the Universe.

*5.4.5.1 Superconducting HEBs*

Superconducting HEBs are thermal detectors with a high TCR and low noise characteristics. They can be considered as direct detectors and, because of their high operational rate, can be used as mixers and for high-rate photon detection. Those devices are attractive, because, being bolometers, they have no parasitic capacitance and, when used in mixers, require low local oscillator (LO) power.

Perhaps, superconducting-normal-superconductor (SNS) junctions were first that were considered as ultrasensitive superconducting detectors. The absorbing element consists of a thin film resistor strip, which is connected to superconducting electrodes (antennas) [221]. The device exploits the Andreev reflection of electrons and weak interaction between electrons and phonons in order to produce a large electron temperature rise. This growth is measured from the temperature dependence of the current-voltage characteristics for a superconductor-insulator-normal metal (SIN) tunnel junction, which is biased at the constant current, when some part of the resistor strip is the normal-metal electrode. The normal-metal absorber is a copper strip. Experimental studies of such HEBs proved the attainability of low NEP values of about $10^{-18}$ W/Hz$^{1/2}$ (see, *e.g.*, Ref. [128]).

The considered NbN-based HEB mixers demonstrate a good noise performance from ~500 GHz up to ~5 THz [222] and, having the critical temperature $T_c \sim 8...11$ K, operate, as a rule, at T < 4 K. The most commonly used materials for THz HEB mixers are NbN and NbTiN thin films (d ~ 5 nm). They have a relatively narrow IF bandwidth ($\leq$ 5 GHz) due to a limited electron-temperature relaxation rate, when their length L satisfies the requirement L << $(12)^{1/2} \times L_{ep}$ [142] in the micro-size bridge design. Here, $L_{ep} = (D\tau_{ep})^{1/2}$, where $\tau_{ep}$ is the time of energy transfer from hot electrons to phonons, and D is the diffusion constant.

HEB mixers with thin (10...20 nm) $MgB_2$ films and $T_c \sim 20$ K have been realized [223], which also can be used as microbolometers. These devices are capable of mixing signals with an IF bandwidth of about 8 GHz at temperatures above 20 K, which means that they can operate with the help of cryogenic-free cooling systems [224].

For practical applications, an important parameter is the required LO power. To achieve low values of this parameter, the area of HEB mixers is reduced to less than 1 $\mu m^2$. The NbN-, NbTiN-, and $MgB_2$-based HEB mixers require an LO pump power of approximately 1...2 $\mu W$ [225–227], which is less than in SIS mixers. Both the diffusion-cooled HEBs (microbridges) and the phonon-cooled HEBs use short superconducting layers connecting two normal metal pads as a mixing element, and generally their operation temperature (T $\sim$ 4...20 K) depends on the superconducting material and the film thickness. Typical double-side band (DSB) noise temperatures of HEB mixers follow the $10 \times h\nu/k_B$ dependence (Fig. 3.16) with approximately 10...15 dB of power conversion losses. The mixer can gain $\sim 5 \cdot h\nu/k_B$ noise temperature at intermediate frequencies lower than 3 GHz with a gain width of about 5.5 GHz in some structures (NbN on GaN membranes) [228].

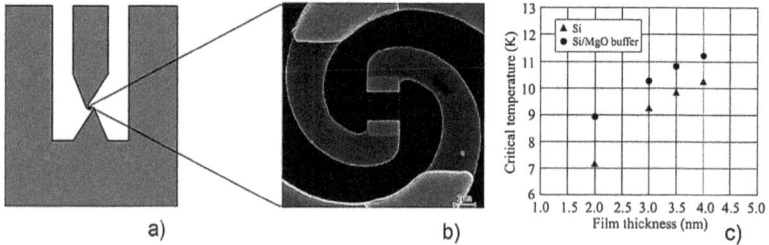

*Figure 5.26. (a) Phonon-cooled NbN-based HEB mixer chip, (b) SEM micrograph of the central area of a mixer with a spiral antenna, (c) dependences $T_c$ versus thickness for NbN films deposited on Si substrates (triangles) and on Si with a MgO buffer layer (circles) [230]. (By permission of SPIE).*

A HEB consists of a thin (several nanometers) superconducting layer with submicron lateral dimensions (Fig. 5.26). Incident radiation absorbed by the antenna heats electrons, which in turn cool either the substrate via the electron-phonon coupling (phonon-cooled HEBs) or the normal-metal pads (diffusion-cooled HEBs). For the efficient detection of radiation, the devices are coupled to suitable normal-metal antenna structures. Superconducting HEBs are typically made from a material with a strong electron-phonon interaction, suc as NbN, NbTiN, MgB$_2$, MoRe, Nb or Al [96, 220, 229].

The superconducting properties of thin films are strongly influenced by their crystalline quality. The lattice constant should match well with that of the substrate. Therefore, the choice of the substrate or the buffer layer is important for the film crystallinity, and the film thickness influences the transition temperature $T_c$, as one can see in Fig. 5.26.c for NbN ultrathin layers.

The superconducting HEBs can operate according to two main mechanisms that allow electrons to exchange their energy with one another faster than they heat phonons:

a) The principle of phonon-cooled HEBs (Fig. 5.27) was considered in Ref. [214], and it was first realized in Ref. [216];

b) The principle of diffusion-cooled HEBs was proposed in Ref. [141], and its first realization seems to be reported in Ref. [217].

Those mechanisms were also considered, *e.g.*, in Refs. [231–233]. The phonon-cooled and diffusion-cooled mechanisms are presented schematically in Fig. 5.27.

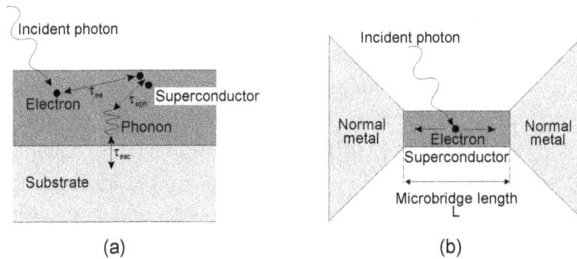

*Figure 5.27. Hot-electron bolometer mechanisms: phonon-cooled (a) and diffusion-cooled (b) principles. [84]. (By permission from Elsevier).*

The thermal dynamics in a superconducting HEB located on a dielectric substrate can be described in terms of four coexisting subsystems: Cooper pairs, quasiparticles (electrons from broken Cooper pairs), phonons in the film, and phonons in the substrate. The general analysis of a nonequilibrium state requires a solution of integral kinetic equations for space- and time-dependent distribution functions. To avoid this complexity, various simplifying assumptions are used to reduce the general problem to analytically solvable rate equations [232]. Below, the schematics of the processes in superconducting HEBs are considered.

In the phonon-cooled HEBs (Fig. 5.27), hot electrons transfer their energy to phonons, and this process is charaterized by the time constant $\tau_{ep}$. In the next step, the excess of phonon energy escapes towards the substrate with the time constant $\tau_{es}$. Several conditions have to be fulfilled to make the phonon-cooled mechanism effective. (i) The electron-electron interaction time ($\tau_{ee}$) must be much shorter than the phonon-electron energy relaxation time $\tau_{ep}$. (ii) The superconducting film must be very thin (a few nm), and the film-to-substrate thermal conductance must be very high ($\tau_{es} << \tau_{ep}$) to obtain an efficient phonon escape from the superconductor to the substrate. (iii) The substrate thermal conductivity must be high and a good thermal contact between the substrate and the cold finger must be ensured.

In HEBs, electrons and phonons have the temperatures $T_e$ and $T_p$, respectively, and are characterized by the time $\tau_{ep}$ of electron energy relaxation via electron-phonon interaction and the time $\tau_{es}$ of phonon escape into the substrate, respectively. The electron and phonon specific heats, $C_e$ and $C_p$, respectively, determine the phonon-electron energy relaxation time $\tau_{pe} = \tau_{ep} \times (C_p/C_e)$. In various superconducting materials, the $C_p/C_e$ ratio, which controls the energy flow from electrons to phonons and the energy backflow due to reabsorption of nonequilibrium phonons by electrons, equals 0.85 (Nb), 6.5 (NbN), and 38 (YBCO). Thus, *e.g.*, in thin (< 10 nm) Nb films, $\tau_{pe} > \tau_{ep}$, and, for a film deposited on a substrate, the effective escape of phonons to the substrate prevails over the energy backflow to electrons. As a result, $\tau_{ep}$ alone governs the response time of thin (< 10 nm) Nb films ($\tau_{es} < \tau_{ep} < \tau_{pe} \sim 10^2$ ns [234]), which is approximately equal to $\tau_{ep} \sim 5$ ns. Thus, these devices, which are sensitive in a wide spectral range, are much faster as compared to bulk semiconductor bolometers operating at T ~ 4 K, and they can reach NEP values of about $3 \times 10^{-13}$ W/Hz$^{1/2}$ [235].

The NbN compound has much shorter $\tau_{ep}$ and $\tau_{pe}$ in comparison with Nb, because of stronger electron-phonon interaction. In ultrathin films with the thickness d ≈3 nm, both $\tau_{ep}$ and $\tau_{pe}$ determine the response time $\tau$ of the detector with NbN sensitive element, which can be about 30 ps [218] near $T_c$ ($\tau_{ep} \sim 10$ ps). The noise-equivalent power can reach NEP values of about $10^{-12}$ W/Hz$^{1/2}$ [215].

Since $C_p/C_e \approx 38$ in high-temperature YBCO superconducting detector layers, these detectors are mainly of the phonon-cooled type, because the energy backflow from phonons to electrons can be neglected, and the thermalization time is about an order of magnitude shorter ($\tau_{ep} \sim 1$ ps) in comparison with that in NbN layers. In YBCO superconducting films excited by fs pulses, non-thermal (hot-electron) and thermal bolometric (phonon) processes are practically decoupled, with the former ones dominating at the early stage of electron relaxation [236]. To decouple electrons from phonons, non-equilibrium phonons should escape from the film (into the substrate) within a time interval shorter than the phonon-electron time $\tau_{pe}$.

When superconducting thin (d $\approx$ 4 nm) small-area (~1 $\mu m^2$) NbN layers are applied as detectors, the time constant can be $\tau \approx 4 \cdot 10^{-11}$ s at T = 4.2 K because of the fast heat transfer into the substrate (see, *e.g.*, Refs. [237–239]). That is a reason why these detectors can be used in coherent systems with a relatively broad bandwidth up to ~ 5 GHz.

In the diffusion-cooled HEBs, hot electrons transfer their energy by diffusion to a normal metal that forms electrical contacts to an external detector readout circuitry and/or the arms of a planar antenna. In this case, the length of the superconducting microbridge must be very short, with the maximum value L $\ll$ L$_{max}$ ~ $2(D_e\tau_{ee})^{1/2}$ ~ 1 $\mu m$, where $D_e$ is the electron diffusivity. The bolometer bandwidth is inversely proportional to the squared microbridge length [142], which falls within the submicron range. The bandwidth of diffusion-cooled bolometers is not limited by $\tau_{ep}$. As a result, larger intermediate frequency $\nu_{IF}$ values, in comparison with phonon-cooled bolometers, can be obtained. In Nb HEBs with L =0.08 $\mu m$, the intermediate frequency $\nu_{IF} \approx$ 8 GHz was achieved at T $\approx$ 4 K.

Typical characteristic times are $\tau_{ee}$ ~ 2 ps for electron-electron interaction, $\tau_{ep}$ ~ 15 ps for electron-phonon interaction, and $\tau_{es}$ ~ 35 ps for phonon-to-substrate escape (at the NbN film thickness d ~ 3.5 nm). Therefore, the cooling mechanism for HEB mixers with distances between the metal contacts shorter than L $\leq$ 1 $\mu m$ (typically L ~ 0.2…1 $\mu m$) is realized through electron diffusion to the normal metal. For larger lengths between the contacts, the phonon-cooling mechanism dominates [142].

The discovery of hot-electron phenomena in thin superconducting films was followed by numerous studies and applications. Unlike other detectors applied in mixers, such as superconductor-insulator-superconductor (SIS) tunnel junctions and Schottky barrier diodes (SBDs), the HEBs did not demonstrate any frequency limitation of the detection mechanism and can operate as incoherent detectors even at $\nu$ > 30 THz [240]. A broadband NbTiN superconducting nanowire detector (T ~ 2.5 K, the thickness d = 8.4 nm) was successfully applied for single-photon detection with a high efficiency (75–92 %) and high detection rates in the spectral range of 1310–1550 nm [241]. These demonstrations are important, *e.g.*, for applications in optical communication and quantum information processing.

HEB mixers are significantly more sensitive than SBD ones, but several times less sensitive than SIS mixers (at $\nu \leq 1$ THz). In the frequency range up to 2.5 THz, the noise temperature closely follows the linear $10 \times h\nu/k_B$ frequency dependence. Above this frequency, the sensitivity becomes somewhat worse, which is caused by increasing losses in the optical components, a lower efficiency of antenna, and skin-effect contributions in the superconducting bridge. Some characteristics of typical coherent detectors are shown in Table 5.6.

*Table 5.6. Some characteristics of typical coherent detectors.*

| Parameter | SIS | HEB | SBD |
|---|---|---|---|
| THz range, $\nu$ | $\leq 1.3 \times$ THz | $\leq 5 \times$ THz | $\leq 3 \times$THz |
| IF, GHz | Up to 12* | $\leq 5...10**$ | $> 10$ |
| $T_{min} = h\nu/k_B$ | See Fig. 3.16 | See Fig. 3.16 | See Fig. 3.16 |
| LO power | $\sim 1$ µW | $\leq(1...2)$ µW | $\sim 1$ mW |
| Operating T, K | $\leq 4$ | $\leq(4...20)*$ | $\sim 20$ to 300 |

* Ref. [242], ** depending on the superconductor and the superconductor film thickness.

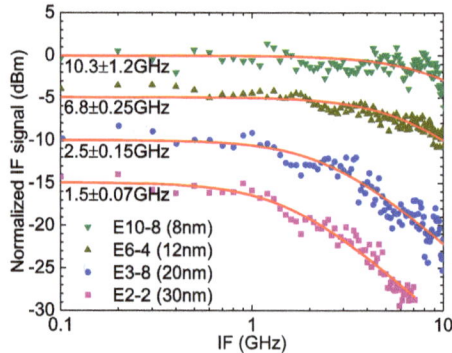

*Figure 5.28. Normalized IF signal of four MgB$_2$ HEB mixers measured at 0.1 THz (E2-2), 0.4 THz (E3-8), and 0.69 THz (E6-4, E10-8) [246]. T$_c$ increases with the film thickness growth. (By permission of E. Novoselov).*

The successful operation of practical instruments (the Heinrich Hertz Telescope, the Receiver Lab Telescope, APEX, SOFIA, Hershel) ensures the importance of the HEB technology despite the lack of rigorous theoretical routine for predicting the performance

[233]. Incoherent HEBs can be used in quick THz imaging and weak THz ultra-short pulse signal detection.

For NbN and NbTiN HEB mixers, the receiver noise temperature increases towards higher IF values, resulting in the operation intermediate frequencies less than 3 GHz (see, e.g., Ref. [230]). Therefore, the number of scientific problems in sub-mm wave astronomy which can be tackled with the help of HEB mixers is limited [243]. For $MgB_2$ hot-electron bolometer mixers, the IF bandwidth can be 2 to 3 times broader as compared to NbN HEBs [244, 245].

*5.4.5.2 Superconducting TES bolometers*

Superconducting TES bolometers are based on thin superconducting films, which $T_c$ is held within a temperature interval of a few mK, where dR/dT strongly changes from the superconducting to the normal state (see Fig. 5.29).

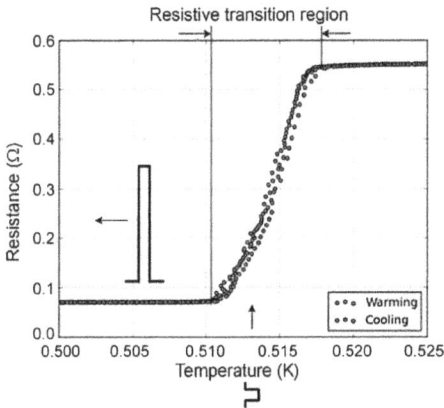

*Figure 5.29. Superconducting-to-normal resistive transition profile of square Ti (200 nm)/Au (20 nm) bilayer TES (adapted from Ref. [247]). (By permission of IOP Publishing).*

In the transition region, the superconducting film has a stable steep dependence of resistance on the temperature. Changes in the transition temperature can be induced by using a bilayer film consisting of a normal material and a layer of superconductor. Such a design enables diffusion of Cooper pairs from the superconductor into the normal metal and makes the latter weakly superconducting; this process is called the proximity effect. As a result, the transition temperature becomes lower in comparison with that in the pure superconducting film, and the bilayer acts as a single superconductor with a tunable $T_c$. The superconducting film is weakly coupled to a hit sink at the temperature $T_0 \sim T_c/2$.

The TES arrays are used for THz and X-ray space science applications. The TES bolometers seem to be among the most widespread superconducting direct detectors and arrays (see Fig. 4.3) with a high sensitivity (see, *e.g.*, Refs. [128, 213, 248]), as it occurs in all thermal superconducting detectors mainly depending on the $G_{th}$ parameter. Due to the DC bias heating, the TES has excess noise and limited saturation power [249].

There is considerable interest in developing superconducting TESs for astronomy and space science. For various purposes (ground-based photometric measurements, space-based instrumentation for Earth observation, and cooled-aperture telescopes at FIR wavelengths), NEP values of about $10^{-17}...10^{-19}$ W/Hz$^{1/2}$ are necessary [247, 250, 251].

Various pairs of superconducting metal films (bilayers) can be used, such as Mo/Au, Mo/Cu, Ti/Au, Al/Mn, and others, but TES bolometers are mainly based on low-temperature superconductors and operate within the temperature interval $T \sim 100...300$ mK. Two metals behave as a single film with a transition temperature between ~800 mK (for Mo) and 0 K (for Au). The transition temperature can be controlled within this interval. The low temperature ($T < 200$ mK) is needed because the energy resolution of these devices scales with the temperature. The heat sink (cold bath) is maintained by a refrigerator with a temperature between 50 and 300 mK.

Another approach to the direct photon detection using superconductivity is to operate far below $T_c$, when most of electrons are bound into Cooper pairs. Absorbed photons can break the Cooper pairs to produce single-electron quasi-particles, similarly to the decay of electron-hole pairs in semiconductors. However, it is difficult to distinguish quasi-particles from Cooper pairs.

The operation of TES bolometers is quite similar to that of HEBs. In the case of HEBs, a high response speed is achieved by allowing the radiation power to be directly absorbed by electrons in the superconductor. In TES bolometers, however, a separate radiation absorber is used that allows the energy to flow to the superconducting TES via phonons. The response time of those devices is considerably longer as compared to that of superconducting HEBs (tens of picoseconds), being in the range from tenths of μs to ms [252, 253].

Among incoherent THz detectors, TESs can have almost ultimate performance characteristics in a broad (from THz up to mm) spectral range with NEP values of $(0.4...3) \times 10^{-19}$ W/Hz$^{1/2}$ at the operation temperatures $T \sim (100...300)$ mK [213, 251, 253, 254]. Modern sub-orbital experiments largely rely on TES bolometers [255–257].

To achieve such low NEPs, low values of the thermal conductance between the sensitive element and the heat bath, $G_{th} \sim (0.1...10)$ pW/K, should be obtained. The lower NEP, the lower $G_{th}$ is required, which leads to some shortcomings connected with a necessity to design long thin legs. However, the larger $G_{th}$, the smaller the effective thermal constant $\tau_{eff}$ (see Fig. 5.30)

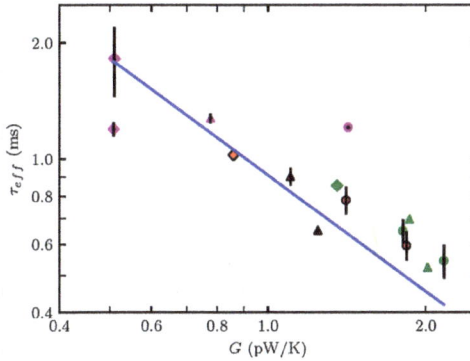

*Figure 5.30. Effective thermal time constant $\tau_{eff}$ versus thermal conductance $G = G_{th}$ for all devices. Circles, triangles, and diamonds represent legs with one, two and three filters, respectively. Error bars correspond to the standard error for the mean of multiple measurements of $\tau_{eff}$ at different bias points. The blue line shows the fitted model $\tau_{eff} \sim 1/G$ [248] (By permission of AIP Publishing).*

An important feature of TES detectors is that they can operate in a wide spectral range between radio and gamma rays [96, 258, 259]. These detectors can be applied for THz photons counting as well. Because of their potentially considerable applicability to CMB ground-based and space-based measurements (see, *e.g.*, Refs. [250, 257, 260, 261]), there are plenty of researches on the TES design and properties.

When a bolometer is biased with a constant voltage, the incoming radiation heats it up and increases its resistance. This increase, in turn, reduces the dissipated radiation power and, consequently, lowers the device temperature. Such a feedback can keep the TES operation temperature almost constant for the radiation power within a certain interval. Nevertheless, a voltage applied to the TES sensor provides its self-heating and that is why the TES has an excess noise and a limited saturation power [249].

The technologies used for the fabrication of TES bolometers are flexible. The latter can be produced with the help of photolithographic processes, and, therefore, specialized detectors and large arrays of TES sensors (see, *e.g.*, Refs. [262–266]) can be designed to meet the needs of specific observations.

The impedance of a TES is low (see Fig. 5.29), so it can deliver a significant power only to low-input impedance amplifiers, which rules out junction gate field-effect transistors and metal oxide semiconductor field-effect transistors (MOSFETs). Instead, the signals are fed into superconducting quantum interference devices (SQUIDs), which are a basis for a growing family of electronic devices that operate by a superconductive mechanism. Therefore, TES sensors are usually interfaced to SQUID amplifiers, which are effective ultra-low-noise ammeters that can operate at the same temperature as the TES thermometers do. In this case, the TES becomes a transformer coupled to the SQUID through an input coil. A current-biased shunt resistor is used to provide a constant bias voltage to the TES. When the shunt resistor is operated close to the detector temperature, a negligible Johnson noise is given from the bias network.

The SQUID readout has a number of advantages. In particular, it operates near the bolometer temperature, has a very low power dissipation and a large noise margin, as well as a low sensitivity to microphonic pickup. In addition, lithographic and other processes used at the fabrication of both SQUID readouts and TES bolometers are similar, which helps their integration on the same chip [267–270].

Proposed space missions using TES bolometer arrays include, *e.g.*, SPIRIT [271]; SPICA-BLISS [272]; SPICA-SAFARI [253, 265, 273], and LiteBIRD [274] (see also Fig. 4.3).

*5.4.5.3 SIN (NIS) detectors*

Superconductor-insulator-normal (SIN) or normal-insulator-superconductor (NIS) detectors are superconducting sensors that are similar to SIS detectors, except that one of the metals is a normal metal to exclude the Josephson current [275–277]. The development of the SIN tunnel junction as a thermometer seems to have started after Ref. [278]. Sensitive HEB Al-based NIS junctions were considered as detectors in Ref. [279].

The SIN devices are fabricated from the following superconductors: Nb [280], $NbN_x$ [281], $TaN_x$ [282], and $TiN_x$ [283] with various critical temperatures varying from $T_c \sim$ 1.35 K ($TiN_x$) to $T_c \sim 10.8$ K ($NbN_x$) in thin layers.

The proposition of SIN junction was connected with the presence of capacity in SIS heterojunctions, which is responsible for the Josephson current in them. To exclude this factor, one of the superconductors was proposed to be substituted by a normal metal [275]. In SIN structures, the I-V characteristics have weaker nonlinear characteristics as compared with SIS structures. As a result, the sensitivity of SIN junctions is lower, but the influence of Josephson effect is absent. To force electron tunnelling from the normal metal into the superconductor, the energy of electrons should not be less than $(\Delta - qV_b)$ above the Fermi level [96], where $V_b$ is the junction bias voltage. The junction current probes the tail of the electron Fermi distribution in the normal metal and is exponentially sensitive to the electron temperature $T_e \sim \exp[-(\Delta - qV_b)/k_B T_e]$.

In the SIN tunnel junction, the electron temperature changes in the normal-metal contact of the junction. Since the SIN junctions operate at sub-K temperatures, the electron-phonon coupling in the normal metal is week, which results in high thermal isolation of electrons. The applied planar antennas concentrate the sub-millimeter radiation energy within a small volume of normal metal. The SIN junction makes up an Andreev mirrors that traps hot electrons in the metal. However, those electrons can tunnel across the SIN junction, resulting in the appearance of a high-sensitive detector [279]. Thereby, the SIN junction becomes a thermometer for measuring the electron temperature in the normal metal. The NEP of SIN detectors can achieve values close to $10^{-16} \ldots 10^{-17}$ W/Hz$^{1/2}$ at T $\approx$ 300 mK [284].

An advantage of SIN tunnel junctions is that they operate in a wider range of temperature than phase transition thermometers do. Their other advantage is that the NEP does not increase much with the increasing absorber volume. These devices promise a higher dynamic range for a given sensitivity in comparison with TES sensors, because they do not have a fixed upper limit of signal response and are capable of removing a background power. They can be read out with the help of SQUIDS, junction-FET amplifiers, or CMOS amplifiers [127]. However, high-quality tunnel junctions are difficult to manufacture [285], and the noise level in them can be rather high.

As in any thermal detector, the NEP can be minimized by minimizing the thermal conductance $G_{th}$. The power absorbed by electrons in the normal metal can be dissipated through several possible channels. The first channel could be electrons flowing out of the active region into the contact pads. Since the contacts are made of a superconductor, the electrons are confined by means of the Andreev reflection. The most important noise source exists due to energy losses spent for producing phonons. At T $\sim$ 100 mK, it can be $G_{th} \sim 2 \times 10^{-13}$ W/K and, then, only for thermal noise, the NEP is proportional to $(4k_B \cdot G_{th} \cdot T^2)^{1/2}$ and amounts to about $3 \times 10^{-19}$ W/Hz$^{1/2}$.

The SIN structures and their variants, *e.g.*, superconductor-insulator-normal metal-insulator-superconductor (SINIS) junctions [286, 287], have found various applications not only as THz detectors. For instance, a SIN microwave source was developed, which is compatible with low-temperature electronics and offers convenient electrical control

over the incoherent photon emission rate with a predetermined frequency [288] for a
quantum-based standard of electric current [289, 290].

### 5.4.5.4 Cold-electron bolometers

The concept of cold-electron bolometers (CEBs) is based on a combination of (i) the RF
capacitive coupling between the absorber and the antenna through the capacitance of the
SIN tunnel junctions [291] and (ii) direct electron cooling of the absorber by the same
SIN tunnel junction [292]. The noise properties of this device are improved by decreasing
the electron temperature. Direct electron cooling also leads to an increase of saturation
power due to the removal of incoming power from the sensitive nanoabsorber.

This concept was implemented as an incoherent detector with a NEP of about $10^{-18}$
W/Hz$^{1/2}$ [293, 294]. Two-dimensional arrays [295] with up to 72 × 4 detectors with NEP
$\approx 10^{-16}$ W/Hz$^{1/2}$ at 350 GHz were fabricated for the OLIMPO telescope [296]. A NEP
value of 1.1 × $10^{-16}$ W/Hz$^{1/2}$ was obtained for a strained silicon CEB at T = 350 mK
[297].

The CEB and the processes running in it are shown schematically in Fig. 5.31.

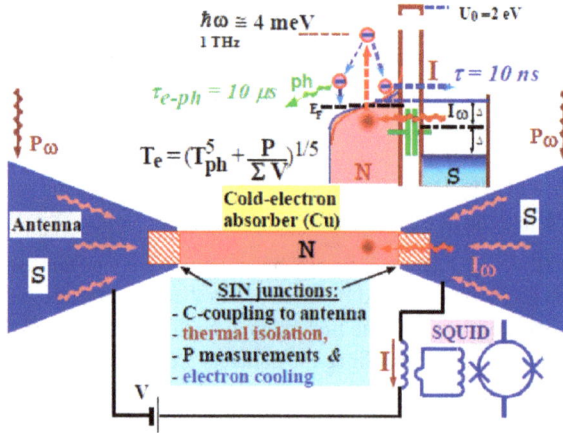

*Figure 5.31. Capacitively coupled CEB with nanoabsorber and SIN tunnel junctions for
direct electron cooling and power measurements. The signal power is supplied to the
sensor through the capacitance of tunnel junctions, dissipated in the nanoabsorber, and
removed back from the absorber as hot electrons by the same SIN junctions. The electron
cooling serves as strong negative electrothermal feedback improving all characteristics
of the CEB: time constant, responsivity, and NEP [249].*

## 5.5 Rectification detectors

For many day-to-day applications, preferable are uncooled detectors operating in a wide spectral range, possessing an acceptable sensitivity (a low NEP) and a wide dynamic range, and fast, thus allowing real-time operation. Rectification detectors seem to comply well with these requirements in many aspects. Detectors based on the rectification processes of electromagnetic signals in structures with a certain kind of non-linear static current–voltage characteristics (see Fig. 3.12) can operate in wide temperature and spectral intervals. No quantum (interband, intersubband, pair-breaking, *etc.*) or thermal processes are involved in this class of detectors.

An important feature of these detectors is that they can detect signals at radiation frequencies higher than their operating ("cut-off") frequencies, the latter, as a rule, being well below 1 THz. The sub-THz/THz broad-band radiation detection, *e.g.*, in long-channel FETs occurs due to the rectification of the high-frequency signal at a short distance from the source (if the receiving antenna is connected to the gate-source, Fig. 5.32), though these FETs cannot operate as signal amplifiers in this frequency range. The rectification phenomenon is related with the fact that the amplitude of an alternating THz signal decreases exponentially along the FET channel and is self-rectified. In SBDs, rectification processes give a possibility to detect signals at radiation frequencies higher than their "cut-off" frequencies $v_{co}$ (*e.g.*, the shunt or series cut-off frequency [298]), especially if SBDs are designed as nano-scale air-bridge devices with a small junction area to reduce the parasitic capacitance.

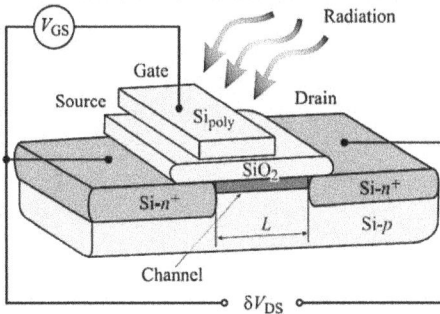

*Figure 5.32. Simplified FET representation. $V_{GS}$ is a gate-source voltage (e.g., from the antenna and the external voltage), and $\delta V_{DS}$ is a dc drain voltage appearing under radiation.*

The governing factor in FET or SBD rectification performance above the cut-off frequencies $v_{co}$ is the presence of parasitic elements (resistors, capacitnaces, and inductances, including those of antenna) [299, 300]. Above $v_{co}$, rectification still takes place. However, the intensity of rectified signals drops by the factor $v^{-2}...v^{-4}$.

These detectors can be used both for incoherent and coherent radiation detection. Deeply cooled STJ, SIS, KID, HEB, and other detectors (see above) can be classified into this group of devices. Nowadays, among those uncooled or not deeply cooled detectors, the most studied ones are SBDs and various kinds of FET detectors and arrays, *e.g.*, Si-MOSFETs, III-V HEMTs (GaAs/InGaAs, GaN/AlGaN *etc.*). As direct detectors, they can operate in wide spectral ranges (*e.g.,* $v \leq 9$ THz for Si MOSFET [301] and $v \leq 10$ THz for SBDs [302]). Important for applications are uncooled rectification detectors. However, modern uncooled rectification detectors are, as a rule, still not enough sensitive, *e.g.*, for passive imaging.

The SBD mixers were used in early astronomical receivers [3, 105]. Now they are almost impractical for modern arrays with SBD mixers because of their limited sensitivity and their high local oscillator power requirements in the 100...1000 μW range per pixel. However, they are widely used as rectification detectors.

The FETs and HEMTs can be used as mixers in heterodyne systems too. The heterodyne FET mixing at THz frequencies was first studied theoretically in Ref. [303]. The heterodyne and subharmonic mixing at THz frequencies using FETs was analyzed for various coupling schemes [304].

Signal rectification and mixing in uncooled THz detectors comprise an advantageous technique for fast detection of radiation. This technique makes it possible to fabricate low-cost sensors with a sufficient sensitivity and extends the dynamic range [305]. This paves the way for the realization of effective, compact, and relatively inexpensive THz cameras for active imaging.

The availability of technologies for manufacturing SBD and FET (HEMT) detectors and arrays with NEP values of $10^{-10}...10^{-12}$ W/Hz$^{1/2}$ (see Table 5.7), which can operate at room temperature, make them favorable for applications in the cost-effective direct active vision and spectroscopy systems. The principal factor that limits the performance of SBD or FET (HEMT) detectors is the presence of parasitic effects (resistors, capacitance, *etc.*) and the requirement of matching their impedance with that of printed antenna used because of small areas of sensitive elements ($\leq 0.5$ μm$^2$).

*Table 5.7. Some typical parameters of uncooled rectification SBD, FET and HEMT THz detectors.*

| Detectors | Operation radiation frequency ν, THz | NEP, W/Hz$^{1/2}$ | Refs |
|---|---|---|---|
| Zero bias SBDs, AlAs/InGaAs/InAs | ≤0.1, ~0.4 | ≈$4.5 \times 10^{-12}$, $8 \times 10^{-12}$ | [309] |
| Zero bias SBDs | 0.15, 0.3, 0.4 | ~$(5...20) \times 10^{-12}$ | [308] |
| Zero bias SBDs, ErAs/InAlGaAs/InP | 0.104 | $1.4 \times 10^{-12}$ | [298] |
| InGaAs SBD | 0.25 | $1.06 \times 10^{-10}$ (300 K) | [307] |
| Zero bias SBDs, InGaAs/InP | ~$(0.3...0.7)$ | ≈$5 \times 10^{-10}$ | [310] |
| Zero bias SBDs, InGaAs/InP (waveguide) | 0.57; 0.63 | $1.6 \times 10^{-11}$; $1.4 \times 10^{-11}$ | [315] |
| InAs SBD | 0.02...0.315 | $1.5 \times 10^{-10}$ (50 GHz) | [324] |
| Si FET | 0.65 | $3 \times 10^{-10}$ | [312] |
| Si n-MOSFET | ~0.32 | $3.2 \times 10^{-10}$ | [311] |
| Si FET | 0.295 | ≈$10^{-11}$ | [313] |
| Si CMOS FET | 0.595; 2.91 | $4.2 \times 10^{-11}$; $4.87 \times 10^{-10}$ | [314] |
| Si MOSFET | 0.63 | $10^{-11}$ | [317] |
| Si MOSFET | 0.5 | $2.9 \times 10^{-11}$ | [320] |
| Si CMOS | 0.86 | $10^{-10}$ | [321] |
| Ge-core/α-Si shell nanowire FET | 1.63 | $6.29 \times 10^{-10}$ | [319] |
| AlGaN/GaN FET | 0.5 | $2.6 \times 10^{-11}$ | [317] |
| Graphene FET | 0.01...0.67 | $(0.6...1) \times 10^{-11}$ | [316] |
| Al$_{0.25}$Ga$_{0.75}$N/GaN HEMT | 0.14 | $0.58 \times 10^{-12}$ (~300 K) $10^{-11}$ (~500 K) | [306] |
| GaN HEMT | 0.15 | ~$10^{-11}$ | [318] |
| InAlAs/InGaAs/InP HEMT | 0.2 | $4.8 \times 10^{-13}$ | [322] |
| Si Schottky gated MODFETs | 0.15; 0.30 | $8.1 \times 10^{-11}$; $7.0 \times 10^{-11}$ | [323] |
| Graphene FET | 0.4 | $1.3 \times 10^{-10}$ | [325] |

| Graphene FET | 0.6 | $5.15 \times 10^{-10}$ | [326] |
|---|---|---|---|
| Graphene FET, bilayer | 0.29...0.38 | $2 \times 10^{-9}$ (T = 300 K) | [330] |
| Graphene ballistic nanorectifier | *)190 Hz (AC) | Up to *)$6.4 \times 10^{-13}$ (T = 300 K) | [329] |
| Graphene ballistic rectifier | 0.07...0.45 | $3.4 \times 10^{-11}$ (antenna effective area A=$\lambda^2/4\pi$) | [328] |
| p–n junction in a graphene channel | 1.8...4.2 | $8 \times 10^{-11}$ | [327] |

*) Assuming the device to be a broadband detector.

At low radiation power levels, these detectors are square-law ones, in which the DC voltage response is proportional to the incoming signal power. The deviation from the linear behaviour of SBD and rectification FET THz detectors can be exploited for nonlinear autocorrelation measurements [331]. The FET detectors can be manufactured at a wafer level as arrays [321, 332] and cameras [321, 333–335].

All described rectification-type detectors are fast (the response time $\tau \leq 10^{-9}$ s in Si-CMOS FETs [301]. In GaAs-based uncooled FETs, $\tau \leq 10^{-11}$ s [336]. In SBDs, $\tau \leq 10^{-11}$ s [308]. The build-up time of the intrinsic rectification signal in a patch-antenna-coupled Si-CMOS detector changes from 20 ps in the deep sub-threshold voltage regime to below 12 ps in the vicinity of the threshold voltage [337]. With these direct detectors, the operation rate in THz vision systems is limited only by parasitic elements and read-out electronics.

### 5.5.1 SBD detectors

The SBD uncooled rectification detectors, which are two-terminal devices, have long been used since 1940s for microwave detection and mixing because of their high sensitivity and ability to operate at ambient and cryogenic temperatures. Together with FET detectors, SBD ones have an advantage in their fast response time as compared with other room-temperature detectors (see Table 5.4), such as Golay cells, pyroelectric detectors, or various kind of uncooled bolometers. The SBDs have an advantage over p-n-junction diodes for high-frequency applications (up to the radiation frequency $\nu \sim 10$ THz [302]) due to the avoidance of minority-carrier storage effects.

Currently, the SBD single incoherent detectors seem to be among the most efficient uncooled ones, especially in the low-frequency section of the THz range, where NEP values of about $5 \times 10^{-13}...10^{-11}$ W/Hz/Hz$^{1/2}$ (see Fig. 5.33) can be achieved. Commercial

SBDs are usually employed as sub-THz (microwave) sensors up to 100 GHz. To extend their application onto higher frequencies, the process of their fabrication has to be changed to reduce the junction area down to sub-micron sizes in order to reduce the capacitance. A smaller junction area and the operation in the THz band can be achieved by modernizing the fabrication processes, in which high-resolution electron beam lithography and air-bridge technologies should be applied. Smaller devices will possess lower junction capacitances, but higher series and differential resistances. Therefore, the increase of the high cut-off frequency for smaller-area SBDs will not be linearly proportional to the junction area.

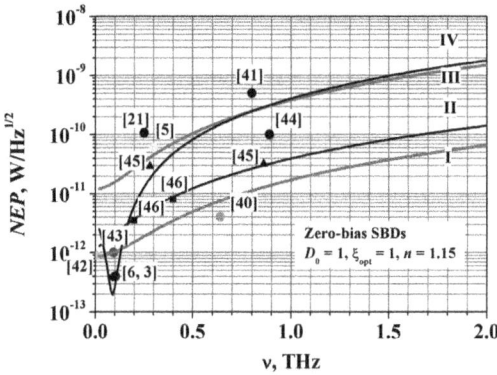

*Figure 5.33. Experimental $NEP_{opt}$ data (points, triangles, squares) for SBDs and the results of calculations (curves I–IV) [338]. The numbers near the experimental data denote the corresponding references in this paper. When calculating curve IV, besides the resistance and capacitance, the inductance was also taken into account, leading to a resonant impedance matching at $v \approx 89$ GHz, but, at the same time, resulting in poor $NEP_{opt}$ values at higher radiation frequencies. (By permission of SPQEO).*

SBDs are used in the THz frequency range as direct detectors and mixers. The SBD operates by sensing a change in the I-V characteristics when the THz radiation voltage is applied. Since the SBDs are small ($< 1$ $\mu m^2$) in comparison with the wavelength in the THz frequency range, antennas should be used to input the radiation, which leads to the frequency dependence of their sensitivity.

SBD mixers operate at frequencies well beyond 5 THz [339]. One of the major advantages of SBD mixers as compared to SIS and HEB ones is that they operate at room temperature, although the optimum performance is achieved at 20 K or below. Schottky mixers require high local oscillator pump power, approximately in the 1 mW range. Typical DSB noise temperatures for room-temperature Schottky mixers are about 1800 K at 500 GHz with approximately 8 dB of conversion losses. However, their noise temperature improves when cooled, *e.g.*, reaching approximately 1200 K (DSB) at 77 K. It was also shown that Schottky mixers can operate with a reduced LO power at the expense of the higher mixer noise temperature [340, 341].

The dominant current transport mechanisms in SBDs are the emission of electrons over the barrier and the quantum-mechanical tunnelling process of electrons through the barrier. A widely recognized model that takes into account both of those effects is the thermionic-field emission model.

Generalized I-V characteristics of an ideal metal-semiconductor junction in either the forward or reverse bias regime can be written in the form

$$I(V) = I_s \left[ \exp\left( \frac{V}{n\varphi_t} \right) - 1 \right]$$

(5.26)

where $I(V)$ is the current through the SBD, $I_S$ is the reverse-bias saturation current, n is the ideality factor (n is approximately equal from 1 to 2, depending on the fabrication process and the design). The thermal voltage $\varphi_t = k_B \cdot T_j/q \approx 25.85$ mV at $T_j = 300$ K, $k_B$ is the Boltzmann constant, q is the elementary charge, and $T_j$ is the temperature of the junction.

The temperature dependence of the reverse-bias saturation current can be expressed as

$$I_S = SA^* T_j^2 \times \exp\left( -\frac{\varphi_b(0)}{n\varphi_t} \right),$$

(5.27)

where $\phi_b(0)$ is the equilibrium Schottky potential barrier, S is the junction area, the Richardson constant $A^* = 4\pi \cdot q \cdot k_B^2 \cdot m_0/h^3 = 120$ A/(cm$^2 \cdot$K$^2$), and $m_0$ is the free electron mass. In semiconductors, $m_0$ should be changed to the effective electron mass m*, so that the Richardson constant will be m*/m$_0$ times lower (this means that the thermo-ionic current will decrease by the same factor). Depending of the reverse bias, the modified Richardson constant can change from A*=30 A/(cm$^2 \cdot$K$^2$) in moderate electric fields to A*

$\approx 7$ A/(cm$^2$·K$^2$) at biases U $\sim$ 1 V [342]. In GaAs, as a rule, it accepts the value A* $\approx$ 8.2 A/(cm$^2$·K$^2$).

With $\phi_b(0) \approx 0.7$ V at T = 300 K ($\phi_b(0)$ changes from 0.6 to 1.0 V for more than 40 metals in GaAs SBDs [343]) at the reverse bias voltage. From Eq. (5.27), in which A* = 8 A/(cm$^2$·K$^2$) and n = 1, it follows that $I_S \sim 1.5 \cdot 10^{-14}$ A for the junction area S $\approx$ 1 µm$^2$, and the junction resistance at zero bias equals

$$R_0 = \left( \frac{dI_S}{dV} |_{V=0} \right)^{-1} = \frac{n\varphi_t}{SA*T_j^2} \times \exp\left( \frac{q\varphi_b(0)}{nk_BT_j} \right) = \frac{n\varphi_t}{I_S} \tag{5.28}$$

so that $R_0 \approx 1.8 \times 10^{12}$ $\Omega$.

Thus, the SBD (with $\phi_b(0) \sim 0.7$ V) operating as a sub-THz detector should be forward biased, which leads to additional (larger by several times and even orders of magnitude) noise in the SBD detector that is conditioned with the 1/f noise component [344–346]. Therefore, it is important to lower the barrier height for the SBD to operate at zero bias in order to suppress the 1/f noise. The lower noise, the less amplification is required to reach the low-noise amplifier (LNA) noise level dominating in the system noise of heterodyne instrumentation.

By lowering the effective barrier in the SBD junction, the differential resistance can be significantly reduced. With the value $\phi_b(0) \approx 0.2$ V for, *e.g.*, Al/GaAs δ-doped SBDs [347], InGaAs/InP [348] or InGaAs [349], $I_S \sim (1.5...10) \times 10^{-6}$ A and lower, depending on the junction area and the zero-bias resistance $R_0$. Therefore, the SBDs with $\phi_b(0) \sim$ (0.2...0.3) V can be used at the zero bias. A qualitative comparison of I-V characteristics for InGaAs and GaAs SBDs is shown in Fig. 5.34. The properties and the parameters of zero-bias SBDs are considered, *e.g.*, in Ref. [350].

One of disadvantages of zero-bias III-V SBDs that are mainly based on ternary alloys is a difficulty of their monolithic integration into arrays with silicon read-out electronics. Another limitation of SBDs consists in that their performance exhibits a strong temperature dependence, if it is the thermionic current that flows through the barrier ($I_S$ changes by approximately a factor of 2 every 20 K). In spite of the fact that SBD direct detectors are well engineered and can hardly be improved, these devices require a high-quality metal and semiconductor interfaces, especially in arrays, as well as a precise control of the interface composition during the deposition process. So far, THz SBD arrays have been demonstrated with a number of elements up to 240 [307, 351–354]. Some parameters of SBD THz detectors are presented in Table 5.7.

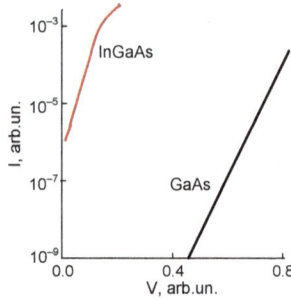

*Figure 5.34. Qualitative comparison of I-V characteristics for AlGaAs and GaAs SBDs with similar junction areas.*

### 5.5.2   FET detectors

Besides two-terminal SBDs, three-terminal devices are also explored as detectors. These are FETs, HEMTs, *etc*. A relatively low noise (when the drain-to-source bias equals zero ($V_{ds} = 0$) [355, 356]) and the availability of technologies make those detectors favorable for sub-THz/THz wave detection as both single devices and arrays.

The operation of a square-law MOSFET detector circuit in the low interval of THz frequencies can be analyzed in the framework of the quasi-static transistor model. In a direct detector, the latter can be realized by coupling the radiation through the antenna to both the gate and the drain simultaneously, resulting in a square-law dependence of the output rectified current [312].

The first usage of FET nonlinear devices in small arrays for detection and getting pictures in the sub-THz region (at the radiation frequency $\nu = 63$ GHz) seems to be reported in Refs. [357, 358].

To a great extent, the studies of FETs as rectifying THz detectors were initiated by the Dyakonov-Shur plasma-wave theory (the plasmonic instability effect) [359] in the hydrodynamic and drift current approximations at strong inversion biases, when the charge density waves in the channel play an increasingly important role [317]. The impact of the channel length, recombination times, and radiation frequencies on the FET detector response was considered, *e.g.*, in Refs. [360, 361].

Following the consideration in Ref. [359], the sub-THz/THz detection was demonstrated in III–V HEMTs [362], later in Si CMOS [355, 360], and more recently, in graphene [363–365]. As compared to many other uncooled detectors, the devices based on CMOS processes can be manufactured at the foundry level, as their technological readiness is

Detectors and Sources for THz and IR                        Materials Research Forum LLC
Materials Research Foundations **72** (2020)                  https://doi.org/10.21741/9781644900758

high. THz imaging with CMOS FETs were obtained in Refs. [313, 366]. The state of the art and the analysis of FET THz detectors up to the year 2015 is given, *e.g.*, in Ref. [367].

Important, for applications, features of HEMT or Si MOSFET direct THz detectors are their power and radiation-frequency dependences. The FETs can operate as linear THz detectors within a wide THz radiation power interval. A linear response to the radiation power, which can be used, *e.g.*, for detector calibration, was observed in InGaAs HEMTs and Si-MOSFETs in a wide range of radiation power (~120 dB) at radiation frequencies of 0.2…3 THz [368, 369]. The interval of linear response is narrower by about an order of magnitude for HEMT THz detectors. HEMT detectors have a higher responsivity, but, when being used in arrays, they have a relatively low integration level. At the same time, detectors manufactured using silicon-based technologies enable the system-on-chip integration and are promising for high production volumes of cost-effective uncooled thermal imaging arrays.

In Ref. [300], it was shown that the frequency response for such detectors depends substantially on the radiation frequency. In the framework of a relatively simple model taking into account parasitic resistances, capacitances, and antennas, it was demonstrated that response decreases with the frequency growth as $R \sim v^{-n}$ with the power exponent n $\approx 2…4$. Such radiation frequency dependences with the indicated n-values were observed in a number of papers (see, *e.g.*, Refs. [369–371]. Similar radiation frequency dependences can be observed in SBD THz detectors as well [302, 372]. The highest radiation frequencies for Si MOSFET ($v$ = 9 THz) and III-V FETs (11.8 THz) were obtained in Refs. [301] and [370], respectively.

When designing FETs as detectors and predicting their corresponding parameters, models for simulating the FET circuits are needed, because the plasma-wave theory does not satisfy well a large number of available experimental data. For instance, this theory predicts resonant detection for high-mobility devices, which is not supported to the full extent by the experiments with Si-MOSFET and III-V HEMT detectors [373–375]. In addition, the plasma-wave theory predicts the increase of the intrinsic responsivity with the frequency growth and its independence of the gate length [316]. Since the optimum signals are observed in the region where the drift and diffusion mechanisms overlap, the diffusion current should also be taken into account [300]. The data obtained [376] make it possilbe to conclude that THz radiation from biased AlGaN/GaN HEMTs cannot be directly attributed to the plasmonic instability effect. Two other emission mechanisms can be assigned to a combination of thermal emission from the heated material and from thermally excited plasmons and trap states.

While sweeping the gate-source voltage $V_{gs}$ of FET detector from the subthreshold one ($V_{th} \sim 0.6...\ 0.7$ V for Si MOSFET) to the strong-inversion operation region, a model taking into account the parasitic parameters of the device and the matching with the antenna should be applied [377]. The dependences of the rectified current (the signal response is proportional to it) strongly depend on the radiation frequency and the FET geometry parameters. The plasma wave theory predicts an enhancement of intrinsic responsivity with radiation frequency growth.

A simplified circuit model of rectifying THz detector was applied to Si MOSFETs with various radiation frequencies, channel lengths, and channel widths. In this model, only the basic extrinsic parasitic components (the resistances and capacitancies, were taken into account. A more complicated FET model [316] requires more parameters to be considered. The intrinsic and extrinsic parameters are bias dependent for both SBDs and FETs. However, even within the simple model, the estimations show strong dependence of the responsivity on the model parameters including the radiation frequency (see Fig. 5.35), which is in qualitative agreement with the experiment [377].

Though there is no quantitative coincidence, a qualitative fit with the experiment is observed. The dependences of the rectified current (the response signal is proportional to it) considerably vary with the radiation frequency and the geometrical FET parameters. The latter are responsible fot the impedance matching between the antenna and the channel [317, 376]. Even the simple model described above predicts a distinct NEP worsening with the growing radiation frequency, which is in agreement with experimental data [301, 314].

A difficulty in determining the component values for a small-signal model of equivalent circuit (the square-law detectors) comes from an ambiguity related to a set of scattering parameters (S-parameters). The latter could stem from different equivalent circuits or multiple instantiations of the equivalent circuit with used different sets of the component values. Therefore, the FET geometry and the thickness of dielectric layers are important for matching the FET and antenna impedances, with the indicated parameters significantly affecting the performance of FETs as detectors. In general, smaller sizes (the width and length) of transistors are favorable [317]. However, depending on the design rules, THz radiation frequencies, a particular antenna type, *etc.*, the FET channels with small lengths L and widths W can exhibit impedances $Z_{det}$ that do not match the antenna impedance $Z_A$. In this case, there appears a strong dependence of the NEP on L and W. For example, the corresponding W-dependences calculated for a simplied circuit with the parasitic elements described in Fig. 5.35.b are shown in Fig. 5.36.

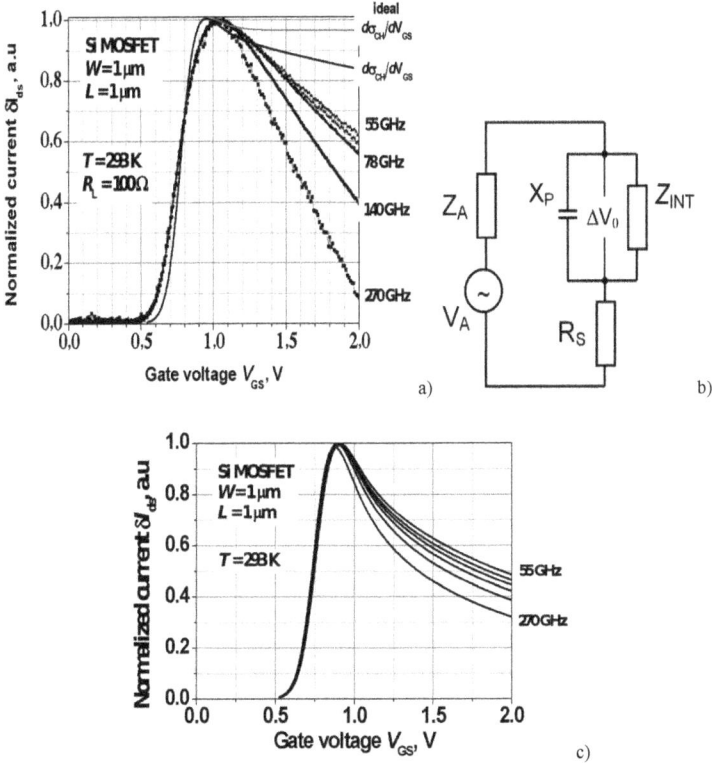

*Figure 5.35. (a) Rectifying current vs gate voltage dependences at various radiation frequencies for a Si-MOSFET THz detector with the indicated sizes. (b) Simplified schematic representation of the rectifying THz detector circuit taking into account the basic extrinsic parasitic components. Here, $V_A$ is the voltage signal, $Z_A$ is the frequency-dependent antenna impedance, $X_P = -j/(vC_P)$ is a parasitic component connected with the parasitic shunting capacity ,$C_P$, and $Z_{INT}$ is the internal source-gate impedance. (c) Estimated tendencies of the normalized current $\delta I_{ds}$ vs the gate-source $V_{GS}$ voltage for an FET transistor with the channel sizes W (width) = L (length) = 1 μm calculated with taking into account the FET and antenna frequency-dependent parameters. (By courtesy of A. Golenkov).*

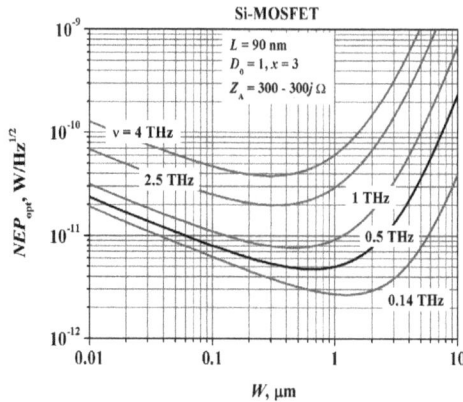

*Figure 5.36. NEP$_{opt}$ dependences on the channel width W for Si-MOSFET THz detectors operating at T = 300 K and various radiation frequencies v [338]. The parameters of FETs were taken from the BSIM 3.3 and BSIM 4 models. (By permission of SPQEO).*

Some experimental data for NEP values in various FETs and HEMTS are presented in Table 5.7.

### 5.6    Graphene-based detectors

Graphene (a single atomic layer of carbon, two-dimensional graphite) is a material, in which the electron transport is mainly governed by Dirac's relativistic equation. Electrons in graphene are constrained to move in two spatial dimensions. Dirac fermions have now been identified in other systems, such as topological insulators (TIs) or quasi-2D organic materials. Recently, graphene has been extensively researched in fundamental science and engineering fields. In the last decade, this material has been well developed for IR and THz detector applications due to its optic, electronic, and thermal properties. There occurs a rapid rise of research in this field (see, *e.g.*, Refs. [365, 378–381]). In realistic TI detectors, the photorespose is usually hindered by high dark currents through the bulk states.

Graphene is a viable alternative to conventional optoelectronic, plasmonic, and nanophotonic materials. It has decisive advantages, such as wavelength-independent absorption, tunable optical properties via electrostatic doping, large charge-carrier concentrations, low dissipation rates, and the ability to confine electromagnetic energy in small volumes [382].

Graphene has optical and electronic characteristics that lead to a broadband performance and high speed when used as a photodetector. It can absorb energy within a broad spectral band from ultraviolet wavelengths through the visible light to THz ones. One suspended layer of graphene has the optical absorption $\pi\alpha$ = 2.3% [383, 284], which depends only on the fine-structure constant $\alpha = e^2/hc \approx 7.299 \cdot 10^{-3}$. Here, c is the speed of light. This parameter determines the binding of light and relativistic electrons. The transmitance of white light through graphene layers decreases by 2.3% with each next layer.

At room temperature, charge carriers in graphene can have the mobility $\mu$ much higher than that in other semiconductors or semimetals. Graphene is highly thermally conductive, $\alpha_{th} >\sim 4 \cdot 10^3$ W·m$^{-1}$·K$^{-1}$, which is several times higher in comparison with natural diamond, and has a high Seebek coefficient (see, *e.g.*, Ref. [385] and references therein), which makes such properties favorable for thermally induced photoresponse mechanisms.

Concerning IR detectors, graphene can have preferences over conventional semiconductor photodetector materials (silicon, indium gallium arsenide, mercury cadmium telluride, and so on). Those materials are widely applied in photonics, but their sensitivity is appreciable only within a small portion of the IR spectrum.

The field of graphene experimental and theoretical research has been developed rapidly since its first isolation and identification by mechanical exfoliation in 2004 [386]. Owing to its band structure (Fig. 5.37) resulting in its specific optoelectronic properties and a broadband absorption spectrum (from the ultraviolet to the THz spectral range), its dispersion relation remains quasi-linear up to ~±4 eV from the Fermi level [382]), graphene is a promising material for detecting radiation through various operating detection mechanisms, *e.g.*, rectification, bolometric, photo-thermoelectric, and plasma-wave-assisted ones.

Graphene is composed of single sheets of carbon atoms that form a flat hexagonal lattice structure (a 2-dimensional sheet of sp$^2$-hybridized carbon atoms arranged in a honeycomb crystal lattice; see Fig. 5.37). It was used earlier mainly for its mechanical properties, but other real industrial applications are quite limited yet [388]. Graphene electronic properties have been intensively studied throughout the world during the last decade, because it is supposed that this material can play an important role in electronics and, possibly, can replace semiconductors in many applications. An important thing is that graphene is chemically inert. However, when designing devices, this circumstance can hamper making good contacts between the graphene and metal layers (see, *e.g.*, Refs. [389–391].

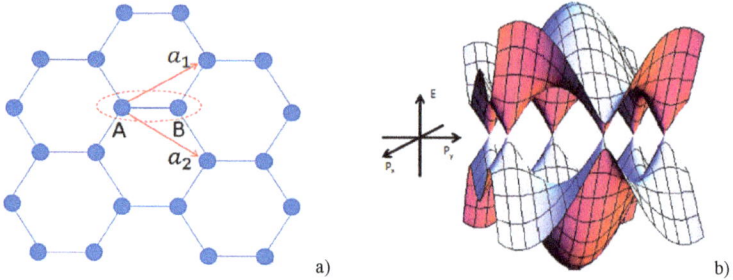

Figure 5.37. a) Hexagonal lattice of carbon atoms in monolayer graphene. Every unit cell of the honeycomb lattice consists of equivalent carbon ions labeled by A and B. The two primitive lattice vectors are labeled as $a_1$ and $a_2$. b) Band structure of graphene. The valence and conduction bands touch each other at six points in the corners of the first Brillouin zone. Near these points, the dispersion relation is approximately linear, resembling that for a massless particle [387]. (By permission of IOP Publishing).

Electrons in graphene have long mean free paths and do not disrupt electron-electron interaction. Therefore, the properties of graphene differ from those of other common metals, semimetals, and semiconductors. In Fig. 5.37.a is shown the two-dimensional hexagonal structure of carbon atoms in graphene [387]. A lattice with a basis of two identical atoms (A and B) in a unit cell forms the honeycomb lattice structure. The unit vectors of this lattice can be expressed as follows:

$$a_1 = a/2(3, \sqrt{3}), \quad a_2 = a/2(3, -\sqrt{3}), \tag{5.29}$$

where the distance between two carbon atoms is approximately equal to 1.42 Å. The reciprocal-lattice vectors are

$$b_1 = 2\pi/3a(1, \sqrt{3}), \quad b_2 = 2\pi/3a(1, -\sqrt{3}). \tag{5.30}$$

Two points K and K' located at the edge of the graphene Brillouin zone (BZ) and called the Dirac points are essential for the physics of graphene. Their positions in the momentum space can be expressed as follows:

$$K = (2\pi/(3a), 2\pi/(3\sqrt{3}a)), \quad K' = (2\pi/(3a), -2\pi/(3\sqrt{3}a)). \tag{5.31}$$

Therefore, the valence and conduction bands touch each other at six points in the corners of the first Brillouin zone (see Fig. 5.37.b) due to the honeycomb lattice structure.

Graphene has a gapless energy spectrum. However, it is not a semiconductor in a literal sense, but rather a semimetal. Therefore, such graphene properties as, *e.g.*, its conductivity type and the amperage can be controlled. Such graphene properties as a high carrier mobility, a small heat capacity (due to the low density of states), weak electron-phonon coupling (the electron temperature can be significantly higher as compared to the phonon one [392]), and a low resistance have promoted extensive research of graphene-based THz optoelectronic devices including detectors possessing a fast response.

In the last decade, the number of papers on the subject concerned grew tremendously. Various laboratory device structures, such as thermal detectors [140, 393–395], various FET designs [325, 326, 330, 363, 365], thermoelectric devices [197, 396, 397], and single graphene THz detectors were used for getting imaging with the help of raster scanning instrumentation [328].

The response time of graphene-based detectors can be very short, and the device speed is high (up to tens of GHz; see, *e.g.*, Refs. [393, 398, 399]). The supercurrent (in mesoscopic junctions consisting of a graphene layer contacting with two closely spaced superconducting electrodes) and superconductivity (with $T_c \approx 1.7$ K) [400, 401] were observed.

Some foreseeable high-performance properties of a number of graphene optoelectronic devices were considered, *e.g.*, in Refs. [381, 402–405].

When fabricating graphene for THz and IR detectors and electronic devices, several substrates can be used: $SiO_2$, mica, hexagonal boron nitride (h-BN), cleaved graphite, *etc.* However, multilayer substrates are used to obtain a good sensitivity (see, *e.g.*, Fig. 5.38). Graphene on the h-BN substrate, because of their close crystal structures, exhibits the rms roughness ($\approx 30$ pm) about an order of magnitude smaller than graphene on $SiO_2$ does ($\approx 225$ pm) [387], being similar to graphene on highly ordered pyrolic graphite (HOPG). The low surface roughness of graphene on the h-BN or HOPG substrate promotes the manifestation of electronic effects and high graphene responsivity.

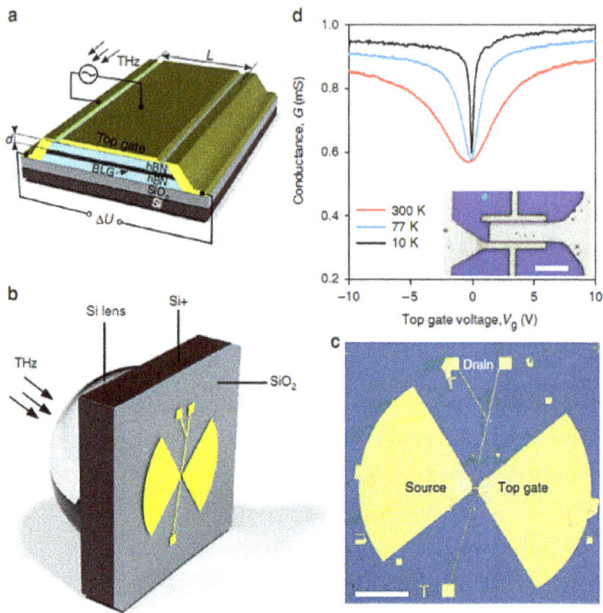

*Figure 5.38. Graphene-based THz detectors. a) Schematic representation of the encapsulated BLG FET. b) 3D image of resonant photodetector. THz radiation is focused on a broadband bow-tie antenna by means of a hemispherical silicon lens yielding modulation of the gate-to-source voltage, as indicated in panel a. c) Optical photograph of a photodetector. The scale bar is 200 μm. d) Conductance of a BLG FET as a function of the gate voltage Vg, measured at a few selected temperatures. Inset: zoomed-in photograph of panel c showing a two-terminal FET with gate and source terminals connected to the antenna. The scale bar is 10 μm [365]. (By permission of AIP Publishing).*

Several mechanisms can be responsible for the photoresponse in photodetectors on the basis of graphene: photovoltaic, bolometric photothermoelectric, and plasma-wave-assisted (the Dyakonov-Shur model) ones, which are schematically shown in Fig. 5.39.

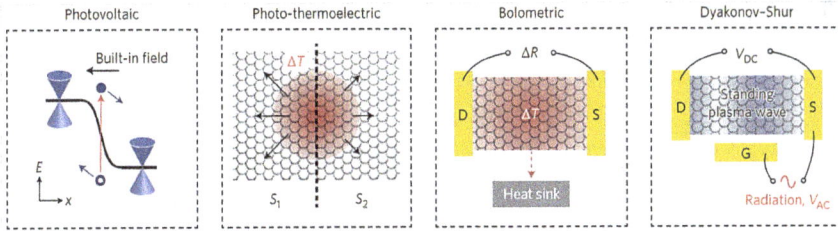

*Figure. 5.39. Schematic representation of four photocurrent generation mechanisms. First panel: electron–hole (solid and open circle) separation by an internal electric field. Second and third panels: the red-shaded area indicates the elevated electron temperature. ΔT is the temperature gradient, ΔR is the resistance across the channel; and S1 and S2 are the Seebeck coefficient in graphene areas with different doping. Third and fourth panels: S, D and G indicate the source, the drain, and the gate, respectively; $V_{DC}$ is the photogenerated D.C. voltage; and $V_{AC}$ is the a.c. voltage applied to the gate [378]. (By permission of Springer Nature Publisher)*

## 5.6.1 Graphene thermal detectors

Graphene is suitable for designing bolometers because of the large temperature-induced changes in its electrical conductivity due to the low electron heat capacity, weak electron-phonon coupling, its unique small thickness, and long electron free path [329], which makes it possible to achieve the room-temperature carrier mobility up to about $(1...2) \times 10^5$ cm$^2$/V·s that is close to the theoretical limit. Those factors enable a carrier mean free path longer than 1 μm and strong Drude absorption of THz radiation, which leads to large light-induced changes in the electron temperature [140, 393, 394]. In Ref. [406], a comprehensive analysis is given of the expected responsivity of a HEB made from a flake that can be embedded into either a planar antenna or a waveguide, and which has NbN or NbTiN superconducting contacts with this flake. The NEP is expected to be less than $10^{-20}$ W/Hz$^{1/2}$ if the device operates at T ~50 mK. Thus, the graphene HEB detector is a radiation background limited device (cosmic background), when installed in facilities with cryogenically cooled mirrors.

Graphene QD detector parameters were considered in Ref. [407]. It was shown that the responsivity decreases with the radiation power increase (from 1 pW to 0.4 nW), but it remains higher than $10^8$ V/W. Some NEP values of graphene thermal detectors are presented in Table 5.8. In Ref. [364], it was shown that, in the FET graphene detectors, both the bolometric and plasmonic mechanisms are responsible for the THz response within a range from 1.63 to 3.11 THz.

*Table 5.8. Some typical parameters of graphene detectors.*

| Detectors | Operation radiation frequency ν, THz | NEP, W/Hz$^{1/2}$ | Refs |
|---|---|---|---|
| Bilayer HEB | | $3.3 \times 10^{-14}$ (T = 5 K) | [393] |
| Hybrid graphene/PbS QD (phototransistor) | λ = 0.3…2 μm | $D^* = 7 \times 10^{13}$ cm·Hz$^{1/2}$/W | [408] |
| Integrated THz antenna | λ = 8…220 μm | S = 5 nA/W (T = 300 K) | [399] |
| Antenna capacitive coupling, effective thermocouple to read out T$_e$ | 0.094 | $<2 \times 10^{-10}$ (T = 50 K) | [397] |
| Photothermoelectric | 1.02 | $1.1 \times 10^{-9}$ (T = 300 K) | [409] |
| FET | 0.3 | $<3 \times 10^{-8}$ (T = 300 K) | [363] |
| FET | 0.4 | $1.3 \times 10^{-10}$ (T = 300 K) | [325] |
| FET | 0.13 | $6 \times 10^{-10}$ (T = 300 K) | [365] |

*T$_e$ is the electron temperature in graphene.
(For more references with graphene detector parameters, see Ref. [382].

In Ref. [410], a graphene-exciton-polariton-based bolometer for the detection of THz radiation was considered. It was predicted that using a single-layer graphene, a THz bolometer can be realized as a simple exciton-polariton ring interferometer device having the sensitivity close to that in presently existing THz bolometers.

In Ref. [411], the cited authors reported about a fabricated photodetector (see Fig. 5.40) with gold-patched graphene nanostripes, which showed ultrafast photodetection and responsivity in a wide spectral range from visible (800 nm) to long IR (20 μm) waves, with a responsivity varying from 0.6 A/W to 8 A/W. The photodetector showed an ultrafast photodetection speed exceeding 50 GHz.

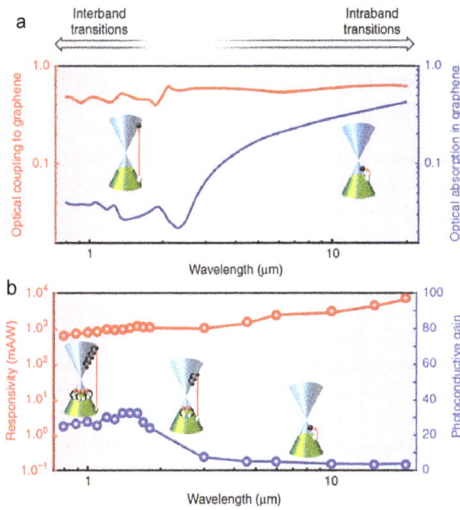

*Figure 5.40. Responsivity and, schematic diagrams of optical transitions and broadband photodetection via gold-patched graphene nano-stripes. a) Numerical estimation of optical coupling (red, upper curve) and optical absorption (blue, lower curve) in graphene nanostripes as a function of wavelength λ. b) The measured responsivity (red symbols, upper curve) and photoconductive gain (blue symbols, lower curve) of the fabricated photodetector at an optical power of 2.5 μW, gate voltage of 22 V, and bias voltage of 20 mV [411].*

## 5.6.2 Graphene FET detectors

Possessing a zero band gap, graphene was stated to be difficult for application in transistor technologies, because the resulting devices would have a low on/off ratio [386, 412]. However, the implementation of the FET graphene THz detectors allowed the designing of high-sensitive and fast room-temperature-operating THz detectors. Their designs are compatible with the CMOS processes. Recently, several THz graphene FET detectors have been proposed (see, *e.g.*, Refs. [325, 363, 365, 413–416]. A design of one of the graphene FET detectors is shown in Fig. 5.38.

In the FET considered in Ref. [365], several different rectification photoresponse mechanisms can be distinguished according to their different temperature dependences. By comparing the photoresponses at various temperatures, it was found that the opposite sign of the Seebeck coefficients in the p-doped graphene channel and at the n-doped graphene-metal interface resulted in a significant photoelectric effect owing to the rectification of high-frequency radiation. Also it was shown that a p–n junction formed in the channel provides an additional rectification mechanism, which allows the enhancement of the responsivity of graphene-based THz detectors.

The authors of Ref. [417] have demonstrated the plasmon-assisted resonant detection of THz radiation with antenna-coupled graphene transistors (see Fig. 5.41), which acted as both a plasmonic Fabry-Perot cavity and a rectifying device. It was quantitatively argued that the peaks observed in the photoresponse emerge as a result of plasmon resonance in the FET channel.

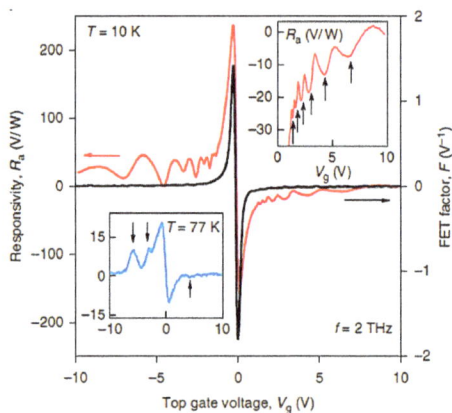

*Figure 5.41. Plasmon-assisted THz photodetection. Gate-voltage dependence of responsivity recorded under 2 THz radiation. The upper inset shows a zoomed-in region of the photovoltage for electron doping. Resonances are indicated by black arrows. Lower inset: resonant responsivity at liquid-nitrogen temperature [417].*

A high-mobility, gate-tunable, h-BN-encapsulated graphene detector, where the incoming THz radiation is focudes on a small photoactive area of graphene, was designed in Ref. [327]. This detector exploits the photothermoelectric effect and is based on a design that employs a dual-gated dipolar antenna with a gap of ~100 nm and operating in an antenna-limited radiation frequency band of 1.8…4.2 THz. The obtained NEP value of about $8 \times 10^{-11}$ W/Hz$^{1/2}$ testifies to its potential applicability.

Still, weak optical absorption of graphene restricts the possible ultimate performance of the planar 2D back-gated graphene field-effect transistor. To avoid this difficuly, a self-rolled-up method was considered in Ref. [416] to transform a 2D buried-gate graphene FET into a 3D tubular one. The 3D grapheme photodetectors demonstrated room-temperature photodetection in the ultraviolet, visible, mid-IR, and THz ranges, with the ultraviolet and the visible-light responsivity S > 1 A/W and responsivity S = 0.232 A/W at a frequency of 3.11 THz.

However, despite that some graphene THz detectors were predicted to have a high responsivity and a low NEP (see Tables 5.7 and 5.8), it is difficult to preview their expansion in the case of mass-production.

In spite of the fact that the photorespose in TIs is usually hindered by the strong bulk dark current, a high mm and THz response of metal-TI-metal structures (on the base of $Bi_2Te_3$ flakes) at room temperature has been demonstrated recently [418]. The noise equivalent power up to NEP = $3.6 \times 10^{-13}$ W/Hz$^{1/2}$ with the response time $\tau$ < 60 ms was attained. The response decreased as the radiation frequency grew in the spectral range $\nu \approx 50...300$ GHz.

## 5.7   THz sensors with metamaterials

Recently, advances in engineered materials [419, 420] called metamaterials (MM) have enabled a new direction in the development of terahertz technologies for creating a new class of THz detectors (see, *e.g.*, Refs. [421–424]). Being composed of periodically arranged unit cells with dimensions of about a wavelength and a specific geometry, metamaterials gave a possibility to confine the radiation flow into a detector array to a certain spectral THz interval using a high-impedance surface approach.

The design flexibility associated with MMs in the microwave and THz ranges provides promising perspective in using next-generation integrated THz arrays and components with the desired spectral functionalities and intended to be applied in multispectral vision, communication, spectroscopy, biomedicine, *etc.* [425–427].

As effective media, MMs can be characterized by a complex permittivity $\varepsilon = \varepsilon_1 + i\varepsilon_2$, where the negative real part $\varepsilon_1$ is related to the phase velocity, and the imaginary part $\varepsilon_2$ determines losses. A large amount of MM research was focused on applications dealing with the creation of materials with requested $\varepsilon_1$. However, manipulating the losses through the design of $\varepsilon_2$ found applications in the THz detectors by using MMs as perfect absorbers with tunable absorption wavelengths [421, 425, 428, 429]. The MM technology gives a possibility to design thin multispectral absorbing elements, which can lead to the enhanced functionality of THz detectors.

The performance of electromagnetic absorbers with a high absorption coefficient in their thin absorbing layers is associated with the shift of the electric field antinode onto the lossy surface of the absorbing layer by the spatial distribution of electromagnetic field. It can be achieved by using the high-impedance surface (HIS) approach [428, 430, 431].

An electrically thin HIS is a subwavelength single-layer frequency-selective meta-surface of a capacitive kind that is placed over a thin grounded dielectric slab with the thickness d $\ll \lambda$ (see the inset in Fig. 5.42.a). At the HIS resonance frequency, the surface impedance tends to infinity, yielding a zero reflection phase. In this case, the tangential component of the magnetic field formed by the incident and reflected waves vanishes upon the HIS surface when the electric component is maximum, and the regime of high absorption can be realized.

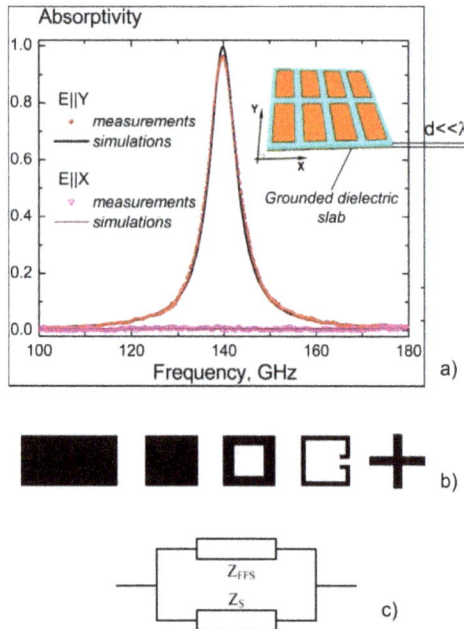

Figure 5.42. a) Spectral performance of the resonant absorber utilizing the rectangular unit-cell array. b) Some of the most common unit cell patterns, which can be used in metamaterial resonators. c) Equivalent circuit of the absorber: $Z_{FFS}$ and $Z_S$ are the surface impedances of the meta-FSS and the thin dielectric substrate, respectively (adapted from Ref. [427]).

A HIS combined with a lossy resistive sheet was demonstrated to be a promising absorbing layer with a subwavelength thickness [419, 427]. In its basic design, the HIS is realized as a single-layer frequency-selective surface (FSS), or meta-surface, of the capacitive type, which is placed above a thin grounded dielectric slab. At the HIS resonance frequency, the surface impedance Z of the structure is considered as a parallel connection (Fig. 5.42.c) of the meta-FSS impedance $Z_{FSS}$ and the impedance $Z_S$ of the grounded dielectric slab [428],

$$Z = \frac{Z_{FFS} \cdot Z_S}{Z_{FSS} + Z_S}$$

(5.32)

It tends to infinity, yielding a zero reflection phase and thereby maximizing the surface electric field. The resonant E-field enhancement enables total absorption of the incident wave even without introducing an auxiliary resistive sheet, but due to properly accounted Ohmic or dielectric losses in the FSS or the carrying substrate [428, 432, 433]. By changing the stuff of the metamaterial array, its thickness, the pitch of unit cells, their geometry and dimensions, one can control the resonant frequency of the detector response or the broad-band response [421, 424], which can be used when developing multicolor THz FPAs [425].

In Refs. [423, 428], a combination of meta-surface THz absorbers with the IR high-emissivity layers for the conversion of THz radiation into IR one was proposed and realized, which allowed THz radiation to be detected in a spectral range of 8…14 μm making use of IR detectors and conventional cameras.

The application of thin metamaterial layers in order to minimize the thickness d ($\lambda$/d >> 1) of absorbing layers covering the bolometric-type detectors [428, 434] used for direct measurements of THz radiation have prospects from the viewpoint of widening the spectral range of uncooled matrix arrays on the basis of $VO_x$ or $\alpha$-Si to lower THz frequencies in comparison with standard ones. The condition $\lambda$/d >> 1 is essential for decreasing the absorber heat capacity and, as a result, enables the achievement of a high sensitivity and low response time in bolometric detectors. In Ref. [427], a reasonable NEP value of $2 \times 10^{-9}$ W//$Hz^{1/2}$ was realized at $\nu \approx 140$ GHz in small (3 × 4 and 3 × 6) pyroelectric arrays with resonant patches in the FSS.

In the biomaterial cantilever detectors with thin metamaterial layers, a NEP of $10^{-8}$ W//$Hz^{1/2}$ was obtained at $\nu$ = 93 GHz and $\nu$ = 693 GHz [421]. Experimental characterization of metamaterial absorbers showed 93…99% of resonant absorptivity in the frequency range of 0.1...0.4 THz, and the IR emissivity was around 93% if a graphitized coating 10 μm in thickness was deposited on the outer surface of the grounded surface layer [428].

In all THz detectors with metamaterials, their response drastically depends on the radiation polarization (see Fig.5.42.a).

### 5.8 Photoconductive antenna (PCA) detectors

Among a wide variety of THz devices for spectroscopy and imaging, TDS-THz ones using PCAs attract attention as emitters and detectors in pulsed and CW THz facilities [435]. The PCA detection of broadband THz radiation is based on antenna structures similar to those used for the generation of pulsed broadband THz radiation emission spectra ($v \sim 0.1...6$ THz [436–438]), e.g., in TDS experiments or imaging (for references up to 2010, see, e.g., Ref. [84]).

The most widely used systems for THz spectroscopy and imaging are based on fs pulses generated by mode-locked lasers (mostly Ti:Sapphire lasers) and a PCA switch (as a rule, based on LTG GaAs) embedded in the antenna structure. To provide the detection of THz radiation, a fs probe pulse is focused, besides the PCA switch, onto the detector, generating electron–hole pairs. The invention of THz TDS was made in Ref. [439]. Since then, photoconductive THz devices have been significantly developed [440]. Today, there exist a number of commercially available PCA-TDS THz instrumentations based on the pump-probe approach.

PCA emission radiation with broadband spectra ($v \sim 0.1...6$ THz [438, 441]) illuminate detectors, which are also excited by fs laser pulses [442–444] with photon energies higher than the band gap in semiconductors with a short ($\sim 10^{-10}...10^{-11}$ s) free-carrier relaxation time. For the excitation of such a photoconductive switch, as a rule, mode-locked Ti:sapphire lasers operating at $\lambda \approx 800$ nm (pulses up to 10 fs) or fs fiber lasers (pulses from $\sim 250$ to 50 fs) operating in the spectral range of 780...1950 nm are used.

In the PCA emitter, a short fs optical (probe) pulse incident between the PCA electrodes excites free charge carriers in the semiconductor. The voltage applied between the photoconductive switch electrodes (Fig. 5.43.a) drifts the charge carriers generated by the fs laser pulse toward the electrodes. The arising photocurrent drives a terahertz antenna connected to the photoconductor through contact electrodes. The generated photocurrent follows the waveform of the optical pump. Both LTG GaAs and narrow-gap InGaAs are used in semiconducting PCA emitters and detectors for their excitation at $\lambda < 1$ μm and $\lambda \approx 1.5$ μm, respectively [444–447].

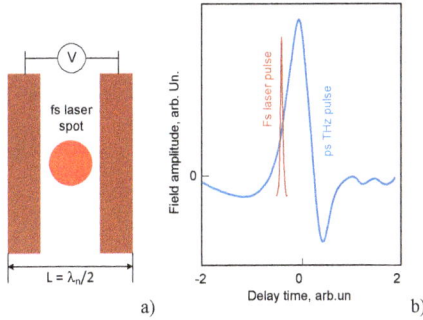

*Figure 5.43. (a) PCA emitter and (b) THz waveform mapped by varying the time-delay between gating and THz pulse.*

The resonance condition is L = m × (λ/2n), where m is an integer, and n is the semiconductor refractive index. Since radiation emission and detection are launched by the same fs laser pulse, the THz signal is coherent.

Some narrow-gap semiconductors, *e.g.*, on the basis of HgCdTe and p-InAs compounds, as well as InGaAs/InAlAs heterostructures, can also be exploited [448]. In addition, more effective LTG heterostructures [449–451], optical nanoantennas [452], as well as photoconductive terahertz emitters based on plasmonic contact electrodes (gratings), can be used [453]. Graphene-based materials [378] and semiconductor nanowire structures [454, 455] are also used for designing PCA detectors.

During a fs laser pulse (from the same fs laser) at the PCA detector (fabricated from the same material as for the photoconductive switch or the wide-bandwidth SBD [456]), the excited carriers are accelerated by the time-dependent electrical field E(t) of the incident THz pulse. A short fs laser pulse superimposed on a much longer THz pulse is sampled with a time-delayed copy of the short laser pulse. The current signal, which arises in the outer circuit, can be analyzed using the inverse Fourier transform procedure. The current is proportional to the amplitude of the incoming THz radiation.

The PCA emitter can be considered as a dipole, which is in resonance with the radiation wavelength inside the semiconductor (Fig. 5.43.a). For the efficient operation at THz frequencies, the transport time of photo-generated carriers to the photoconductor contact electrodes should be a fraction of the required oscillation period of radiation [457].

Since both the THz and probe pulses are present during the delay time in PCA active area, the photoinduced carriers move under the THz-field action, giving rise to a current between two electrodes. Therefore, it is the field created by THz pulses rather than their intensity that is registered at the THz pulse detection.

The delay time between the fs laser and THz pulses coming to the PCA detector is controlled, *e.g.*, by a mechanical stage in the delay cell (see Ch. 6) or some other translation mechanism. That is why the TDS-THz instrumentation looks like a Mach-Zehnder interferometer, where the transformation from optical to THz frequencies occurs in one of its arms.

The time domain current I(τ), where τ is the delay time between the probe and the THz pulse, is a convolution of the THz pulse electric field E(t) and the detector response D(t):

$$I(\tau) = \frac{1}{T}\int_0^T E(t)D(t-\tau)dt.$$

(5.33)

Since the laser pulse is short as compared to the time duration of the THz pulse, the latter acts as a gated sampling signal (Fig. 5.43.b).

PCA detectors are widely used in THz spectroscopy and imaging. For instance, in biomedicine, the cancer diagnostics by means of pulsed THz imaging demonstrated a potential of its application to non-invasive processes for determining regions of skin cancer with the help of reflection-geometry systems [458, 459]. THz TDS can be applicable even to test the quality of white wine [460]. Pulsed systems can provide a broader range of information, including frequency domain or time domain information, and they can be applied to obtain depth-related information and information on the nature of scattering objects [461–463].

Important for the sensitivity of the TDS THz instrumentation based on Ti-sapphire or fiber femtosecond lasers generating radiation with spectra from 0.1 to 5.0 THz, is a dynamic range of 60 dB and even larger when operating at the peak of the THz signal (ν ~ 0.5…1.0 THz). This is conditioned by the fact that a detector senses radiation only when excited by a fs laser. Therefore, the noise between two sequential pulses is not registered, and the dynamic range can be large.

Imaging with pulsed time-delay spectroscopy TDS yields a broadband spectrum, but it is difficult to realize PCA arrays in such a system. Still, the first multichannel THz-TDS system operating with a photoconductive antenna array of 15 detection channels was presented in Ref. [464]. The spectrum of THz pulses within the range from 0.1 to 0.8 THz could be obtained 15 times faster than in a single-detector TDS system with a spatial resolution of about 1 mm. The photoconductive emitter array was considered in Ref. [465].

PCAs can also be used for coherent detection in the THz spectral range in both the pulse and CW modes when they are illuminated by optical lasers with the radiation frequencies $v_1$ and $v_2$ (photomixing), provided that their difference is in the THz range (see, *e.g.*, Refs. [466, 467]); as a rule, $v_1 - v_2 \sim 0.5...2.5$ THz [438]. Photomixing (optical heterodyning) involves the use of two IR or visible lasers to generate THz radiation at their difference (beat) frequency.

High absorption in the Earth atmosphere makes the standard TDS-THz method insufficient for a reliable substance identification even under laboratory conditions [468]. Therefore, this feature of the standard TDS-THz method applied at relatively long distances leads to a large number of false positives. However, this is not the case if applying this method to biomedical science (short distances).

Pulsed THz techniques are widely used for obtaining images of biomedical objects. Because of relatively low-power ($\sim 0.1...100$ μW) THz radiation [438, 449, 469], the peculiarities of optical setups with a single detector result in time-consuming experiments, since the raster technique has to be applied. Even a small-area image requires long time, because the reconstruction of an N-pixel image requires N measurements. For example, processing the area of $10 \times 10$ mm$^2$ with an x-y stepper takes 8 min to get a picture composed of $0.25 \times 0.25$ mm$^2$ pixels [470]. For large-area objects, even hours are demanded for obtaining a picture [471].

## 5.9  Summary

In this chapter, recent progress and trends in the evolution of basic IR and THz detectors was briefly discussed. The NEP values for many contemporary detectors were pointed out in the tables. Nowadays, the THz detector technologies are among the promising ones in high-speed and broadband communications, biomedicine, high-resolution spectroscopy, food control, security screening, *etc.*, to say nothing of space research. The application of THz detectors with the highest possible performance parameters plays one of the most important roles in those technologies. Some technological advancement in practical applications owing to the rapid current development of THz detector technologies can be expected.

For plenty of applications aimed at extracting more information, THz or IR arrays with a large number of sensitive detectors are wishful for fast sensing in different parts of the spectrum (multicolor operation). They are most desirable for contemporary IR systems. The array sizes seem to continue growing, but probably at rates that fall below the observed Moore's Law, though their size is technically feasible. Large optoelectronic arrays can be and are used, *e.g.*, in astronomy instead of earlier exploited photographic films. Since large arrays significantly increase the data output of telescope systems, the development of large format THz and IR sensors for ground-based astronomy is a goal of many observatories. However, the market demand for larger arrays is not as strong as earlier.

**References to Ch 5**

[1] T. Endoh, S. Tohyama, T. Yamazaki, Yutaka Tanaka, *et al.*, Uncooled infrared detector with 12 μm pixel pitch video graphics array, Proc. SPIE. 8704 (2013) 87041G. https://doi.org/10.1117/12.2013690

[2] D. Lohrmann, R. Littleton, C. Reese, D. Murphy, J. Vizgaitis, Uncooled long-wave infrared small pixel focal plane array and system challenges, Opt. Eng. 52 (2013) 061305. https://doi.org/10.1117/1.OE.52.6.061305

[3] G. Chattopadhyay, Submillimeter-wave coherent and incoherent sensors for space applications, in: S.C. Mukhopadhyay, R.Y.M. Huang (Eds.), Sensors, Springer-Verlag, Berlin, Heidelberg, 2008, pp. 387-414. https://doi.org/10.1007/978-3-540-69033-7_19

[4] Huber M.C.E., Pauluhn A., Culhane J.L., Timothy J.G., Wilhelm K., Zehnder A. (Eds.), 2013. Observing Photons in Space. A Guide to Experimental Space Astronomy. New York, Heidelberg, Dordrecht, London: Springer. https://doi.org/10.1007/978-1-4614-7804-1

[5] M.A. Kinch, State-of-the-Art Infrared Detector Technology, SPIE Press Book, Bellingham, 2014. https://doi.org/10.1117/3.1002766

[6] A. Rogalski, Next decade in infrared detectors, Proc. SPIE. 10433 (2017) 104330L. https://doi.org/10.1117/12.2300779

[7] A. Betz, R. Boreiko, Y. Zhou, J. Zhao, *et al.*, HgCdTe photoconductive mixers for 3-15 terahertz, 14th International Symposium on Space Terahertz Technology. April 22-24, 2003 Loews Ventana Canyon Resort Tucson, Arizona, pp. 102-111.

[8] B. Starr, L. Mears, C. Fulk, J. Getty, *et al.*, RVS large format arrays for astronomy, Proc. SPIE. 9915 (2016) 9915. https://doi.org/10.1117/12.2233033

[9] D. Lee, M. Carmody, E. Piquette, P. Dreiske, *et al.*, High-operating temperature HgCdTe: A vision for the near future, J. Electr. Mater. 45 (2016) 4587-4595. https://doi.org/10.1007/s11664-016-4566-6

[10] D. Figer, J. Lee, E. Corrales, J. Getty, L. Mears, HgCdTe detectors grown on silicon substrates for observational, Proc. SPIE. 10709 (2018) 1070926.

[11] R. Mills, E. Beuville, E. Corrales, A. Hoffman, *et al.*, Evolution of large format

impurity band conductor focal plane arrays for astronomy applications, Proc. SPIE. 8154 (2011) 81540R. https://doi.org/10.1117/12.897292

[12] R.J. Nicholas, K. von Klitzing, R A. Stradling, An observation by photoconductivity of strain splitting of shallow bulk donors located near to the surface in silicon MOS devices, Solid State Commun. 20 (1976) 77-80. https://doi.org/10.1016/0038-1098(76)91703-8

[13] A.G. Kazanskii, P.L. Richards, E.E. Haller, Far-infrared photoconductivity of uniaxially stressed germanium, Appl. Phys. Lett. 31 (1977) 496-497. https://doi.org/10.1063/1.89755

[14] G.A. Thomas, M. Capizzi, F. DeRosa, R.N. Bhatt, T.M. Rice, Optical study of interacting donors in semiconductors, Phys. Rev. B. 23 (1981) 5472. https://doi.org/10.1103/PhysRevB.23.5472

[15] H. Hogue, E. Atkins, D. Reynolds, M. Salcido, L Dawson, D. Molyneux, and M. Muzilla, Update on blocked impurity band detector technology from DRS, Proc. SPIE. 7780 (2010) 778004. https://doi.org/10.1117/12.862708

[16] E.E. Haller, J.W. Beeman, Far infrared photoconductors: Recent advances and future prospects, Far-IR, Sub-mm and Mm Detector Technology Workshop, 1-3 April, Monterey, CA, 2002 (http://www.sofia.usra.edu/det_workshop, id.13).

[17] L.A. Reichertz, J.W. Beeman, B.L. Cardozo, N.M. Haegel, *et al.*, GaAs BIB photodetector development for far-infrared astronomy, Proc. SPIE. 5543 (2004) 231-238. https://doi.org/10.1117/12.560291

[18] N. Deßmann, S.G. Pavlov, V.N. Shastin, R.Kh. Zhukavin, *et al.*, Time-resolved electronic capture in n-type germanium doped with antimony, Phys. Rev. B. 89 (2014) 035205. https://doi.org/10.1103/PhysRevB.89.035205

[19] K.S. Liao, N. Li, C. Wang, L. Li, *et al.*, Extended mode in blocked impurity band detectors for terahertz radiation detection, Appl. Phys. Lett. 105 (2014) 143501. https://doi.org/10.1063/1.4897275

[20] J.E. Huffman, A.G. Crouse, B.L. Halleck, T.V. Downes, T.L. Herter, Si:Sb blocked impurity band detectors for infrared astronomy, J. Appl. Phys. 72 (1992) 273-275. https://doi.org/10.1063/1.352127

[21] A. Rogalski, Progress in focal plane array technologies, Progr. Quant. Electr. 36 (2012) 342-473. https://doi.org/10.1016/j.pquantelec.2012.07.001

[22] E.E. Haller, W.L. Hansen, F.S. Goulding, Photothermal ionization spectroscopy in semiconductors, IEEE Trans. Nucl. Sci. NS-22 (1975) 127-134. https://doi.org/10.1109/TNS.1975.4327629

[23] M.D. Petroff, M.G. Stapelbroek, U.S. Patent 4,568,960. Filled Oct. 23 (1980),

granted Feb. 4 (1986).

[24] S.B. Stetson, D.B. Reynolds, M.G. Stapelbroek, R.L. Stermer, Design and performance of blocked-impurity-band detector focal plane arrays, Proc SPIE. 686 (1986) 48-65. https://doi.org/10.1117/12.936525

[25] F. Szmulowicz, F.L. Madarsz, Blocked impurity band detectors: an analytical model - figures of merit, J Appl Phys. 62 (1987) 2533-2540. https://doi.org/10.1063/1.339466

[26] W. Raab, Semiconductors for low energies: incoherent infrared/sub-millimetre detectors, in: M.C.E. Huber, A. Pauluhn, J.L. Culhane, J.G. Timothy, K. Wilhelm, A. Zehnder (Eds.), Observing Photons in Space. A Guide to Experimental Space Astronomy, Springer, New York, Heidelberg, Dordrecht, London, 2013, pp. 525-542. https://doi.org/10.1007/978-1-4614-7804-1_30

[27] D.M. Watson, J.E. Huffman, Germanium blocked-impurity-band far-infrared detectors, Appl. Phys. Lett. 52 (1988) 1602-1604. https://doi.org/10.1063/1.99094

[28] J. Zhu, H. Zhu, H. Xu, Z. Weng, H. Wu, Ge-based mid-infrared blocked-impurity-band photodetectors, Infr. Phys. Techn. 92 (2018) 13-17. https://doi.org/10.1016/j.infrared.2018.04.015

[29] A.K. Mainzer, J. Hong, M.G. Stapelbroek, H. Hogue, *et al.*, A new large well 1024×1024 Si:As detector for mid-infrared, Proc. SPIE. 5881 (2005) 58810Y. https://doi.org/10.1117/12.633492

[30] L.A. Reichertz, J.W. Beeman, B.L. Cardozo, G. Jakob, *et al.*, Development of a GaAs-based BIB detector for sub-mm wavelengths, Proc. SPIE. 6275 (2006) 62751S. https://doi.org/10.1117/12.673039

[31] L.A. Reichertz, B.L. Cardozo, J.W. Beeman, D.I. Larsen, *et al.*, First results on GaAs blocked impurity band (BIB) structures for far-infrared detector arrays, Proc. SPIE. 5883 (2005) 58830Q. https://doi.org/10.1117/12.620156

[32] X. Wang, B. Wang, Y. Chen, L. Hou, *et al.*, Spectral response characteristics of novel ion-implanted planar GaAs blocked-impurity-band detectors in the terahertz domain, Opt. Quant. Electron. 48 (2016) 518. https://doi.org/10.1007/s11082-016-0778-5

[33] V. Khalap, H. Hogue, Antimony doped silicon blocked impurity band (BIB) arrays for low flux applications, Proc. SPIE. 8512 (2012) 85120O. https://doi.org/10.1117/12.930769

[34] H. Zhu, Z. Weng, J. Zhu, J. Xu, *et al.*, Surface plasmon enhanced Si-based BIB terahertz detectors, Appl. Phys. Lett. 111 (2017) 053102. https://doi.org/10.1063/1.4996496

[35] H.H. Hogue, M.G. Mlynczak, M.N. Abedin, S.A. Masterjohn, J.E. Huffman, Far-

infrared detector development for space-based Earth observation, Proc. SPIE. 7082 (2008) 70820E. https://doi.org/10.1117/12.797078

[36] A. Poglitsch, R.O. Katterloher, R. Hoenle, J.W. Beeman, *et al.*, Far-infrared photoconductors for Herschel and SOFIA, Proc. SPIE. 4855 (2003) 115-128. https://doi.org/10.1117/12.459184

[37] M. Hanaoka, H. Kaneda, S. Oyabu, M. Yamagishi, *et al.*, Development of blocked-impurity-band type Ge detectors fabricated with the surface-activated wafer bonding method for far-infrared astronomy, J. Low Temper. Phys. 184 (2016) 225-230. https://doi.org/10.1007/s10909-016-1484-1

[38] T. Wada, Y. Arai, S. Baba, M. Hanaoka, *et al.*, Development for germanium blocked impurity band far-infrared image sensors with fully-depleted silicon-on-insulator CMOS readout integrated circuit, J. Low Temper. Phys. 184 (2016) 217-224. https://doi.org/10.1007/s10909-016-1522-z

[39] A. Poglitsch, C. Waelkens, O. H. Bauer, J. Cepa, *et al.*, The photodetector array camera and spectrometer (PACS) for the Herschel space observatory, Proc. SPIE. 7010 (2008) 701005, in: Space Telescopes and Instrumentation 2008: Optical, Infrared, and Millimeter, Eds. J.M. Oschmann, M.W. de Graauw, H.A. MacEwen. https://doi.org/10.1117/12.790016

[40] V.V. Rumyantsev, D.V. Kozlov, S.V. Morozov, M.A. Fadeev, *et al.*, Terahertz photoconductivity of double acceptors in narrow gap HgCdTe epitaxial films grown by molecular beam epitaxy on GaAs(013) and Si(013) substrates, Semicond. Sci. Techn. 32 (2017) 095007. https://doi.org/10.1088/1361-6641/aa76a0

[41] D.E. Dolzhenko, A.V. Nicorici, L.I. Ryabova, D.R. Khokhlov, A new type of sensitive semiconductor detectors of terahertz radiation, Proc. SPIE. 8431 (2012) 843126. https://doi.org/10.1117/12.922173

[42] A. Poglitsch, C. Waelkens, N. Geis, H. Feuchtgruber, *et al.*, The photodetector array camera and spectrometer (PACS) on the Herschel space observatory, Astronomy and Astrophysics. 518 (2010) L2, 12 pp.

[43] J.W. Beeman, S. Goyal, L.A. Reichertz, E.E. Haller, Ion-implanted Ge:B far-infrared blocked-impurity-band detectors, Infr. Phys. Technol. 51 (2007) 60-65. https://doi.org/10.1016/j.infrared.2006.12.001

[44] L.C. West, S.J. Eglash, First observation of an extremely large-dipole infrared transition within the conduction band of a GaAs quantum well, Appl. Phys. Lett. 46 (1985) 1156-1158. https://doi.org/10.1063/1.95742

[45] H. Schneider, H.C Liu, Quantum Well Infrared Photodetectors: Physics and Applications. Springer, Heidelberg, 2007.

[46] F. Sizov, Semiconductor superlattice and quantum well detectors, in: A. Rogalski (Ed.), Infrared Photon Detectors, SPIE Opt. Eng. Press, Bellingham, 1995, pp. 561-623.

[47] A. Rogalski, Infrared Detectors, second ed., CRC Press, Boca Raton, 2010. https://doi.org/10.1201/b10319

[48] B.F. Levine, A. Zussman, S.D. Gunapala, M.T. Asom, *et al.*, Photoexcited escape probability, optical gain, and noise in quantum well infrared photodetectors, J. Appl. Phys. 72 (1992) 4429-4443. https://doi.org/10.1063/1.352210

[49] F. Klappenberger, A.A. Ignatov, S. Winner, E. Schomburg, *et al.*, Broadband semiconductor detector for THz radiation, Appl. Phys. Lett. 78 (2001) 1673-1675. https://doi.org/10.1063/1.1352669

[50] F. Wu, W. Tian, W.Y. Yan, J. Zhang, *et al.*, Terahertz intersubband transition in GaN/AlGaN step quantum well, J. Appl. Phys. 113 (2013) 154505. https://doi.org/10.1063/1.4802496

[51] D. Feezell, Y. Sharma, S. Krishna, Optical properties of nonpolar III-nitrides for intersubband photodetectors, J. Appl. Phys. 113 (2013) 133103. https://doi.org/10.1063/1.4798353

[52] D. Morozov, P. Mauskopf, I. Bacchus, M. Elliott, *et al.*, THz direct detector with 2D electron gas periodic structure absorber, 18th Int. Symp. Space Terahertz Technol., 21-23 March, 2007, Pasadena, California. Ed. A. Karpov, pp. 123-127.

[53] R. Paiella, F. Capasso, C. Gmachl, D. L. Sivco, *et al.*, Self-mode-locking of semiconductor lasers with giant ultrafast optical nonlinearities, Science. 290 (2000) 1739-1742. https://doi.org/10.1126/science.290.5497.1739

[54] S. Winner, E. Schomburg, S. Brandl, F. Klappenberger, *et al.*, A superlattice detector as a fast direct detector and autocorrelator for terahertz radiation, Proc. SPIE. 3795 (1999) 116-124. https://doi.org/10.1117/12.370155

[55] Z. Chen, Z.Y. Tan, Y.J. Han, R. Zhang, *et al.*, Wireless communication demonstration at 4.1 THz using quantum cascade laser and quantum well photodetector, Electr. Lett. 47 (2011) 1002-1004. https://doi.org/10.1049/el.2011.1407

[56] P.D. Grant, S.R. Laframboise, R. Dudek, M. Graf, *et al.*, Terahertz free space communications demonstration with quantum cascade laser and quantum well photodetector, Electr. Lett. 45 (2009) 952-954. https://doi.org/10.1049/el.2009.1586

[57] T. Zhou, R. Zhang, X.G. Guo, Z.Y. Tan, *et al.*, Terahertz imaging with quantum-well photodetectors, IEEE Photonics Technol. Lett. 24 (2012) 1109-1111. https://doi.org/10.1109/LPT.2012.2196033

[58] C. Kadow, A.W. Jackson, A.C. Gossard, S. Matsuura, G.A. Blake, Self-assembled

ErAs islands in GaAs for optical-heterodyne THz generation, Appl. Phys. Lett. 76 (2000) 3510-3512. https://doi.org/10.1063/1.126690

[59] H.C. Liu, C.Y. Song, A.J. SpringThorpe, J.C. Cao, Terahertz quantum-well photodetector, Appl. Phys. Lett. 84 (2004) 4068-4070. https://doi.org/10.1063/1.1751620

[60] M. Graf, G. Scalari, D. Hofstetter, J. Faist, *et al.*, Terahertz range quantum well infrared photodetector, Appl. Phys. Lett. 84 (2004) 475-477. https://doi.org/10.1063/1.1641165

[61] H. Luo, H.C. Liu, C.Y. Song, Z.R. Wasilewski, Background-limited terahertz quantum-well photodetector, Appl. Phys. Lett. 86 (2005) 231103. https://doi.org/10.1063/1.1947377

[62] M. Patrashin, I. Hosako, Terahertz frontside-illuminated quantum-well photodetector, Opt Lett. 33 (2008) 168-70. https://doi.org/10.1364/OL.33.000168

[63] X.G. Guo, J.C. Cao, R. Zhang, Z.Y. Tan, H.C. Liu, Recent progress in terahertz quantum-well photodetectors, IEEE J. Selected Topics Qunt. Electr. 19 (2013) 8500508. https://doi.org/10.1109/JSTQE.2012.2201136

[64] F. Castellano, Quantum well photodetectors, in: M. Perenzoni, D.J. Paul (Eds.), Physics and Applications of Terahertz Radiation, Springer, Dordrecht, Heidelberg, New York, London, 2014, pp. 3-34. https://doi.org/10.1007/978-94-007-3837-9_1

[65] F. Gouider, Yu.B. Vasilyev, J. Könemann, P.D. Buckle, *et al.*, Detection of THz radiation with devices made from wafers with HgTe and InSb quantum wells, Physics of Semiconductors, AIP Conf. Proc., 1399 (2011) 1019-1020. https://doi.org/10.1063/1.3666725

[66] C. Zoth, P. Olbrich, P. Vierling, K.-M. Dantscher, *et al.*, Quantum oscillations of photocurrents in HgTe quantum wells with Dirac and parabolic dispersions, Phys. Rev. B. 90 (2014) 205415. https://doi.org/10.1103/PhysRevB.90.205415

[67] R. Bonk, C. Stellmach, Yu.B. Vasilyev, C.R. Becker, *et al.*, Fast and tunable photodetectors for terahertz radiation based on the Landau quantization at HgTe/HgCdTe focal plane arrays, AIP Conf. Proc. 893 (2007) 491. https://doi.org/10.1063/1.2729980

[68] S. Komiyama, O. Astafiev, V. Antonov, T. Kutsuwa, and H. Hirai, A single photon detector in the far-infrared range, Nature. 403 (2000) 405-407. https://doi.org/10.1038/35000166

[69] P. Kleinschmidt, S.P. Giblin, V. Antonov, H. Hashiba, *et al.*, A highly sensitive detector for radiation in the terahertz region, IEEE Trans. Instr. Measur. 56 (2007) 463-467. https://doi.org/10.1109/TIM.2007.891146

[70] O. Astafiev, S. Komiyama, T. Kutsuwa, V. Antonov, *et al.*, Single photon detector in the microwave range, Appl. Phys. Lett. 80 (2002) 4520-4522. https://doi.org/10.1063/1.1482787

[71] P. Martyniuk, A. Rogalski, Quantum-dot infrared photodetectors: Status and outlook, Progr. Quant. Electr. 32 (2008) 89-120. https://doi.org/10.1016/j.pquantelec.2008.07.001

[72] X. H. Su, J. Yang, and P. Bhattacharya, Terahertz detection with tunneling quantum dot intersublevel photodetector, Appl. Phys. Lett. 89 (2006) 031117. https://doi.org/10.1063/1.2233808

[73] H. Hashiha, V. Antonov, L. Kulik, A. Tzalenchuk, and S. Komiyama, Sensing individual terahertz photon, Nanotechn. 21 (2010) 165203. https://doi.org/10.1088/0957-4484/21/16/165203

[74] J.-H. Dai, Jh.-H. Lee, Y.-L. Lin, and S.-Ch. Lee, In(Ga)As quantum rings for terahertz detection, Jap. J. Appl. Phys. 47 (2008) 2924-2926. https://doi.org/10.1143/JJAP.47.2924

[75] M. Patrashin, and H. Iwao, Terahertz frontside-illuminated quantum well photodetector, J. Nation. Inst. Information and Com. 55 (2008) 29-35. https://doi.org/10.1117/12.797653

[76] G. Huang, J. Yang, P. Bhattacharya, G. Ariyawansa, A.G.U. Perera, A multicolor quantum dot intersublevel detector with photoresponse in the terahertz range, Appl. Phys. Lett. 92 (2008) 011117. https://doi.org/10.1063/1.2830994

[77] Information on http://www.ccas-web.org/superconductivity (Superconductivity: Properties, History, Applications and Challenges. CCAS/IEEE CSC Outreach (2008).

[78] C.P. Poole, Superconductivity. Acad. Press, New York, 2007. https://doi.org/10.1016/B978-012088761-3/50034-X

[79] D.J. Thoen, B.G. C. Bos, E A. F. Haalebos, T.M. Klapwijk, *et al.*, Superconducting NbTiN thin films with highly uniform properties over a 100 mm wafer, IEEE Trans. Appl. Supercond. 27 (2016) 1500505. https://doi.org/10.1109/TASC.2016.2631948

[80] J. Nagamatsu, N. Nakagawa, T. Muranaka, Y. Zenitani, J. Akimitsu, Superconductivity at 39 K in magnesium diboride, Nature. 410 (2001) 63-64. https://doi.org/10.1038/35065039

[81] C. Zhang, Y. Wang, D. Wang, Y. Zhang, *et al.*, Suppression of superconductivity in epitaxial MgB2 ultrathin films, J. Appl. Phys. 114 (2013) 023903. https://doi.org/10.1063/1.4812738

[82] D.D.E. Martin, P. Verhoeve, Superconducting tunnel junctions, in: M.C.E. Huber, A. Pauluhn, J.L. Culhane, J.G. Timothy, K. Wilhelm, A. Zehnder (Eds.), Observing

Photons in Space. A Guide to Experimental Space Astronomy, Springer, New York, Heidelberg, Dordrecht, London, 2013, pp. 441-457.

[83] B. Josephson, Possible new effects in superconductive tunnelling, Phys. Lett. 1 (1962) 251-253. https://doi.org/10.1016/0031-9163(62)91369-0

[84] F. Sizov, A. Rogalski, THz detectors, Progr. Quant. Electr. 34 (2010) 278-347. https://doi.org/10.1016/j.pquantelec.2010.06.002

[85] E. Burstein, D.N. Langenberg, B.N. Taylor, Superconductors as quantum detectors for microwave and sub-millimeter radiation, Phys. Rev. Lett. 6 (1961) 92-94. https://doi.org/10.1103/PhysRevLett.6.92

[86] S. Friedrich, Superconducting tunnel junction photon detectors: Theory and applications, J. Low Temp. Phys. 151 (2008) 277-286. https://doi.org/10.1007/s10909-007-9697-y

[87] V.V. Samedov, B.M. Tulinov, Relationship between the heat flow and the current through a superconducting tunnel junction, Physics Procedia. 36 (2012) 312-317. https://doi.org/10.1016/j.phpro.2012.06.165

[88] D.D.E. Martin, P. Verhoeve, Superconductive tunnel junctions, in: M.C.E. Huber, A. Pauluhn, J.L. Culhane, J.G. Timothy, K. Wilhelm, A. Zehnder (Eds.), Observing Photons in Space. A Guide to Experimental Space Astronomy, Springer, New York, Heidelberg, Dordrecht, London, 2013, pp. 479-496.

[89] A. Peacock, P. Verhoeve, N. Rando, A. van Dordrecht, *et al.*, Single optical photon detection with a superconducting tunnel junction, Nature. 381 (1996) 135-137. https://doi.org/10.1038/381135a0

[90] S. Ariyoshi, T. Taino, A. Dobroiu, H. Sato, *et al.*, Terahertz detector based on a superconducting tunnel junction coupled to a thin superconductor film, Appl. Phys. Lett. 95 (2009) 193504. https://doi.org/10.1063/1.3263711

[91] F.P. Mena, J.W. Kooi, A.M. Baryshev, C.F.J. Lodewijk, *et al.*, Design and performance of a 600-720-GHz sideband-separating receiver using AlOx and AlN SIS junctions, IEEE Trans. Microw. Theory Techn. 59 (2011) 166-177. https://doi.org/10.1109/TMTT.2010.2090417

[92] C. Wilson, L. Frunzio, D. Prober, Time-resolved measurements of thermodynamic fluctuations of the particle number in a nondegenerate Fermi gas, Phys. Rev. Lett. 87 (2001) 067004. https://doi.org/10.1103/PhysRevLett.87.067004

[93] C.A. Mears, Q. Hu, P.L. Richards, A.H. Worsham, *et al.*, Quantum limited heterodyne detection of millimeter waves using super conducting tantalum tunnel junctions, Appl. Phys. Lett. 57 (1990) 2487-2489. https://doi.org/10.1063/1.104111

[94] J.R. Tucker, M.J. Feldman, Quantum detection at millimeter wavelengths, Rev.

Mod. Phys. 57 (1985) 1055-1113. https://doi.org/10.1103/RevModPhys.57.1055

[95] G.H. Rieke, Detection of Light: From the Ultraviolet to the Submillimeter, Cambridge University Press, Cambridge, 2003. https://doi.org/10.1017/CBO9780511606496

[96] J. Zmuidzinas, P.L. Richards, Superconducting detectors and mixers for sub-mm astrophysics, Proc. IEEE. 92 (2004) 1597-1616. https://doi.org/10.1109/JPROC.2004.833670

[97] C. Otani, S. Ariyoshi, H. Matsuo, T. Morishima, *et al.*, Terahertz direct detector using superconducting tunnel junctions, Proc. SPIE. 5354 (2004) 86-93. https://doi.org/10.1117/12.528886

[98] P. Verhoeve, N. Rando, A. Peacock, D. Martin, R. den Hartog, Superconducting tunnel junctions as photoncounting imaging spectrometers from the optical to the X-ray band, Opt. Eng. 41 (2002) 1170-1184. https://doi.org/10.1117/1.1475738

[99] R. Schoelkopf, S. Moseley, C. Stachle, P. Wahlgren, P. Delsing, A concept for a sub-millimeter-wave single-photon counter, Trans. Appl. Supercond. 9 (1999) 2935-2939 https://doi.org/10.1109/77.783645

[100] S. Ariyoshi, C. Otani, A. Dobroiu, H. Sato, *et al.*, Terahertz imaging with a direct detector based on superconducting tunnel junctions, Appl. Phys. Lett. 88 (2006) 203503. https://doi.org/10.1063/1.2204842

[101] W. Wild, Coherent far-infrared/sub-millimeter detectors, in: M.C.E. Huber, A. Pauluhn, J.L. Culhane, J.G. Timothy, K. Wilhelm, A. Zehnder (Eds.), Observing Photons in Space. A Guide to Experimental Space Astronomy, Springer, New York, Heidelberg, Dordrecht, London, 2013, pp. 543-553.

[102] Y. Kozuki, H. Ishida, Y. Hasegawa, K Kuroiwa, *et al.*, Development for a wideband 100 GHz SIS mixer, 26th Intern. Symp. Space Terahertz Technol., 16-18 March, Cambridge, MA, 2015, P-233.

[103] S. Selig, M.P. Westig, K. Jacobs, M. Schultz, N. Honingh, Heat transfer coefficient saturation in superconducting Nb tunnel junctions contacted to a NbTiN circuit and an Au energy relaxation layer, IEEE Trans. Appl. Supercond. 25 (2015) 2400705. https://doi.org/10.1109/TASC.2014.2378054

[104] J. Zmuidzinas, J.W. Kooi, J. Kawamura, G. Chattopadhyay, *et al.*, Development of SIS mixers for 1 THz, Proc. SPIE. 3357 (1998) 53-61. https://doi.org/10.1117/12.317386

[105] U.U. Graf, C.E. Honingh, K. Jacobs, J. Stutzki, Terahertz heterodyne array receivers for astronomy, Journal of Infrared, Millimeter, and Terahertz Waves 36 (2015) 896-921. https://doi.org/10.1007/s10762-015-0171-7

[106] C. Groppi, C. Wheeler, H. Mani, S. Weinreb, *et al.*, The kilopixel array pathfinder project (KAPPa): A 16 pixel 660 GHz pathfinder instrument with an integrated heterodyne focal plane detector, in: Proc. 22nd Int. Symp. Space THz Techn, April 25-28, Loews Ventema Canyon Resort, Tuscon, AZ USA, 2011, pp. 164-170.

[107] C.E. Groppi, C.H. Wheeler, H. Mani, P. McGarey, *et al.*, The kilopixel array pathfinder project (KAPPa), a 16 pixel integrated heterodyne focal plane array, Proc. SPIE. 8452 (2012) 8452OY. https://doi.org/10.1117/12.927358

[108] G.H. Wheeler, M. Neric, C.E. Groppi, M. Underhill, *et al.*, Results of using permanent magnets to suppress Josephson noise in the KAPPs SIS receiver, Proc. SPIE. 9914 (2016) 99141W. https://doi.org/10.1117/12.2231358

[109] P.K. Day, H.G. Leduc, B.A. Mazin, A. Vayonakis, J. Zmuidzinas, A broadband superconducting detector suitable for use in large arrays, Nature. 425 (2003) 817-821. https://doi.org/10.1038/nature02037

[110] B.A. Mazin, Microwave kinetic inductance detectors: The first decade, The 13th International Workshop on Low Temperature Detectors-LTD13. AIP Conf. Proceed. 1185(1) (2009) 135-142.

[111] J.J. A. Baselmans, Kinetic inductance detectors, J. Low Temp. Phys. 167 (2011) 292-304. https://doi.org/10.1007/s10909-011-0448-8

[112] J. Zmuidzinas, Superconducting microresonators: Physics and applications, Annual Rev. Condensed Matter Phys. 3 (2012) 169-214. https://doi.org/10.1146/annurev-conmatphys-020911-125022

[113] R.M.J. Janssen, J.J.A. Baselmans, A. Endo, L. Ferrari, *et al.*, High optical efficiency and photon noise limited sensitivity of microwave kinetic inductance detectors using phase readout, Appl. Phys. Lett. 103 (2013) 203503. https://doi.org/10.1063/1.4829657

[114] R.M.J. Janssen, J.J.A. Baselmans, A. Endo, L. Ferrari, *et al.*, Performance of hybrid NbTiN-Al microwave kinetic inductance detectors as direct detectors for sub-millimeter astronomy, Proc. SPIE. 9153 (2014) 91530T. https://doi.org/10.1117/12.2055537

[115] D.J. Goldie, S. Withington, Non-equilibrium superconductivity in quantum-sensing superconducting resonators, Supercond. Sci. Technol. 26 (2013) 015004. https://doi.org/10.1088/0953-2048/26/1/015004

[116] T. Guruswamy, D.J. Goldie, S. Withington, Quasiparticle effective temperature in superconducting thin films illuminated at THz frequencies, 26th Int. Symposium on Space Terahertz Technol., W3-2, Cambridge, MA, 16-18 March (2015).

[117] J.J. A. Baselmans, J. Bueno, S.J. C. Yates, O. Yurduseven, *et al.*, A kilo-pixel

imaging system for future space based far-infrared observatories using microwave kinetic inductance detectors, J. Astron. Astrophys. 601 (2017) A89. (Free Access). https://doi.org/10.1051/0004-6361/201629653

[118] B.A. Mazin, B. Bumble, S.R. Meeker, K. O'Brien, *et al.*, A superconducting focal plane array for ultraviolet, optical, and near-infrared astrophysics, Opt. Express. 20 (2012) 1503-15111. (Open Access) https://doi.org/10.1364/OE.20.001503

[119] P.D. Mauskopf, S. Doyle, P. Barry, S. Rowe, *et al.*, Photon-noise limited performance in aluminum LEKIDs, J. Low Temp. Phys., 176 (2014) 545-552; https://doi.org/10.1007/s10909-013-1069-1

[120] A. Catalano, A. Benoit, O. Bourrion, M. Calvo, *et al.*, Maturity of lumped element kinetic inductance detectors for space-borne instruments in the range between 80 and 180 GHz, Astron. Astrophys. 592 (2016) A26. https://doi.org/10.1051/0004-6361/201527715

[121] J. Hubmayr, J. Beall, D. Becker, H.-M. Cho, *et al.*, Photon-noise limited sensitivity in titanium nitride kinetic inductance detectors, Appl. Phys. Lett. 106 (2015) 073505. https://doi.org/10.1063/1.4913418

[122] J. Baselmans, J. Bueno, O. Yurduseven, S. Yates, *et al.*, Performance of a 961 pixel kinetic inductance detector system for future space borne observatories, IRMMW-THz 41st Intern. Conf., 25−30 Sept. 2016, Copenhagen, Denmark; doi: 10.1109/IRMMW-THz.2016.7758393. https://doi.org/10.1109/IRMMW-THz.2016.7758393

[123] R. Barends, N. Vercruyssen, A. Endo, P.J. de Visser, *et al.*, Reduced frequency noise in superconducting resonators, Appl. Phys. Lett. 97 (2010) 033507. https://doi.org/10.1063/1.3467052

[124] A. Bruno, G. de Lange, S. Asaad, K.L. van der Enden, *et al.*, Reducing intrinsic loss in superconducting resonators by surface treatment and deep etching of silicon substrates, Appl. Phys. Lett. 106 (2015) 182601. https://doi.org/10.1063/1.4919761

[125] R. Barends, S. van Vliet, J.J.A. Baselmans, S.J.C., Yates, *et al.*, Enhancement of quasiparticle recombination in Ta and Al superconductors by implantation of magnetic and nonmagnetic atoms, Phys. Rev. B. 79 (2009) 020509. https://doi.org/10.1103/PhysRevB.79.020509

[126] A. Pauluhn, Detector types used in space, in: M.C.E. Huber, A. Pauluhn, J.L. Culhane, J.G. Timothy, K. Wilhelm, A. Zehnder (Eds.), Observing Photons in Space. A Guide to Experimental Space Astronomy, Springer, New York, Heidelberg, Dordrecht, London, 2013, pp. 363-366. https://doi.org/10.1007/978-1-4614-7804-1_20

[127] Y.A.C. Eaton, Infrared imaging bolometers, in: M.C.E. Huber, A. Pauluhn, J.L. Culhane, J.G. Timothy, K. Wilhelm, A. Zehnder (Eds.), Observing Photons in Space.

A Guide to Experimental Space Astronomy, Springer, New York, Heidelberg, Dordrecht, London, 2013, pp. 515-524

[128] B.S. Karasik, A.V. Sergeev, D.E. Prober, Nanobolometers for THz photon detection, EEE Trans. Terahertz Sci. Techn. 1 (2011) 97-111. https://doi.org/10.1109/TTHZ.2011.2159560

[129] R. Müller, B. Gutschwager, J. Hollandt, M. Kehrt, *et al.*, Characterization of a large-area pyroelectric detector from 300 GHz to 30 THz, J. IR, Mm, THz Waves. 36 (2015) 654-661. (Open Access). https://doi.org/10.1007/s10762-015-0163-7

[130] N. Oda, Technology trend in real-time, uncooled image sensors for sub-THz and THz wave detection, Proc. SPIE. 9836 (2016) 98362P. In: Micro- and Nanotechnology Sensors, Systems, and Applications VIII, Eds. Thomas George, Achyut K. Dutta, M. Saif Islam (2016).

[131] D. Dufour, L. Marchese, M, Terroux, H. Oulachgar, *et al.*, Review of terahertz technology development at INO, Journal of Infrared, Millimeter, and Terahertz Waves. 36 (2015) 922-946. https://doi.org/10.1007/s10762-015-0181-5

[132] N. Oda, S. Kurashina, M. Miyoshi, K. Doi., *et al.*, Journal of Infrared, Millimeter, and Terahertz Waves. 36 (2015) 947-960. https://doi.org/10.1007/s10762-015-0184-2

[133] M. Hoefle, K. Haehnsen, I. Oprea, *et al.*, Compact and sensitive millimetre wave detectors based on low barrier Schottky diodes on impedance matched planar antennas, Journal of Infrared, Millimeter, and Terahertz Waves. 35 (2014) 891-908. https://doi.org/10.1007/s10762-014-0090-z

[134] Information on http://www.ophiropt.com/

[135] J. Clarke, G. I. Hoffer, P. L. Richards, and N. H. Yeh, Superconductive bolometers for sub-millimeter wavelengths, J. Appl. Phys. 48 (1977) 4865-4879. https://doi.org/10.1063/1.323612

[136] A. van der Ziel, Noise in Measurements, Wiley-Interscience, New York, 1976.

[137] P.W. Kruse, Uncooled Thermal Imaging, SPIE Press, Bellingham, 2001.

[138] C.M. Hanson, Uncooled IR detector performance limits and barriers. Proc. SPIE. 4028 (2000) 2-11. https://doi.org/10.1117/12.391717

[139] E.L. Dereniak, G.D. Boreman, Infared Detectors and Systems, Wiley-Interscience, New-York, 1996.

[140] C.B. McKitterick, D.E. Prober, B.S. Karasik, Performance of graphene thermal photon detectors. J. Appl. Phys. 113 (2013) 044512. https://doi.org/10.1063/1.4789360

[141] D.E. Prober, Superconducting terahertz mixer using a transition-edge microbolometer. Appl. Phys. Lett. 62 (1993) 2119-2121.

https://doi.org/10.1063/1.109445

[142] P.J. Burke, R.J. Schoelkopf, D.E. Prober, A. Skalare, *et al.* Length scaling of bandwidth and noise in hot electron superconducting mixers, Appl. Phys. Lett. 68 (1996) 3344-3346. https://doi.org/10.1063/1.116052

[143] J.C. Mather, Bolometer noise: nonequilibrium theory, Appl. Opt. 21 (1982) 1125-1129. https://doi.org/10.1364/AO.21.001125

[144] J.A. Ratches, Current and future trends in military night vision applications, Ferroelectrics. 342 (2006) 183-192. https://doi.org/10.1080/00150190600946351

[145] M. Kohin, N. Butler, Performance limits of uncooled VOx microbolometer focal-plane arrays, Proc. SPIE. 5406 (2004) 447-453. https://doi.org/10.1117/12.542482

[146] J. Bock, The future of far-infrared detectors for space, Far-Infrared Astronomy from Space, Pasadena, USA, 28-30 May (2008), https://www.ipac.caltech.edu/irspace/pres/bock.pdf.

[147] P.L. Marasco, E.L. Dereniak, Uncooled infrared sensor performance, Proc. SPIE. 2020 (1993) 363-378. https://doi.org/10.1117/12.160557

[148] M.J.E. Golay, A pneumatic infrared detector, Rev. Sci. Instr. 18 (1947) 357-362. https://doi.org/10.1063/1.1740949

[149] G. Shol, Y. Marfaing, M. Munch, P. Thorel, and P. Combette, Les Detecteurs de Rayonnement Infra-Rouge, Dunod, Paris, 1966.

[150] P.H. Siegel, THz technology: an overview. Intern. J. High Speed Electr. Systems. 13 (2003) 351-394. https://doi.org/10.1142/S0129156403001776

[151] D. Denison, M. Knotts, M. McConney, V. Tsukruk, Experimental characterization of mm-wave detection by a micro-array of Golay cells, Proc SPIE. 7309 (2009) 73090J. https://doi.org/10.1117/12.818387

[152] V. Desmaris, H. Rashid, A. Pavolotsky, V. Belitsky, Design, simulations and optimization of micromachined Golay-cell based THz sensors operating at room temperature, Proceed. Chemistry. 1 (2009) 1175-1178. https://doi.org/10.1016/j.proche.2009.07.293

[153] N. Karpowicz, H. Zhong, J. Xu, K.-I. Lin, *et al.*, Non-destructive sub-THz CW imaging, Proc. SPIE. 5727 (2005) 132-142. https://doi.org/10.1117/12.590539

[154] D.D. Eden, Antenna-coupled microbolometer millimeter wave focal plane array technology, Proc. SPIE. 4719 (2002) 370-381. https://doi.org/10.1117/12.477473

[155] C. Vedel, J.-L. Martin, J.-L.O. Buffet, J.-L. Tissot, *et al.*, Amorphous silicon based uncooled microbolometer IR FPA, Proc. SPIE. 3698 (1999) 276-283. https://doi.org/10.1117/12.354529

[156] A.W.M. Lee, Q. Hu, Real-time, continuous-wave terahertz imaging by use of a microbolometer focal-plane array. Opt. Lett. 30 (2005) 2563-2565. https://doi.org/10.1364/OL.30.002563

[157] A.W.M. Lee, B.S. Williams, S. Kumar, Q. Hu, Real-time imaging using a 4.3-THz quantum cascade laser and a 320×240 microbolometer focal-plane array, IEEE Phot. Tech. Lett. 18 (2006) 1415-1417. https://doi.org/10.1109/LPT.2006.877220

[158] O. Oda, Detection of terahertz radiation from quantum cascade laser, using vanadium oxide microbolometer focal plane arrays, Proc. SPIE. 6940 (2008) 69402Y.

[159] F. Simoens, T. Durand, J. Meilhan, P. Gellie, *et al.*, Terahertz imaging with a quantum cascade laser and amorphous-silicon microbolometer array, Proc. SPIE. 7485 (2009) 74850M. https://doi.org/10.1117/12.830350

[160] F. Simoens, J. Martyrs, Terahertz real-time imaging uncooled array based on antenna- and cavity-coupled bolometers, Philos. Trans. R. Soc. A372 (2014) 20130111. https://doi.org/10.1098/rsta.2013.0111

[161] J.L. Tissot, C. Trouilleau, B. Fieque, A. Crastes, O. Legras, Uncooled microbolometer detector: recent developments at ULIS, Proc. SPIE. 5957 (2005) 5957OM, in: Infrared Photoelectronics, Congress on Optics and Optoelectronics, 2005, Warsaw, Poland, Eds. A. Rogalski; E.L. Dereniak; F.F. Sizov. https://doi.org/10.1117/12.621884

[162] F. Simoens, J. Meilhan, J.-A. Nicolas, Terahertz real-time imaging uncooled arrays based on antenna-coupled bolometers or FET developed at CEA-Leti, J. IR. Mm, THz Waves. 36 (2015) 961-985. https://doi.org/10.1007/s10762-015-0197-x

[163] J. Gou, Q. Niu, K. Liang, J. Wang, Y. Jiang, Frequency modulation and absorption improvement of THz micro-bolometer with micro-bridge structure by spiral-type antennas, Nanoscale Res. Lett. 13 (2018) 74. https://doi.org/10.1186/s11671-018-2484-7

[164] M.V.S. Ramakrishna, G. Karunasiri, P. Neuzil, U. Sridhar, W.J. Zeng, Highly sensitive infrared temperature sensor using self-heating compensated microbolometers, Sens. Actuators. A 79 (2000) 122-127. https://doi.org/10.1016/S0924-4247(99)00280-0

[165] F. Niklau, C. Vieider, H. Jakobsen, MEMS/MOEMS technologies and applications III, Proc. SPIE. 6836 (2007) 68360D. https://doi.org/10.1117/12.755128

[166] F. Simoens, The bolometer detectors, in: M. Perenzoni, D.J. Paul (Eds.), Physics and Applications of Terahertz Radiation, Springer, Dordrecht-Heidelberg, 2014, pp. 35-76. https://doi.org/10.1007/978-94-007-3837-9_2

[167] F. Sizov, V. Reva, A. Golenkov, V. Zabudsky. Uncooled detector challenges for

THz/sub-THz arrays imaging. Journal of Infrared, Millimeter, and Terahertz Waves. 32 (2011) 1192-1206. https://doi.org/10.1007/s10762-011-9789-2

[168] A. Rogalski, F. Sizov, Terahertz detectors and focal plane arrays, Opto-Electron. Rev. 19 (2011) 346-404. https://doi.org/10.2478/s11772-011-0033-3

[169] G. Hyseni, N. Caka, K.H. Hyseni, Infrared thermal detectors parameters: Semiconductor bolometers versus pyroelectrics, WSEAS Trans. on Circuits and Systems. 9 (2010) 238-247.

[170] C. Middleton, G. Zummo, A. Weeks, A. Pergande, *et al.*, Passive millimeter-wave focal plane array, Joint 29th Int. Conf. Infrared and Millimeter Waves, 12th Int. Conf. Terahertz Electronics (2004), pp. 745-746.

[171] E.N. Grossman, A.J. Miller, Active millimeter-wave imaging for concealed weapons detection, Proc. SPIE. 5077 (2003) 62-70. https://doi.org/10.1117/12.488198

[172] I. Kašalynas, R. Venckevicius, L. Minkevicius, A. Sešek, *et al.*, Spectroscopic terahertz imaging at room temperature employing microbolometer terahertz sensors and its application to the study of carcinoma tissues, Sensors. 16 (2016) 432 (Open Access). https://doi.org/10.3390/s16040432

[173] J. Trontelj, G. Valušis, R. Venckevicius, I. Kašalynas, *et al.*, A high performance room temperature THz sensor, Proc. SPIE. 91990 (2014) 91990K. https://doi.org/10.1117/12.2060692

[174] W. Li, J. Wang, J. Gou, Z. Huang, and Y. Jiang, Fabrication and characterization of linear terahertz detector arrays based on lithium tantalate crystal, Journal of Infrared, Millimeter, and Terahertz Waves. 36 (2015) 42-48. https://doi.org/10.1007/s10762-014-0115-7

[175] S. Bevilacqua, S. Cherednichenko, Low noise nanometer scaleroom-temperature $YBa_2Cu_3O_{7-x}$ bolometers for THz direct detection, IEEE Trans. Terahertz Sci. Technol. 4 (2014) 653-660. https://doi.org/10.1109/TTHZ.2014.2344435

[176] F. Simoens, J. Meilhan, Terahertz real-time imaging uncooled array based on antenna- and cavity-coupled bolometers, Phil. Trans. R. Soc. A. 372 (2014) 20130111. https://doi.org/10.1098/rsta.2013.0111

[177] T. Xue-Cou, K. Lin, L. Xin-Hua, M. Qing-Kai, *et al.*, $Nb_5N_6$ microbolometer arrays for terahertz detection, Chin. Phys. B. 22 (2013) 040701.

[178] F. Voltolina, A. Tredicucci, P.H. Bolivar, Low cost thermopile detectors for THz imaging and sensing, IRMMW-THz 2008, 33rd International Conf., Pasadena, CA, USA,15-19 Sept. 2008, DOI:10.1109/ICIMW.2008.4665525. https://doi.org/10.1109/ICIMW.2008.4665525

[179] A. Luukanen, V.-P. Viitanen, Terahertz imaging system based on antenna-coupled

microbolometers, Proc. SPIE. 3378 (1998) 34-44. https://doi.org/10.1117/12.319403

[180] A.K. Huhn, G. Spickermann, A. Ihring, U. Schinkel, *et al.*, Uncooled antenna-coupled THz detectors with 22 μs response time based on BiSb/Sb thermocouples, App. Phys. Lett. 102 (2013) 121102. https://doi.org/10.1063/1.4798369

[181] J. Gou, J. Wang, X. Zheng, D. Gu, *et al.*, Detection of terahertz radiation from 2.52 THz $CO_2$ laser using a 320×240 vanadium oxide microbolometer focal plane array, RSC Adv. 5 (2015) 84252. https://doi.org/10.1039/C5RA15049C

[182] D. Rutledge, S. Schwarz, Planar multimode detector arrays for infrared and millimeter-wave applications, IEEE J. Quant. Electr. 17 (1981) 407-414. https://doi.org/10.1109/JQE.1981.1071097

[183] I. Kasalynas, A.J.L. Adam, T.O. Klaassen, N.J. Hovenier, *et al.*, Some properties of a room temperature THz detection array, Proc. SPIE. 6596 (2007) 65960J. https://doi.org/10.1117/12.726404

[184] Z. Jiang, L. Men, C. Wan, P. Xiao, *et al.*, Low-noise readout integrated circuit for terahertz array detector, IEEE Trans. Terahertz Sci. Technol. 8 (2018) 350-356. https://doi.org/10.1109/TTHZ.2018.2819502

[185] A. Banerjee, H. Satoh, Y. Sharma, N. Hiromoto, H. Inokawa, Characterization of platinum and titanium thermistors for terahertz antenna-coupled bolometer applications, Sensors and Actuators. A. 273 (2018) 49-57. https://doi.org/10.1016/j.sna.2018.02.014

[186] D.P. Neikirk, D.B. Rutledge, W. Lam, Far-infrared microbolometer detectors, Int. J. Infr. Milli Waves. 5 (1984) 245-277. https://doi.org/10.1007/BF01009656

[187] A. Sesek, A. Zemva, J. Trontelj, A microbolometer system for radiation detection in the THz frequency range with a resonating cavity fabrication in the CMOS technology, Recent Patents on Nanotechnology. 12 (2018) 34-44. https://doi.org/10.2174/1872210511666170704103627

[188] A. Sešek, I. Kašalynas, A. Zemva, J. Trontelj, Antenna-coupled Ti-microbolometers for high-sensitivity terahertz Imaging, November, Sensors and Actuators A: Physical. 268 (2017) 133-140. https://doi.org/10.1016/j.sna.2017.11.029

[189] Y.-S. Lee, Principles of Terahertz Science and Technology, Springer, Berlin, 2009.

[190] F.L. Bakker, J. Flipse, B. van Wees, Nanoscale temperature sensing using the Seebeck effect, J. Appl. Phys. 111 (2012) 084306. https://doi.org/10.1063/1.3703675

[191] D.P. Neikirk, D.B. Rutledge, Self-heated thermocouples for far-infrared detection, Appl. Phys. Lett. 41 (1982) 400-402. https://doi.org/10.1063/1.93554

[192] S. Ben Mbarek, S. Euphrasie, T. Baron, L. Thiery, *et al.*, Room temperature

thermopile THz sensor, Sensors and Actuators A. 193 (2013) 155-160. https://doi.org/10.1016/j.sna.2013.01.014

[193] J.A. Russer, C. Jirauschek, G.P. Szakmany, M. Schmidt, *et al.*, High-speed antenna-coupled terahertz thermocouple detectors and mixers, IEEE Trans. Microw. Theory Techn. 63 (2015) 4236-4246. https://doi.org/10.1109/TMTT.2015.2496379

[194] J.A. Cox, R. Higashi, F. Nusseibeh, K. Newstrom-Peitso, C. Zins, Uncooled MEMS-based detector arrays for THz imaging applications, Proc. SPIE. 7311 (2009) 73110R.

[195] U. Schinkel, E. Kessler, A. Ihring, U. Dillner, *et al.*, Uncooled thermocouple air-bridge structure for a THz imaging system, AMA Conferences 2013 - SENSOR 2013, OPTO 2013, 14-16 May, 2013, Nurnberg, Germany.

[196] A. Ihring, E. Kessler, U. Dillner, F. Haenschke, *et al.*, High performance uncooled THz sensing structures based on antenna-coupled air-bridges, Microelectr. Eng. 98 (2012) 512–515. https://doi.org/10.1016/j.mee.2012.07.074

[197] A.L. Hsu, P.K. Herring, N.M. Gabor, S. Ha, *et al.*, Graphene-based thermopile for thermal imaging applications, Nano Lett. 15 (2015) 7211–7216. https://doi.org/10.1021/acs.nanolett.5b01755

[198] W.S. Boyle, K.F. Rodgers, Performance characteristics of a new low-temperature bolometer, J. Opt. Soc. Amer. 49 (1959) 66–69. https://doi.org/10.1364/JOSA.49.000066

[199] F.J. Low, Low-temperature germanium bolometer, J. Opt. Soc. Amer. 31 (1961) 130–134.

[200] M.A. Kinch, Compensated silicon-impurity conduction bolometer, J. App. Phys. 42 (1971) 5861–5863. https://doi.org/10.1063/1.1660027

[201] Detectors needs for long wavelength astrophysics, A report by the infrared, submillimeter and millimeter detector working group, http://safir.gsfc.nasa.gov/docs/ISMDWG_final.pdf (2002).

[202] V. Goudon, A. Aliane, W. Rabaud, C. Vialle, *et al.*, Design and fabrication of cooled silicon bolometers for mm wave detection, Nuclear Instruments and Methods in Physics Research Section A: Accelerators, Spectrometers, Detectors and Associated Equipment. 912 (2018) 78–81. https://doi.org/10.1016/j.nima.2017.10.057

[203] K.M. Itoh, E.E. Haller, J.W. Beeman, W.L. Hansen, *et al.*, Hopping conduction and metal-insulator transition in isotopically enriched neutron-transmutation-doped 70Ge:Ga, Phys. Rev. Lett. 77 (1996) 4058–4061. https://doi.org/10.1103/PhysRevLett.77.4058

[204] W. Holmes, J.J. Bock, and A.E. Lange, Heat capacity of neutron transmutation

doped Ge type 18, AIP Conf. Proc. 1185 (2009) 103-107.
https://doi.org/10.1063/1.3292293

[205] P.D. Mauskopf, J.J. Bock, H. Del Castillo, W.L. Holzapfel, A.E. Lange, Composite infrared bolometers with $Si_3N_4$ micromesh absorbers, Appl. Opt. 36 (1997) 765-771. https://doi.org/10.1364/AO.36.000765

[206] A. L. Woodcraft, R.V. Sudiwal, E. Wakui, and C. Paine, Hopping conduction in NTD germanium: comparison between measurement and theory, J. Low Temp. Phys. 134 (2004) 925-944. https://doi.org/10.1023/B:JOLT.0000013209.08494.01

[207] D.C. Alsop, C. Inman, A.E. Lange, and T. Wilbanks, Design and construction of high sensitivity infrared bolometers for operation at 300 mK, Appl. Opt. 31 (1992) 6610-6615. https://doi.org/10.1364/AO.31.006610

[208] S.T. Tanaka, A. Clapp, M. Devlin, M. Fisher, *et al.*, A 100 mK bolometric receiver for low background astronomy, Proc. SPIE. 1946 (1993) 110-115. https://doi.org/10.1117/12.158665

[209] J.W. Kooi, Advanced receivers for submillimeter and far infrared astronomy, Print Partners Ipskamp B.V., Enschede, The Netherlands, 2008 (ISBN 978-90-367-3653-4).

[210] A. Orduna, C.G. Trevino, A. Torres, R. Delgado, M.A. Dominguez, Micromachined and characterization of cooled a-Si:B:H microbolometer array in the terahertz region, in: R.P. Campos, A.C. Cuevas, R.A.E. Munoz (Eds.), Characterization of Metals and Alloys, Springer, Switzerland, 2017, pp 191-199. https://doi.org/10.1007/978-3-319-31694-9_16

[211] K. Farooqui, J.O. Gundersen, P.T. Timbie, G.W. Wilson, *et al.*, The monolithic silicon bolometer as an ultrasensitive detector for millimeter and sub-millimeter wavelength, 8th Intern. Symposium on Space Terahertz Technol., Harvard University, March (1997), pp. 546-555.

[212] P.L. Richards, Bolometers for infrared and millimeter waves, J. Appl. Phys. 76 (1994) 1–24. https://doi.org/10.1063/1.357128

[213] B.S. Karasik, R. Cantor, Demonstration of high optical sensitivity in far-infrared hot-electron bolometer, Appl. Phys. Lett. 98 (2011) 193503. https://doi.org/10.1063/1.3589367

[214] E.M. Gershenzon, G.N. Gol'tsman, I.G. Gogdize, Y.P. Gusev, *et al.*, Millimeter and submillimeter range mixer based on electronic heating of superconducting films in the resistive state, Sov. Phys. Supercond. 3 (1990) 1582-1597.

[215] Yu.P. Gousev, G.N. Gol'tsman, A.D. Semenov, E M. Gershenzon, Broadband ultrafast superconducting NbN detector for electromagnetic radiation, J. Appl. Phys. 75 (1994) 3695. https://doi.org/10.1063/1.356060

[216] B. Karasik, G.N. Gol'tsman, B.M. Voronov, S.I. Svechnikov, *et al.*, Hot electron quasioptical NbN superconducting mixer, IEEE Trans. Appl. Supercond. 5 (1995) 2232-2235. https://doi.org/10.1109/77.403029

[217] A. Skalare, W.R. McGrath, B. Bumble, H.G. LeDuc, *et al.*, Large bandwidth and low noise in a diffusion-cooled hot-electron bolometer mixer, Appl. Phys. Lett. 68 (1996) 1558-1560. https://doi.org/10.1063/1.115698

[218] K.S. Il'in, M. Lindgren, M. Currie, A.D. Semenov, *et al.*, Picosecond hot-electron energy relaxation in NbN superconducting photodetectors, Appl. Phys. Lett. 76 (2000) 2752-2754. https://doi.org/10.1063/1.126480

[219] J. Chen, S. Jiang, L. Kang, P.H. Wu, Terahertz direct detectors based on superconducting hot electron, Proc. SPIE. 10209 (2017) 1020909. https://doi.org/10.1117/12.2264878

[220] S. Cherednichenko, S. Bevilacqua, E. Novoselov, THz Hot-Electron Bolometer Mixers, 39th International Conference on Infrared, Millimeter, and Terahertz Waves, 14-19 Sept., 2014, Tucson, AZ, USA (DOI: 10.1109/IRMMW-THz.2014.6956174). https://doi.org/10.1109/IRMMW-THz.2014.6956174

[221] M. Nahum, J.M. Martinis, Novel hot-electron microbolometer, Physica B. 194 (1994) 109-110. https://doi.org/10.1016/0921-4526(94)90384-0

[222] J.L. Kloosterman, D.J. Hayton, Y. Ren, T.Y. Kao, *et al.*, Hot electron bolometer heterodyne receiver with a 4.7-THz quantum cascade laser as a local oscillator, Appl. Phys. Lett. 102 (2013) 011123. https://doi.org/10.1063/1.4774085

[223] S.C. Cherednichenko, V. Drakinsky, Terahertz mixing in MgB2 microbolometers, Appl. Phys. Lett. 90 (2007) 023507. https://doi.org/10.1063/1.2430928

[224] D. Cunnane, J. Kawamura, B.S. Karasik, M.A. Wolak, X.X. Xi, Development of hot-electron THz bolometric mixers using MgB2 thin films, Proc. SPIE. 9153 (2014) 91531Q. https://doi.org/10.1117/12.2054607

[225] S. Cherednichenko, M. Kroug, H. Merkel, E. Kollberg, *et al.*, Local oscillator power requirement and saturation effects in NbN HEB mixers, in: Proc. 12th int. Symp. Space THz Technol. (ISSTT), San Diego, CA, USA, 2001, pp. 273-285.

[226] J.J.A. Baselmans, M. Hajenius, J.R. Gao, A. Baryshev, *et al.*, NbN hot electron bolometer mixers: sensitivity, LO power, direct detection and stability, IEEE Trans. Appl. Supercond. 15 (2005) 484-489. https://doi.org/10.1109/TASC.2005.849884

[227] E. Novoselov, S. Bevilacqua, S. Cherednichenko, H. Shibata, Y. Tokura, Effect of the critical and operational temperatures on the sensitivity of MgB2 HEB mixers, IEEE Trans. Terahertz Sci. Technol. 6 (2016) 238-244. https://doi.org/10.1109/TTHZ.2016.2520659

[228] S. Krause, D. Meledin, V. Desmaris, A. Pavolotsky, *et al.*, Noise and IF gain bandwidth of a balanced waveguide NbN/GaN hot electron bolometer mixer operating at 1.3 THz, IEEE Trans. Terahertz Sci. Techn. 8 (2018) 365-371. https://doi.org/10.1109/TTHZ.2018.2824027

[229] X.X. Xi, A.V. Pogrebnyakov, S.Y. Xu, K. Chen, *et al.* MgB2 thin films by hybrid physical-chemical vapour deposition, Physica C. 456 (2007) 22-37. https://doi.org/10.1016/j.physc.2007.01.029

[230] G.N. Gol'tsman, Yu.B. Vachtomin, S.V. Antipov, M.I. Finkel, *et al.*, NbN phonon-cooled hot-electron bolometer mixer for terahertz heterodyne receivers, Proc. SPIE. 5727 (2005) 95-106 in: Terahertz and Gigahertz Electronics and Photonics, Ed. R. Jennifer, K.J. Linden. https://doi.org/10.1117/12.590490

[231] W.R. McGrath, Novel hot-electron bolometer mixers for submillimeter applications: An overview of recent developments, Proc. URSI Int. Symp. Signals, Systems, and Electronics, 1995, pp. 147-152.

[232] A. Semenov, G. Gol'tsman, R. Sobolewski, Hot-electron effect in superconductors and its applications for radiation sensors, Supercond. Sci. Technol. 15 (2002) R1-R16. https://doi.org/10.1088/0953-2048/15/4/201

[233] A. Shurakov, Y. Lobanov, G. Goltsman, Superconducting hot-electron bolometer: from the discovery of hot-electron phenomena to practical applications, Supercond. Sci. Technol. 29 (2016) 023001. https://doi.org/10.1088/0953-2048/29/2/023001

[234] E.M. Gershenzon, M.E. Gershenzon, G.N. Gol'tsman, A.M. Lyul'kin, *et al.*, Electron-phonon interaction in ultrathin Nb films, Sov. Phys. JETP. 70 (1990) 505-511.

[235] E.M. Gershenson, M.E. Gershenson, G.N. Goltsman, B.S. Karasik, *et al.*, Ultra-fast superconducting electron bolometer, J. Tech. Phys. Lett. 15 (1989) 118-119.

[236] A. Semenov, G.N. Gol'tsman, R. Sobolewski, Hot−electron effect in semiconductors and its applications for radiation sensors, LLE Review. 87 (2001) 134-143.

[237] Y.E. Tong, R. Blundell, D.C. Papa, M. Smith, *et al.*, An all solid-state superconducting heterodyne receiver at terahertz frequencies. IEEE Microwave and Guided Wave Lett. 9 (1999) 366-368. https://doi.org/10.1109/75.790476

[238] A.D. Semenov, H.-W. Hübers, H. Richter, M. Birk, *et al.*, Superconducting hot-electron bolometer mixer for terahertz heterodyne receivers, IEEE Trans. Appl. Supercond. 13 (2003) 168-172. https://doi.org/10.1109/TASC.2003.813672

[239] I. Milostnaya, A. Korneev, O. Minaeva, I. Rubtsova, *et al.*, Superconducting nanostructured detectors capable of single photon counting of mid-infrared optical

radiation, Proc. SPIE. 5957 (2005) 60-68. https://doi.org/10.1117/12.623767

[240] G. N. Gol'tsman, O. Okunev, G. Chulkova, A. Lipatov, *et al.*, Picosecond superconducting single-photon optical detector, Appl. Phys. Lett. 79 (2001) 705-707. https://doi.org/10.1063/1.1388868

[241] I.E. Zadeh, J.W.N. Los, R.B.M. Gourgues, V. Steinmetz, *et al.*, Single-photon detectors combining high efficiency, high detection rates, and ultra-high timing resolution, APL Photonics, 2 (2017) 111301. https://doi.org/10.1063/1.5000001

[242] A.M. Baryshev, R. Hesper, F.P. Mena, T.M. Klapwik, *et al.*, The ALMA band 9 receiver, Astron. Astrophys. 577 (2015) A129.

[243] A. F. Loenen, S. Lord, P. Morris, and T. G. Phillips, Excitation of the molecular gas in the nuclear region of M82, Astron. Astrophys. 521 (2010) L2 https://doi.org/10.1051/0004-6361/201015114

[244] E. Novoselov, S. Cherednichenko, Broadband MgB2 hot-electron bolometer THz mixers operating up to 20 K, IEEE Trans. Appl. Supercond. 27 (2017) 2300504. https://doi.org/10.1109/TASC.2017.2654861

[245] E. Novoselov, S. Cherednichenko, Low noise terahertz MgB2 hot-electron bolometer mixers with an 11 GHz bandwidth, Appl. Phys. Lett. 110 (2017) 032601. https://doi.org/10.1063/1.4974312

[246] E. Novoselov, MgB2 hot-electron bolometer mixers for sub-mm wave astronomy, Chalmers Reproservice, Goteborg, Sweden, 2017, Thesis.

[247] C.M. Posada, P.A.R. Ade, Z. Ahmed, K. Arnold, Fabrication of large dual-polarized multichroic TES bolometer arrays for CMB measurements with the SPT-3G camera, Supercond. Sci. Technol. 28 (2015) 094002.

[248] E.A. Williams, S. Withington, C.N. Thomas, D.J. Goldie, D. Osman, Superconducting transition edge sensors with phononic thermal isolation, J. Appl. Phys. 124 (2018) 144501. https://doi.org/10.1063/1.5041348

[249] L.S. Kuzmin, Cold-electron bolometer, in: A.G.U. Perera (Ed.), Bolometers, InTech, Rijeka, Croatia, 2012, pp. 77-106. (Open Access).

[250] D.J. Goldie, D.M. Glowacka, S. Withington, J. Chen, *et al.*, Performance of horn-coupled transition edge sensors for L- and S-band optical detection on the SAFARI instrument. Proc. SPIE. 9914 (2016) 99140A. https://doi.org/10.1117/12.2232740

[251] J. Hubmayr, J.E. Austermann, J.A. Beall, D.T. Becker, *et al.* Low temperature detectors for CMB imaging arrays, J. Low. Temp. Phys. 193 (2018) 633-647. https://doi.org/10.1007/s10909-018-2029-6

[252] B.S. Karasik, D. Olaya, J. Wei, S. Pereverzev, *et al.*, Record-low NEP in hot-

electron titanium nanobolometers, IEEE Trans. Appl. Supercond. 17 (2007) 293-297. https://doi.org/10.1109/TASC.2007.897167

[253] W.D.J. Goldie, J.R. Gao, D.M. Glowacka, D.K. Griffin, *et al.*, Ultra-low-noise transition edge sensors for the SAFARI L-band on SPICA, Proc. SPIE. 8452 (2012) 84520A. https://doi.org/10.1117/12.925861

[254] M. Kenyon, P.K. Day, C.M. Bradford, J.J. Bock, H.G. Leduc, Progress on background-limited membrane-isolated TES bolometers for far-IR/submillimeter spectroscopy, Proc. SPIE. 6275 (2006) 627508. https://doi.org/10.1117/12.672036

[255] J. Dunkley, A. Amblard, C. Baccigalupi, M. Betoule, D. Chuss, *et al.*, CMBPol mission concept study: Prospects for polarized foreground, American Institute of Physics conference series. 1141 (2009) 222-264. https://doi.org/10.1063/1.3160888

[256] R. O'Brient, P. Ade, K. Arnold, J. Edwards, *et al.*, A dual-polarized broadband planar antenna and channelizing filter bank for millimeter wavelengths, Appl. Phys. Lett. 102 (2013) 063506. https://doi.org/10.1063/1.4791692

[257] A program of technology development and of sub-orbital observations of the cosmic microwave background polarization leading to and including a satellite mission, A Report for the Astro-2010 Decadal Committee on Astrophysics, April (2009).

[258] B. Cabrera, R. Clarke, P. Colling, A. Miller, *et al.*, Detection of single infrared, optical, and ultraviolet photons using superconducting transition edge sensors, Appl. Phys. Lett. 73 (1998) 735-737. https://doi.org/10.1063/1.121984

[259] G.C. Hilton, J.M. Martinis, K.D. Irwin, N.F. Bergren, *et al.*, Microfabricated transition-edge X-ray detectors, IEEE Trans. Appl. Supercond. 11 (2001) 739-742. https://doi.org/10.1109/77.919451

[260] B. Westbrook, A. Cukierman, A. Lee, A. Suzuki, *et al.*, Development of the next generation of multi-chroic antenna-coupled transition edge sensor detectors for CMB polarimetry, J. Low Temp. Phys. 184 (2016) 74-81. https://doi.org/10.1007/s10909-016-1508-x

[261] A. Suzuki, K. Arnold, J. Edwards, G. Engargiola, *et al.*, Multi-chroic dual-polarization bolometric detectors for studies of the cosmic microwave background. J. Low. Temp. Phys. 176 (2014) 650-656. https://doi.org/10.1007/s10909-013-1049-5

[262] J. Gildemeister, A. Lee, P. Richards, Monolithic arrays of absorber-coupled voltage-biased superconducting bolometers, Appl. Phys. Lett. 77 (2000) 4040-4042. https://doi.org/10.1063/1.1326844

[263] T. May, V. Zakosarenko, E. Kreysa, W. Esch, *et al.*, Design, realization, and characteristics of a transition edge bolometer for sub-millimeter wave astronomy, Rev.

Sci. Instrum. 83 (2012) 114502. https://doi.org/10.1063/1.4764581

[264] D. M. Glowacka, M. Crane, D. J. Goldie, S. Withington, A fabrication route for arrays of ultra-low-noise MoAu transition edge sensors on thin silicon nitride for space applications, J. Low Temp. Phys. 167 (2012) 516. https://doi.org/10.1007/s10909-012-0580-0

[265] M.D. Audley, G. de Langeb, J.-R. Gao, P, Khosropanah, *et al.*, Optical performance of prototype horn-coupled TES bolometer arrays for SAFARI, Rev. Sci. Instrum. 87 (2016) 043103. https://doi.org/10.1117/12.2231088

[266] C.M. Posada, P.A.R. Ade, A.J. Anderson, J. Avva, *et al.*, Large arrays of dual-polarized multichroic TES detectors for CMB measurements with the SPT-3G receiver, Proc. SPIE. 9914 (2016) 9914E.

[267] J.A. Chervenak, K.D. Irwin, E.N. Grossman, J.M. Martinis, *et al.*, Superconducting multiplexer for arrays of transition edge sensors, Appl. Phys. Lett. 74 (1999) 4043-4045. https://doi.org/10.1063/1.123255

[268] P.J. Yoon, J.Clarke, J.M. Gildemeister, A.T. Lee, *et al.*, Single superconducting quantum interference device multiplexer for arrays of low-temperature sensors, Appl. Phys. Lett. 78 (2001) 371-373. https://doi.org/10.1063/1.1338963

[269] K.D. Irwin, SQUID multiplexers for transition-edge sensors, Physica C. 368 (2002) 203-210. https://doi.org/10.1016/S0921-4534(01)01167-4

[270] S.M. Stanchfield, P.Ade, J. Aguirre, J. Brevik, *et al.*, Development of a microwave SQUID-multiplexed TES array for MUSTANG-2, J. Low Temp. Phys. 184 (2016) 460. https://doi.org/10.1007/s10909-016-1570-4

[271] D.J. Benford, S.A. Rinehart, D.T. Leisawitz, T.T. Hyde, Cryogenic far-infrared detectors for the Space Infrared Interferometric Telescope (SPIRIT), Proc. SPIE. 6687 (2007) 66870. https://doi.org/10.1117/12.734751

[272] C.M. Bradford, M. Kenyon, W. Holmes, J. Bock, T. Koch, Sensitive far-IR survey spectroscopy: BLISS for SPICA, Proc. SPIE. 7020 (2008) 702010. https://doi.org/10.1117/12.788271

[273] B.D. Jackson, P.A.J. de Korte, J. Van der Kuur, P.D. Mauskopf, *et al.*, The SPICA-SAFARI detector system: TES detector arrays with frequency-division multiplexed SQUID readout, IEEE Trans. THz Sci. Techn. 2 (2012) 12-21. https://doi.org/10.1109/TTHZ.2011.2177705

[274] T. Matsumura, Y. Akiba, J. Borrill, Y. Chinone, *et al.*, Mission design of LiteBIRD, J. Low Temp. Phys. 176 (2014) 733-740. https://doi.org/10.1007/s10909-013-0996-1

[275] R. Blundell, K.H. Gundlach, A quasioptical SIN mixer for 230 GHz frequency

range, Journal of Infrared, Millimeter, and Terahertz Waves. 8 (1987) 1573-1579. https://doi.org/10.1007/BF01012443

[276] M. Nahum, P.L. Richards, C.A. Mears, Design analysis of a novel hot-electron microbolometer, IEEE Trans. Appl. Supercond. 3 (1993) 2124-2127. https://doi.org/10.1109/77.233921

[277] D. Golubev, L. Kuzmin, Nonequilibrium theory of a hot-electron bolometer with normal metal-insulator-superconductor tunnel junction, J. Appl. Phys. 89 (2001) 6464-6472. https://doi.org/10.1063/1.1351002

[278] J.W. Bakker, H. Van Kempen, P. Wyder, Adiabatic demagnetization of ruby and the use of a superconducting tunnel junction thermometer, Phys. Letts. A. 31 (1970) 290-291. https://doi.org/10.1016/0375-9601(70)90861-3

[279] M. Nahum, and J.M. Martinis, Ultrasensitive-hot-electron microbolometer, Appl. Phys. Lett. 63 (1993) 3075-3077. https://doi.org/10.1063/1.110237

[280] M.R. Nevala, S.Chaudhuri, J. Halkosaari, J.T. Karvonen, I.J. Maasilta, Sub-micron normal-metal/insulator/superconductor tunnel junction thermometer and cooler using Nb, Appl. Phys. Lett. 101 (2012) 112601. https://doi.org/10.1063/1.4751355

[281] S. Chaudhuri, M.R. Nevala, I.J. Maasilta, Niobium nitride-based normal metal-insulator-superconductor tunnel junction microthermometer, Appl. Phys. Lett. 102 (2013) 132601. https://doi.org/10.1063/1.4800440

[282] S. Chaudhuri, I.J. Maasilta, Superconducting tantalum nitride-based normal metal-insulator-superconductor tunnel junctions, Appl. Phys. Lett. 104 (2014) 122601. https://doi.org/10.1063/1.4869563

[283] A. Torgovkin, S. Chaudhuri, J. Malm, T. Sajavaara, I. Maasilta, Normal metal-insulator-superconductor tunnel junction with atomic layer deposited titanium nitride as suprconductor, IEEE Trans. Appl. Supercond. 25 (2015) 1101604. https://doi.org/10.1109/TASC.2014.2383914

[284] D.R. Schmidt, K.W. Lehnert, A.M. Clark, W.D. Duncan, *et al.*, A superconductor-insulator-normal metal bolometer with microwave readout suitable for large-format arrays, Appl. Phys. Lett. 86 (2005) 053505. https://doi.org/10.1063/1.1855411

[285] H. Kraus, Superconductive bolometers and calorimeters, Supercond. Sci. Technol. 9 (1996) 827-842. https://doi.org/10.1088/0953-2048/9/10/001

[286] M.A. Tarasov, V.S. Edelman, S. Mahashabde, L. Kuzmin, Powerload and temperature dependence of cold-electrical bolometer optical response at 350 GHz, IEEE Trans. Appl. Supercond. 24 (2014) 2400105. https://doi.org/10.1109/TASC.2014.2316196

[287] S. Lemzyakov, M. Tarasov, S. Mahashabde, R. Yusupov, *et al.*, Experimental study

of a SINIS detector response time at 350 GHz signal frequency, IOP Conf. Series: J. Physics: Conf. Series. 969 (2018) 012081. https://doi.org/10.1088/1742-6596/969/1/012081

[288] S. Masuda, K.Y. Tan, M. Partanen, R.E.Lake, *et al.*, Observation of microwave absorption and emission from incoherent electron tunneling through a normal-metal-insulator-superconductor junction, Sci. Reports. 8 (2018) 3966. https://doi.org/10.1038/s41598-018-21772-5

[289] J.P. Pekola, J.J. Vartiainen, M. Möttönen, M. Meschke, *et al.*, Hybrid single-electron transistor as a source of quantized electric current, Nature Phys. 4 (2008) 120-124. https://doi.org/10.1038/nphys808

[290] A. Kemppinen, V. F. Maisi, O.-P. Saira, S. Kafanov, *et al.*, Radio-frequency transport of single electrons in superconductor-normal-metal tunnel junctions and the quantum metrological triangle, 30th URSI General Assembly and Scientific Symposium, Istanbul, Turkey, 20 October, 2011. https://doi.org/10.1109/URSIGASS.2011.6050291

[291] L. Kuzmin, Capacitively coupled hot electron microbolometer as perspective IR and sub-mm wave sensor, Proc. 9th Int. Symposium on Space Terahertz Techn., Pasadena, 1998, pp. 99-103.

[292] L. Kuzmin, and D. Golubev, On the concept of an optimal hot-electron bolometer with NIS tunnel junctions, Physica C. 378 (2002) 372-376. https://doi.org/10.1016/S0921-4534(02)00704-9

[293] M. Tarasov, L. Kuzmin, Concept of a mixer based on a cold-electron bolometer, JETP Lett. 81 (2005) 538-541. https://doi.org/10.1134/1.1996765

[294] M. Salatino, P. de Bernardis, S. Mahashabde, L. Kuzmin, S. Masi, Cold electron bolometers for future mm and sub-mm sky surveys, Proc. SPIE. 9153 (2014) 91530A. https://doi.org/10.1117/12.2056744

[295] L. Kuzmin, 2D array of cold-electron nanobolometers with double polarization cross-dipole antennae, Nanoscale Res. Lett. 7 (2012) 224. https://doi.org/10.1186/1556-276X-7-224

[296] L.S. Kuzmin, A resonant cold-electron bolometer with a kinetic inductance nanofilter, IEEE Trans. Terahertz Sci. Technol. 4 (2014) 6778093. https://doi.org/10.1109/TTHZ.2014.2311321

[297] T.L.R. Brien, P.A.R. Ade, P.S. Barry, C. Dunscombe, *et al.*, A strained silicon cold electron bolometer using Schottky contacts, Appl. Phys. Lett. 105 (2014) 043509. https://doi.org/10.1063/1.4892069

[298] E.R. Brown, A.C. Young, J. Zimmerman, H. Kazerni, A.C. Gossard, Advances in

Schottky rectifier performance, IEEE Microw. Mag. 8 (2007) 54-59.
https://doi.org/10.1109/MMW.2007.365059

[299] E. Donchev, J.S. Pang, P.M. Gammon, A. Centeno, *et al.*, The rectenna device: from theory to practice (a review), MRS Energy and Sustainability - A Review J. 1 (2014) 1-34. https://doi.org/10.1557/mre.2014.10

[300] M. Sakhno, A. Golenkov, F. Sizov, Uncooled detector challenges: Millimeter wave and terahertz long channel field effect transistor and Schottky barrier diode detectors, J. Appl. Phys. 114 (2013) 164503. https://doi.org/10.1063/1.4826364

[301] A. Lisauskas, M. Bauer, S. Boppel, M. Mundt, *et al.*, Exploration of Terahertz Imaging with Silicon MOSFETs, J. Infrared, Millimeter, and Terahertz Waves. 35 (2014) 63-80. https://doi.org/10.1007/s10762-013-0047-7

[302] A.J.M. Kreisler, Submillimeter wave applications of submicron Schottky diodes, Proc. SPIE. 666 (1986) 51-63. https://doi.org/10.1117/12.938820

[303] B. Gershgorin, V.Y. Kachorovskii, Y. Lvov, M.S. Shur, Field effect transistor as heterodyne terahertz detector, Electron. Lett. 44 (2008) 1036-1037. https://doi.org/10.1049/el:20080737

[304] A. Lisauskas, S. Boppel, M. Mundt, V. Krozer, H.G. Roskos, Subharmonic mixing with field effect transistors: Theory and experiment at 639 GHz high above ft, IEEE Sensors J. 13 (2013) 124-132. https://doi.org/10.1109/JSEN.2012.2223668

[305] S. Boppel, A. Lisauskas, A. Max, V. Krozer, H.G. Roskos, CMOS detector arrays in a virtual 10-kilopixel camera for coherent terahertz real-time imaging, Opt. Lett. 37 (2012) 536-538. https://doi.org/10.1364/OL.37.000536

[306] H.W. Hou, Z. Liu, J.H. Ten, T. Palacios, S.J. Chua, High temperature terahertz detectors realized by a GaN high electron mobility transistor, Scientific Reports. 7 (2017) 46664. https://doi.org/10.1038/srep46664

[307] S.-P. Han, H. Ko, J.-W. Park, N. Kim, *et al.*, InGaAs Schottky barrier diode array detector for a real-time compact terahertz line scanner, Opt. Express. 21 (2013) 25874-25882. https://doi.org/10.1364/OE.21.025874

[308] L. Liu, J.L. Hesler, H. Xu, A.W. Lichtenberger, R.M. Weikle, A broadband quasi-optical terahertz detector utilizing a zero bias Schottky diode, IEEE Microw. Wireless Compon. Lett. 20 (2010) 504-506. https://doi.org/10.1109/LMWC.2010.2055553

[309] D. Schoenherr, C. Bleasdale, T. Goebel, C. Sydlo, *et al.*, Extreamly broadband characterization of a Schottky diode based THz detectors, in: 35th International Conference on Infrared, Millimeter and Terahertz Waves, IEEE, 2010, DOI: 10.1109/ICIMW.2010.5613008. https://doi.org/10.1109/ICIMW.2010.5613008

[310] A. Semenov, O. Cojocari, H.-W. Hübers, F. Song, *et al.*, Application of zero-bias

quasi-optical Schottky-diode detectors for monitoring short-pulse and weak terahertz radiation, IEEE Electron Device Lett. 31 (2010) 674-676. https://doi.org/10.1109/LED.2010.2048192

[311] A. Pleteršek, J. Trontelj, A self-mixing NMOS channel-detector optimized for mm-wave and THz signals, J. Infrared, Millimeter, and Terahertz Waves. 33 (2012) 615-626. https://doi.org/10.1007/s10762-012-9901-2

[312] E. Ojefors, U.R. Pfeiffer, A. Lisauskas, H.G. Roskos, A 0.65 THz focal-plane array in a quarter-micron CMOS process technology, IEEE J. Solid-St. Circ. 44 (2009) 1968-1976. https://doi.org/10.1109/JSSC.2009.2021911

[313] F. Schuster, D. Coquillat, H. Videlier, M. Sakowicz, *et al.*, Broadband terahertz imaging with highly sensitive silicon CMOS detectors, Opt. Express. 19 (2011) 7827-7832. https://doi.org/10.1364/OE.19.007827

[314] S. Boppel, A. Lisauskas, M. Mundt, D. Seliuta, *et al.*, CMOS integrated antenna-coupled field-effect transistors for the detection of radiation from 0.2 to 4.3 THz IEEE Trans. Microw. Theory Techn. 60 (2012) 3834-3843. https://doi.org/10.1109/TMTT.2012.2221732

[315] X.-Y. Liu, Y. Zhang, D.-J. Xia, T.-H. Ren, *et al.*, A high-sensitivity terahertz detector based on a low-barrier Schottky diode, Chin. Phys. Lett. 34 (2017) 070701. https://doi.org/10.1088/0256-307X/34/7/070701

[316] M.A. Andersson, J. Stake, An accurate empirical model based on Volterra series for FET power detectors, IEEE Trans. Microw. Theory Techn. 64 (2016) 1431-1441. https://doi.org/10.1109/TMTT.2016.2532326

[317] M. Bauer, S. Boppel, J. Zhang, A. Ramer., *et al.*, Optimization of the design of terahertz detectors based on Si CMOS and AlGaN/GaN field-effect transistors, Int. J. High Speed Electr. Systems. 25 (2016) 1640013. https://doi.org/10.1142/S0129156416400139

[318] E. Javadi, J.A. Delgado-Notario, N. Masoumi, M. Shahabadi, *et al.*, Continuous wave terahertz sensing using GaN HEMTs, Phys. Stat. Sol. A. (2018) 1700607. https://doi.org/10.1002/pssa.201700607

[319] X. Deng, M. Simanullang, Y. Kawano, Ge-core/α-Si shell nanowire-based field-effect transistor for sensitive terahertz detection, Photonics. 5 (2018) 5020013. https://doi.org/10.3390/photonics5020013

[320] M.I.W. Khan, S. Kim, D.-W. Park, H.-J. Kim, *et al.*, Nonlinear analysis of nonresonant THz response of MOSFET and implementation of a high-responsivity cross-coupled THz detector, IEEE Trans. Terahertz Sci. Technol. 8 (2018) 108-120. https://doi.org/10.1109/TTHZ.2017.2778499

[321] R. Al Hadi, H. Sherry, J. Grzyb, Y. Zhao, *et al.*, A 1 k-pixel video camera for 0.7-1.1 terahertz imaging applications in 65-nm CMOS, IEEE J. Solid-St. Circuits. 47 (2012) 2999-3012. https://doi.org/10.1109/JSSC.2012.2217851

[322] Y. Kurita, G. Ducournau, D. Coquillat, A. Satou, *et al.*, Ultrahigh sensitive sub-terahertz detection by InP-based asymmetric dual-grating-gate high-electron-mobility transistors and their broadband characteristics, Appl. Phys. Lett. 104 (2014) 251114. https://doi.org/10.1063/1.4885499

[323] J.A. Delgado-Notario, J.E. Velazquez-Perez, Y.M. Meziani, K. Fobelets, Sub-THz imaging using non-resonant HEMT detectors, Sensors. 18 (2018) 543. https://doi.org/10.3390/s18020543

[324] A. Westlund, P. Sangar, G, Ducournau, P.-A. Nilsson, *et al.*, Terahertz detection in zero-bias InAs self-switching diodes at room temperature, Appl. Phys. Lett. 103 (2013) 133504. https://doi.org/10.1063/1.4821949

[325] A.A. Generalov, M.A. Andersson, X. Yang, A. Vorobiev, J. Stake, A 400-GHz Graphene FET Detector. IEEE Transactions on Terahertz Science and Technology. 7 (2017) 614-616. https://doi.org/10.1109/TTHZ.2017.2722360

[326] A. Zak, M. Andersson, M. Bauer, J. Matukas, *et al.*, Antenna-integrated 0.6 THz FET direct detectors based on CVD graphene, Nano Lett. 14 (2014) 5834-5838. https://doi.org/10.1021/nl5027309

[327] S. Castilla, B. Terres, M. Autore, L. Viti, *et al.*, Fast and sensitive terahertz detection using an antenna-integrated graphene p-n-junction, Nano Lett. (2019) DOI: 10.1021/acs.nanolett.8b04171. https://doi.org/10.1021/acs.nanolett.8b04171

[328] G. Auton, D.B. But, J. Zhang, E. Hill, *et al.*, Terahertz detection and imaging using graphene ballistic rectifiers. Nano Letters. 17 (2017) 7015-7020. https://doi.org/10.1021/acs.nanolett.7b03625

[329] G. Auton, J. Zhang, R.K. Kumar, H. Wang, *et al.*, Graphene ballistic nano-rectifier with very high responsivity, Nature Commun. 7 (2016) 11670. https://doi.org/10.1038/ncomms11670

[330] D. Spirito, D. Coquillat, S.L. De Bonis, A. Lombardo, *et al.*, High performance bilayer-graphene terahertz detectors, Appl. Phys. Lett. 104 (2014) 061111. https://doi.org/10.1063/1.4864082

[331] S. Preu, M. Mittendorf, S. Winnerl, O. Cojocari, A. Penirschke, THz autocorrelators for ps pulse characterization based on Schottky diodes and rectifying field-effect transistors, IEEE Trans. Terahertz Sci. Technol. 5 (2015) 922-929. https://doi.org/10.1109/TTHZ.2015.2482943

[332] A. Lisauskas, U. Pfeiffer, E. Oejefors, P.H. Bolivar, *et al.*, Rational design of high-

responsivity detectors of terahertz radiation based on distributed self-mixing in silicon field-effect transistors. J. Appl. Phys. 105 (2009) 114511. https://doi.org/10.1063/1.3140611

[333] I. Kukushkin, V. Muravev, G. Tsydynzhapov, U.S. Patent WO2013096805A1 (2013).

[334] J. Zdanevičius, M. Bauer, S. Boppel, V. Palenskis, *et al.*, Camera for high-speed THz imaging, Journal of Infrared, Millimeter, and Terahertz Waves. 36 (2015) 986-997. https://doi.org/10.1007/s10762-015-0169-1

[335] D.Y. Kim, S. Park, R. Han, K.K. O, Design and demonstration of 820-GHz array using diode-connected NMOS transistors in 130-nm CMOS for active imaging, IEEE Trans. THz Sci. Technol. 6 (2016) 306-317. https://doi.org/10.1109/TTHZ.2015.2513061

[336] S. Preu, M. Mittendorf, S. Winnerl, H. Lu, *et al.*, Ultra-fast transistor-based detectors for precise timing of near infrared and THz signals, Opt. Express. 21 (2013) 17941-17950. https://doi.org/10.1364/OE.21.017941

[337] A. Lisauskas, K. Ikamas, S. Massabeau, M. Bauer, *et al.*, Field-effect transistors as electrically controllable nonlinear rectifiers for the characterization of terahertz pulses, APL Photonics. 3 (2018) 051705. https://doi.org/10.1063/1.5011392

[338] A. Golenkov, F. Sizov, Performance limits of terahertz zero biased rectifying detectors for direct detection, Semicond. Phys., Quantum Electr. Optoelectr. 19 (2016) 129-138. https://doi.org/10.15407/spqeo19.02.129

[339] M. Tonouchi, Cutting edge THz technology, Nature Photonics. 1 (2007) 97-105. https://doi.org/10.1038/nphoton.2007.3

[340] J.L. Hesler, W.R. Hall, T.W. Crowe, R.M. Weikle, *et al.*, Fixed tuned submillimeter wavelength waveguide mixers using planar Schottky-barrier diodes, IEEE Trans. Microwave Theory and Tech. 45 (1997) 653-658. https://doi.org/10.1109/22.575581

[341] G. Chattopadhyay, Sensor technology at submillimeter wavelengths for space applications, Proc. 2nd Int. Conf. on Sensing Technol., Palmestron North, New Zealand, November, 26, 2007.

[342] F.D. Shepherd, Infrared detectors: State of the art, Proc. SPIE. 1735 (1992) 250-261. https://doi.org/10.1117/12.138629

[343] G. Myburg, F.D. Auret, W.E. Meyer, C.W. Louw, M.J. van Staden, Summary of Schottky barrier height data on epitaxially grown n- and p-GaAs, Thin Solid Films. 325 (1998) 181-186. https://doi.org/10.1016/S0040-6090(98)00428-3

[344] A.M. Cowley, H.O. Sorensen, Quantitative comparison of solid state microwave

detectors, IEEE Trans. Microw. Theory Techn. 14 (1966) 588-602.
https://doi.org/10.1109/TMTT.1966.1126337

[345] C. Yu, C. Wu, S. Kshattry, Y. Yun, *et al.*, Compact, high impedance and wide
bandwidth detectors for characterization of millimeter wave performance, IEEE J.
Solid-St. Circuits. 47 (2012) 2335-2349. https://doi.org/10.1109/JSSC.2012.2219155

[346] J.J. Lynch, H.P. Moyer, J.H. Schaffner, Y. Royter, *et al.*, Passive millimeter-wave
imaging module with preamplified zero-bias detection, IEEE Trans. Microw. Theory
Techn. 56 (2008) 1592-1600. https://doi.org/10.1109/TMTT.2008.924361

[347] V.I. Shashkin, A.V. Murel, V.M. Daniltsev, O.I. Khrykin, Control of charge
transport mode in the Schottky barrier by δ-doping: Calculation and experiment for
Al/GaAs, Semiconductors. 36 (2002) 505-510. https://doi.org/10.1134/1.1478540

[348] C. Sydlo, O. Cojocari, D. Schönherr, T. Goebel, *et al.*, Fast THz detectors based on
InGaAs Schottky diodes, Frequenz. 62 (2008) 107-110.
https://doi.org/10.1515/FREQ.2008.62.5-6.107

[349] I. Oprea, A. Walber, O. Cojocari, H. Gibson, *et al.*, 183 GHz mixer on InGaAs
Schottky diodes, 21st Intern. Symp. Space Teraherz Techn., Oxford, 23-25 March,
2010, pp. 159-160.

[350] M. Hrobak, Critical mm-Wave Components for Synthetic Automatic Test Systems,
Springer Vieweg, Wiesbaden, 2015. https://doi.org/10.1007/978-3-658-09763-9

[351] H. Liu, J. Yu, P. Huggard, B. Alderman, A multichannel THz detector using
integrated bow-tie antennas, Int. J. Antennas and Propagation. 2013 (2013) 417108.
https://doi.org/10.1155/2013/417108

[352] R. Han, Y. Zhang, Y. Kim, D.Y. Kim, *et al.* Active terahertz imaging using
Schottky diodes in CMOS: array and 860-GHz pixel, IEEE J. Solid-St. Circuits. 48
(2013) 2296-2308. https://doi.org/10.1109/JSSC.2013.2269856

[353] S.P. Han, H. Ko, N. Kim, W.H. Lee, *et al.*, Real-time continuous-wave terahertz
line scanner based on a compact 1×240 InGaAs Schottky barrier diode array detector,
Opt. Exp. 17 (2014) 28977-28983. https://doi.org/10.1364/OE.22.028977

[354] D.W. Park, E.S. Lee, J.-W. Park, H.-S. Kim, *et al.*, InGaAs Schottky barrier diode
array detectors integrated with broadband antenna, Proc. SPIE. 10103 (2017)
10103OF.

[355] R. Tauk, F. Teppe, S. Boubanga, D. Coquillat, *et al.*, Plasma wave detection of
terahertz radiation by silicon field effects transistors: Responsivity and noise
equivalent power, Appl. Phys. Lett. 89 (2006) 253511.
https://doi.org/10.1063/1.2410215

[356] A. Lisauskas, S. Boppel, J. Matukas, V. Palenskis, *et al.*, Terahertz responsivity

and low-frequency noise in biased silicon field-effect transistors, Appl. Phys. Lett. 102 (2013) 153505. https://doi.org/10.1063/1.4802208

[357] W. Chew, H.R. Fetterman, Millimeter-wave imaging using FET detectors integrated with printed circuit antennas, J. Infrared, Millimeter, and Terahertz Waves. 10 (1989) 565-578. https://doi.org/10.1007/BF01043255

[358] W. Chew, H.R. Fetterman, Printed circuit antennas with integrated FET detectors for millimeter-wave quasi optics, IEEE Trans. Microwave Theory Techn. MTT-37 (1989) 593-597. https://doi.org/10.1109/22.21632

[359] M.I. Dyakonov, M.S. Shur, Plasma wave electronics: Novel terahertz devices using two dimensional electron fluid, IEEE Trans. Electr. Devices. 43 (1996) 1640-1645. https://doi.org/10.1109/16.536809

[360] W. Knap, V. Kachorovskii, Y. Deng, S. Rumyantsev, *et al.*, Nonresonant detection of terahertz radiation in field effect transistors, J. Appl. Phys. 91 (2002) 9346-9353. https://doi.org/10.1063/1.1468257

[361] W. Knap, M. Dyakonov, D. Coquillat, F. Teppe, *et al.*, Field effect transistors for terahertz detection: Physics and first imaging applications, J. Infrared, Millimeter, and Terahertz Waves. 30 (2009) 1319-1337. https://doi.org/10.1007/s10762-009-9564-9

[362] J.-Q. Lü, M.S. Shur, J.L. Hesler, L. Sun, and R. Weikle, Terahertz detector utilizing two-dimensional electronic fluid, IEEE Electr. Device Lett. 19 (1998) 373-375. https://doi.org/10.1109/55.720190

[363] L. Vicarelli, M.S. Vitiello, D. Coquillat, A. Lombardo, *et al.*, Graphene field-effect transistors as room-temperature terahertz detectors, Nature Mater. 11 (2012) 865 871. https://doi.org/10.1038/nmat3417

[364] A.V. Muraviev, S.L. Rumyantsev, G. Liu, A.A. Balandin, *et al.*, Plasmonic and bolometric terahertz detection by graphene field-effect transistor, Appl. Phys. Lett. 103 (2013) 181114. https://doi.org/10.1063/1.4826139

[365] D.A. Bandurin, I. Gayduchenko, Y. Cao, M. Moskotin, *et al.*, Dual origin of room temperature sub-terahertz photoresponse in graphene field effect transistors, Appl. Phys. Lett. 112 (2018) 141101. https://doi.org/10.1063/1.5018151

[366] E. Oejefors, A. Lisauskas, D. Glaab, H.G. Roskos, U.R. Pfeiffer, Terahertz imaging detectors in CMOS technology, J. Infrared, Millimeter, and Terahertz Waves. 30 (2009) 1269-1280. https://doi.org/10.1007/s10762-009-9569-4

[367] T. Otsuji, Trends in the research of modern terahertz detectors: Plasmon detectors, IEEE Trans. Terahertz Sci., Technol. 5 (2015) 1110-1120.

[368] D.B. But, C. Drexler, M.V. Sakhno, N. Dyakonova, O. Drachenko, F.F. Sizov, A. Gutin, S.D. Ganichev, W. Knap, Nonlinear photoresponse of field effect transistors

terahertz detectors at high irradiation intensities, J. Appl. Phys. 115 (2014) 164514. https://doi.org/10.1063/1.4872031

[369] P. Zagrajek, S.N. Danilov, J. Marczewski, M. Zaborowski, *et al.*, Time resolution and dynamic range of field effect transistor based terahertz detectors, J. Infrared, Millimeter, and Terahertz Waves. 40 (2019) 703-719. https://doi.org/10.1007/s10762-019-00605-0

[370] S. Regensburger, A.K. Mukherjee, S. Schonhuber, M.A. Kainz, *et al.*, Broadband terahertz detection with zero-bias field-effect transistors between 100 GHz and 11.8 THz with a noise equivalent power of 250 pW/√Hz at 0.6 THz, IEEE Trans. Terahertz Sci. Technol. 8 (2018) 465-471. https://doi.org/10.1109/TTHZ.2018.2843535

[371] P. Hillger, J. Grzyb, R. Jain, U.R. Pfeiffer, Terahertz imaging and sensing applications with silicon-based technologies, IEEE Trans. Terahertz Sci. Technol. 9 (2019) 1-19. https://doi.org/10.1109/TTHZ.2018.2884852

[372] Z. Ahmad, A. Lisauskas, H.G. Roskos, K.K. O, Design and demonstration of antenna coupled Schottky diodes in a foundry complementary metal-oxide semiconductor technology for electronic detection of far-infrared radiation, J. Appl. Phys. 125 (2019) 194501. https://doi.org/10.1063/1.5083689

[373] W. Knap, Y. Deng, S. Rumynantsev, M.S. Shur, Resonant detection of subterahertz and terahertz radiation by plasma waves in submicron field-effect transistors, Appl. Phys. Lett. 81 (2002) 4637-4639. https://doi.org/10.1063/1.1525851

[374] V.I. Gavrilenko, E.V. Demidov, K.V. Marem'yanin, S.V. Morozov, *et al.*, Electron transport and detection of terahertz radiation in a GaN/AlGaN submicrometer field-effect transistor, Semiconductors. 41 (2007) 232-234. https://doi.org/10.1134/S1063782607020224

[375] F. Teppe, A. El Fatimy, S. Boubanga, D. Seliuta, *et al.*, Terahertz resonant detection by plasma waves in nanometric transistors, Acta Phys. Polon. 113 (2008) 815-820. https://doi.org/10.12693/APhysPolA.113.815

[376] A. Lisauskas, A. Rämer, M. Burakevič, S, Chevtchenko, *et al.*, THz radiation from biased AlGaN/GaN high-electron-mobility transistors, J. Appl. Phys. 125 (2019) 151614. https://doi.org/10.1063/1.5083838

[377] A. Golenkov, F. Sizov, I. Lysiuk, Frequency sensitive effect of rectifying FET terahertz detectors, Proc. 46th European Microw. Week Conf., 4-6 October (2016) London, UK, pp. 1107-1110. https://doi.org/10.1109/EuMC.2016.7824541

[378] F.H.L. Koppens, T. Mueller, Ph. Avouris, A.C. Ferrari, *et al.*, Photodetectors based on graphene, other two-dimensional materials and hybrid systems, Nature Nanotechn. 9 (2014) 780-793. https://doi.org/10.1038/nnano.2014.215

[379] X. Li, L. Tao, Z. Chen, H. Fang, *et al.*, Graphene and related two-dimensional materials: Structure-property relationships for electronics and optoelectronics, Appl. Phys. Rev. 4 (2017) 021306. https://doi.org/10.1063/1.4983646

[380] G.X. Ni, A.S. McLeod, Z. Sun, L. Wang, *et al.*, Fundamental limits to graphene plasmonics, Nature. 557 (2018) 530-533. https://doi.org/10.1038/s41586-018-0136-9

[381] R. Wang, L. Jiang, Graphene based functional devices: A short review, Front. Phys. 14 (2019) 13603. https://doi.org/10.1007/s11467-018-0859-y

[382] A.C. Ferrari, F. Bonaccorso, V. Fal'ko, K.S. Novoselov, *et al.*, Science and technology roadmap for graphene, related two-dimensional crystals, and hybrid systems, Nanoscale. 7 (2015) 4598-4810. https://doi.org/10.1039/C4NR01600A

[383] R.R. Nair, P. Blake, A.N. Grigorenko, K.S. Novoselov, *et al.*, Fine structure constant defines visual transparency of grapheme, Science. 320 (2008) 1308. https://doi.org/10.1126/science.1156965

[384] A.K. Geim, K.S. Novoselov, The rise of grapheme, Nat. Mater. 6 (2007) 183-191. https://doi.org/10.1038/nmat1849

[385] M. Sang, J. Shin, K. Kim, K.J. Yu, Electronic and thermal properties of grapheme and recent advances in graphene based electronics applications, Nanomaterials. 9 (2019) 374. https://doi.org/10.3390/nano9030374

[386] K.S. Novoselov, A.K. Geim, S.V. Morozov, D. Jiang, Y. Zhang, *et al.*, Electric field effect in atomically thin carbon films. Science. 306 (2004) 666-669. https://doi.org/10.1126/science.1102896

[387] M. Yankowitz, J. Xue., B.J. LeRoy, Graphene on hexagonal boron nitride, J. Phys.: Condens. Matt. 26 (2014) 303201. https://doi.org/10.1088/0953-8984/26/30/303201

[388] Information on www.sciencedaily.com/releases/2018/02/180206115108.htm

[389] G. Liang, N. Neophytou, M.S. Lundstrom, D.E. Nikonov, Contact effects in graphene nanoribbon transistors, Nano Lett. 8 (2008) 1819-1824. https://doi.org/10.1021/nl080255r

[390] P.A. Khomyakov, G. Giovannetti, P.C. Rusu, G. Brocks, *et al.*, First-principles study of the interaction and charge transfer between graphene and metals, Phys. Rev. B. 79 (2009) 195425. https://doi.org/10.1103/PhysRevB.79.195425

[391] A. Hsu, H. Wang, K.K. Kim, J. Kong, T. Palacious, Impact of graphene interface quality on contact resistance and RF device performance, IEEE Electron Device Lett. 32 (2011) 1008-1010. https://doi.org/10.1109/LED.2011.2155024

[392] C. Betz, S.H. Jhang, E. Pallecchi, R. Ferreira, *et al.*, Supercollision cooling in undoped graphene, Nature Phys. 9 (2013) 109-112. https://doi.org/10.1038/nphys2494

[393] J. Yan, M.-H. Kim, J.A. Elle, A.B. Sushkov, *et al.* Dual-gated bilayer graphene hot-electron bolometer, Nature Nanotechn. 7 (2012) 472-478. https://doi.org/10.1038/nnano.2012.88

[394] B.S., Karasik, C.B. McKitterick, D.E. Prober, Prospective performance of graphene HEB for ultrasensitive detection of sub-mm radiation, J. Low Temp. Phys. 176 (2014) 249-254. https://doi.org/10.1007/s10909-014-1087-7

[395] X. Du, D.E. Prober, H. Vora, C.B. Mckitterick, Graphene based bolometers, Graphene 2D Mater. 1 (2014) 1-22. https://doi.org/10.2478/gpe-2014-0001

[396] A.A. Balandin, Thermal properties of graphene, Nature Mater. 10 (2011) 569-582. https://doi.org/10.1038/nmat3064

[397] G. Skoblin, J. Sun, A. Yurgens, Graphene bolometer with thermoelectric readout and capacitive coupling to an antenna, Appl. Phys. Lett. 112 (2018) 063501. https://doi.org/10.1063/1.5009629

[398] L. Liao, Y.-C. Lin, M. Bao, R. Cheng, *et al.*, High-speed graphene transistors with a self-aligned nanowire gate, Nature. 467 (2010) 305-308. https://doi.org/10.1038/nature09405

[399] M. Mittendorff, S. Winnerl, J. Kamann, J. Eroms, *et al.*, Ultrafast graphene-based broadband THz detector, Appl. Phys. Lett. 103 (2013) 021113 https://doi.org/10.1063/1.4813621

[400] H.B. Heersche, P. Jarillo-Herrero, J.B. Oostinga, L.M.K. Vandersypen, A.F. Morpurgo, Bipolar supercurrent in graphene, Nature. 446 (2007) 56-59. https://doi.org/10.1038/nature05555

[401] Y. Cao, V. Fatemi, S. Fang, K. Watanabe, *et al.*, Unconventional superconductivity in magic-angle graphene superlattices, Nature. 556 (2018) 43-50. https://doi.org/10.1038/nature26160

[402] A.H. Castro Neto, F. Guinea, N.M.R. Peres, K.S. Novoselov, A.K. Geim, The electronic properties of graphene, Rev. Modern Phys. 81 (2009) 109-162. https://doi.org/10.1103/RevModPhys.81.109

[403] C.N.R. Rao, A.K. Sood, R. Voggu, K.S. Subrahmanyam, Some novel attributes of graphene, J. Phys. Chem. Lett. 1 (2010) 572-580. https://doi.org/10.1021/jz9004174

[404] V. Ryzhii, T. Otsuji, M. Ryzhii, N. Ryabova, *et al.*, Graphene terahertz uncooled bolometers, J. Phys. D: Appl. Phys. 46 (2013) 065102. https://doi.org/10.1088/0022-3727/46/6/065102

[405] S. Lin, Y. Lu, J. Xu, S. Feng, J. Li, High performance graphene/semiconductor Van Der Waals heterostructure optoelectronic devices, Nano Energy. 40 (2017) 122-148. https://doi.org/10.1016/j.nanoen.2017.07.036

[406] B.S. Karasik, C.B. McKitterick, D.E. Prober, Monolayer graphene bolometer as a sensitive far-IR detector, Proc. SPIE. 9153 (2014) 915309. https://doi.org/10.1117/12.2055317

[407] A. El Fatimy, A. Nath, B.D. Kong, A.K. Boyd, *et al.*, Ultra-broadband photodetectors based on epitaxial graphene quantum dots, Nanophot. 7 (2018) 20170100. https://doi.org/10.1515/nanoph-2017-0100

[408] G. Konstantatos, M. Badioli, L. Gaudreau, J. Osmond, *et al.* Hybrid graphene-quantum dot phototransistors with ultrahigh gain, Nature Nanotech. 7 (2012) 363-368. https://doi.org/10.1038/nnano.2012.60

[409] X. Cai, A.B. Sushkov, R.J. Suess, M.M. Jadidi, *et al.*, Sensitive room-temperature terahertz detection via the photothermoelectric effect in graphene, Nature Nanotechn. 9 (2014) 814–819. https://doi.org/10.1038/nnano.2014.182

[410] G.G. Paschos, T.C.H. Liew, Z. Hatzopoulos, A.V. Kavokin, *et al.*, An exciton-polariton bolometer for terahertz radiation detection, Scientific Rep. 8 (2018) 10092. https://doi.org/10.1038/s41598-018-28197-0

[411] S. Cakmakyapan, P.K. Lu, A. Navabi, M. Jarrahi, Gold-patched graphene nano-stripes for high-responsivity and ultrafast photodetection from the visible to infrared regime, Light: Sci. Appl. 7 (2018) 1-9. (Open Access). https://doi.org/10.1038/s41377-018-0020-2

[412] K.S. Novoselov, A.K. Geim, S.V. Morozov, D. Jiang, *et al.*, Two-dimensional gas of massless Dirac fermions in graphene, Nature. 438 (2005) 197–200. https://doi.org/10.1038/nature04233

[413] A. Tredicucci, M.S. Vitiello, Device concepts for graphene-based terahertz photonics, IEEE J. Select. Top. Quantum Electr. 20 (2014) 8500109. https://doi.org/10.1109/JSTQE.2013.2271692

[414] J. Tong, M. Muthee, S.-Y. Chen, S.K. Yngvesson, J. Yan, Antenna enhanced graphene THz emitter and detector, Nano Lett. 15 (2015) 5295-5301. https://doi.org/10.1021/acs.nanolett.5b01635

[415] H. Qin, J. Sun, S. Liang, X. Li, *et al.*, Room-temperature, low-impedance and high-sensitivity terahertz direct detector based on bilayer graphene field-effect transistor, Carbon. 116 (2017) 760-765. https://doi.org/10.1016/j.carbon.2017.02.037

[416] T. Deng, Z. Zhang, Y. Liu, Y. Wang, *et al.*, Three-dimensional graphene field-efect transistors as high-performance photodetectors, Nano Lett. 19 (2019) 1494-1503. https://doi.org/10.1021/acs.nanolett.8b04099

[417] D.A. Bandurin, D. Svintsov, I. Gayduchenko, Sh.G. Xu, *et al.*, Resonant terahertz detection using graphene plasmons, Nature Commun. 9 (2018) 5392. (Open Access).

https://doi.org/10.1038/s41467-018-07848-w

[418] W. Tang, A. Politano, Ch. Guo, W. Guo, *et al.*, Ultrasensitive room-temperature terahertz direct detection based on a bismuth selenide topological insulator, Adv. Funct. Mater. 28 (2018) 1801786. https://doi.org/10.1002/adfm.201801786

[419] N. Engheta, R.W. Ziolkowski, Metamaterials: Physics and Engineering Explorations. Wiley-IEEE Press, New York, 2006. https://doi.org/10.1002/0471784192

[420] S. Walia, C.M. Shah, P. Gutruf, H. Nili, *et al.*, Flexible metasurfaces and metamaterials: A review of materials and fabrication processes at micro- and nano-scales, Appl. Phys. Rev. 2 (2015) 011303. https://doi.org/10.1063/1.4913751

[421] H. Tao, E.A. Kadlec, A.C. Strikwerda, K. Fan, Microwave and Terahertz wave sensing with metamaterials, Opt. Express. 19 (2011) 21620-21626. https://doi.org/10.1364/OE.19.021620

[422] F. Alves, D. Grbovic, B. Kearney, N.V. Lavrik, G. Karunasiri, Bi-material terahertz sensors using metamaterial structures, Opt. Express. 20 (2013) 13256-13271. https://doi.org/10.1364/OE.21.013256

[423] S.A. Kuznetsov, A.G. Paulish, A.V. Gelfand, P.A. Lazorskiy, V.N. Fedorinin, Bolometric THz-to-IR converter for terahertz imaging, Appl. Phys. Lett. 99 (2011) 023501. https://doi.org/10.1063/1.3607474

[424] I. Carranza, J. Grant, J. Gough, D. Cumming, Terahertz metamaterial absorbers implemented in CMOS technology for imaging applications: Scaling to large format focal plane arrays, IEEE J. Selected Topics Quant. Electr. 23 (2017) 4700508. https://doi.org/10.1109/JSTQE.2016.2630307

[425] Z. Zhou, T. Zhou, S. Zhang, Z. Shi, *et al.*, Multicolor T-Ray imaging using multispectral metamaterials, Adv. Sci. 5 (2018) 1700982. https://doi.org/10.1002/advs.201700982

[426] K.A. Moldosanov, A.V. Postnikov, V.M. Lelevkin, N.J. Kairyev, Terahertz imaging technique for cancer diagnostics using frequency conversion by gold nano-objects, Ferroelectrics. 509 (2017) 158-166. https://doi.org/10.1080/00150193.2017.1296344

[427] S.A. Kuznetsov, A.G. Paulish, M. Navarro-Cia, A.V. Arzhannikov, Selective pyroelectric detection of millimetre waves using ultra-thin metasurface absorbers, Sci. Rep. 6 (2016) 21079. (Open Access). https://doi.org/10.1038/srep21079

[428] S.A. Kuznetsov, A.G. Paulish, A.V. Gelfand, M. A. Astafiev, *et al.*, Extremely thin metamaterial absorbers for subterahertz waves: from fundamentals towards applications in uncooled bolometric sensors, Proc. SPIE. 8432 (2012) 8423OS.

https://doi.org/10.1117/12.922728

[429] B. Kearney, F. Alves, D. Grbovic, G. Karunasiri, Al/SiOx/Al single and multiband metamaterial absorbers for terahertz sensor applications, Opt. Eng. 52 (2013) 013801. https://doi.org/10.1117/1.OE.52.1.013801

[430] F. Capolino, Theory and Phenomena of Metamaterials, CRC Press, Boca Raton-London-New York, 2009.

[431] V. Savinov, V.A. Fedotov, P.A.J. de Groot, N.I. Zheludev, Radiation-harvesting resonant superconducting sub-THz metamaterial bolometer, Supercond. Sci. Technol. 26 (2013) 084001. https://doi.org/10.1088/0953-2048/26/8/084001

[432] F. Costa, S. Genovesi, A. Monorchio, On the bandwidth of high-impedance frequency selective surfaces, IEEE Antennas Wirel. Propag. Lett. 8 (2009) 1341-1344. https://doi.org/10.1109/LAWP.2009.2038346

[433] F. Costa, S. Genovesi, A. Monorchio, G. Manara, A circuit based model for the interpretation of perfect metamaterial absorbers, IEEE Trans. Antennas and Propag. 61 (2013) 1201-1210. https://doi.org/10.1109/TAP.2012.2227923.

[434] A.G. Paulish, B.N. Novgorodov, S.V. Khryashchev, S.A. Kuznetsov, THz imager based on a THz-to-IR converter, Optoelectr. Instrum. Data Process. 55 (2019) 45-51. https://doi.org/10.3103/S8756699019010084

[435] N.M. Burford, and M.O. El-Shenawee, Review of terahertz photoconductive antenna technology, Opt. Eng. 56 (2017) 010901. https://doi.org/10.1117/1.OE.56.1.010901

[436] R.A. Lewis, A review of terahertz sources, J. Phys. D. Appl. Phys. 47 (2014) 374001. https://doi.org/10.1088/0022-3727/47/37/374001

[437] J.R. Freeman, H.E. Beere, D.A. Ritchie, Generation and detection of terahertz radiation, in: K.-E. Peiponen, J.A. Zeitler, M.Kuwata-Gonokami (Eds.), Teraherz Spectroscopy and Imaging, Springer-Verlag, Berlin, Heidelberg, 2013, pp. 1-27.

[438] D.J. Paul, The progress towards terahertz quantum cascade lasers on silicon substrates, Laser Photon Rev. 4 (2010) 610-632. https://doi.org/10.1002/lpor.200910038

[439] D.H. Auston, K.P. Cheung, P.R. Smith, Picosecond photoconducting Hertzian dipoles, Appl. Phys. Lett. 45 (1984) 284-286. https://doi.org/10.1063/1.95174

[440] E. Castro-Camus, Photoconductive devices for THz time-domain spectroscopy, in: S.S. Dhillon, M.S. Vitiello, E.H. Linfield, *et al.*, The 2017 terahertz science and technology roadmap, J. Phys. D: App. Phys. 50 (2017) 043001.

[441] P.C. Upadhya, W. Fan, A. Burnett, J. Cunningham, *et al.*, Excitation-density-

dependent generation of broadband terahertz radiation in an asymmetrically excited photoconductive antenna, Opt. Lett. 32 (2007) 2297-2299. https://doi.org/10.1364/OL.32.002297

[442] P.U. Jepsen, D.G. Cooke, M. Koch, Terahertz spectroscopy and imaging-modern techniques and applications, Laser Photon. Rev. 5 (2011) 124-166. https://doi.org/10.1002/lpor.201000011

[443] P. Shumyatsky, R.R. Alfano, Terahertz sources, J. Biomedical Optics. 16 (2011) 033001. https://doi.org/10.1117/1.3554742

[444] D.V. Lavrukhin, G.M. Katyba, A.E. Yachmenev, R.R. Galiev, *et al.*, Numerical simulations and experimental study of terahertz photoconductive antennas based on GaAs and its ternary compounds, Proc. SPIE. 10680 (2018) 106801M. https://doi.org/10.1117/12.2306189

[445] H. Page, S. Malik, M. Evans, I. Gregory, *et al.*, Waveguide coupled terahertz photoconductive antennas: Toward integrated photonic terahertz devices, Appl. Phys. Lett. 92 (2008) 163502. https://doi.org/10.1063/1.2909539

[446] M. Tani, K.-S. Lee, and X.-C. Zhang, Detection of terahertz radiation with low-temperature-grown GaAs-based photoconductive antenna using 1.55 μm probe, Appl. Phys. Lett. 77 (2000) 1396-1398. https://doi.org/10.1063/1.1289914

[447] M. Suzukia, M. Tonouchi, Fe-implanted InGaAs photoconductive terahertz detectors triggered by 1.56 μm femtosecond optical pulses, Appl. Phys. Lett. 86 (2005) 163504. https://doi.org/10.1063/1.1901817

[448] A. Krotkus, R. Adomavicius, V.L. Malevich, Terahertz emission from semiconductors excited by ultrafast laser pulses, in: R.E. Miles, X.-C. Zhang, H. Eisele, A. Krotkus (Eds), Terahertz Frequency Detection and Identification of Materials and Objects, Springer, Netherlands, 2007, pp. 3-16. https://doi.org/10.1007/978-1-4020-6503-3_1

[449] R.J.B. Dietz, B. Globisch, M. Gerhard, A. Velauthapillai, *et al.*, 64 μW pulsed terahertz emission from growth optimized InGaAs/InAlAs heterostructures with separated photoconductive and trapping regions, Appl. Phys. Lett. 103 (2013) 061103. https://doi.org/10.1063/1.4817797

[450] R.J.B. Dietz, N. Vieweg, T. Puppe, A. Zach, *et al.*, All fiber-coupled THz-TDS system with kHz measurement rate based on electronically controlled optical sampling, Opt. Lett. 39 (2014) 6482-6485. https://doi.org/10.1364/OL.39.006482

[451] A. Jooshesh, F. Fesharaki, V. Bahrami-Yekta, M. Mahtab, *et al.*, Plasmon-enhanced LT-GaAs/AlAs heterostructure photoconductive antennas for sub-bandgap terahertz generation, Optics Express. 25 (2017) 22140.

https://doi.org/10.1364/OE.25.022140

[452] S. Lepeshov, A. Gorodetsky, A. Krasnok, E. Rafailov, P. Belov, Enhancement of terahertz photoconductive antenna operation by optical nanoantennas, Laser Photon. Rev. 11 (2016) 1600199. https://doi.org/10.1002/lpor.201600199

[453] C.W. Berry, M. Jarrahi, Terahertz generation using plasmonic photoconductive gratings, New J. Phys. 14 (2012) 105029. https://doi.org/10.1088/1367-2630/14/10/105029

[454] K. Peng, P. Parkinson, L. Fu, Q. Gao, *et al.*, Single nanowire photoconductive terahertz detectors, NanoLett. 15 (2015) 206-210. https://doi.org/10.1021/nl5033843

[455] K. Peng, P. Parkinson, J.L. Boland, Q. Gao, *et al.*, Broadband phase-sensitive single InP nanowire photoconductive terahertz detector, NanoLett. 16 (2016) 4925-4931. https://doi.org/10.1021/acs.nanolett.6b01528

[456] F. Rettich, N. Vieweg, O. Cojocari, A. Deninger, Field intensity detection of individual terahertz pulses at 80 MHz repetition rate, Journal of Infrared, Millimeter, and Terahertz Waves. 36 (2015) 607-612. https://doi.org/10.1007/s10762-015-0162-8

[457] S. Preu, G.H. Dohler, S. Malzer, L.J. Wang, A.C. Gossard, Tunable, continuous-wave terahertz photomixer sources and applications, J. Appl. Phys. 109 (2011) 061301. https://doi.org/10.1063/1.3552291

[458] R.M. Woodward, V.P. Wallace, R.J. Pye, B.E. Cole, *et al.*, Terahertz pulse imaging of ex vivo basal cell carcinoma, J. Invest. Dermatol. 120 (2003) 72-78. https://doi.org/10.1046/j.1523-1747.2003.12013.x

[459] V. Wallace, A. Fitzgerald, S. Shankar, N. Flanagan, *et al.*, Terahertz pulsed imaging of basal cell carcinoma ex vivo and in vivo, Br. J. Dermatol. 151 (2004) 424-432. https://doi.org/10.1111/j.1365-2133.2004.06129.x

[460] S. Ye-xin, L. Jiu-sheng, Terahertz spectrum analysis of various white wine, The Intern. Photon. Optoelectr. Meeting, 2017, OSA Technical Digest, 2017, paper AS3A.21. https://doi.org/10.1364/ASA.2017.AS3A.21

[461] N. Karpowicz, H. Zhong, J. Xu, K.-I. Lin, *et al.*, Comparison between pulsed terahertz time-domain imaging and continuous wave terahertz imaging, Semicond. Sci. Technol. 20 (2005) S293-S299. https://doi.org/10.1088/0268-1242/20/7/021

[462] Yi. Sun, M.Y. Sy, Yi-X. J. Wang, A.T. Ahuja, *et al.*, A promising diagnostic method: Terahertz pulsed imaging and spectroscopy, World J. Radiol. 3 (2011) 55-65. https://doi.org/10.4329/wjr.v3.i3.55

[463] M.R. Grootendorst, A.J. Fitzgerald, S.G. Brouwer de Koning, A. Santaolalla, *et al.*, Use of a handheld terahertz pulsed imaging device to differentiate benign and malignant breast tissue, Biomed. Opt. Exp. 8 (2017) 2932-2945.

https://doi.org/10.1364/BOE.8.002932

[464] A. Brahm, A. Wilms, R. Dietz, T. Gobel, *et al.*, Multichannel terahertz time-domain spectroscopy system at 1030 nm excitation wavelength, Opt. Exp. 22 (2014) 12982-12993. https://doi.org/10.1364/OE.22.012982

[465] C. Berry, M. Hashemi, M. Jarrahi, Generation of high power pulsed terahertz radiation using a plasmonic photoconductive emitter array with logarithmic spiral antennas, App. Phys. Let. 104 (2014) 081122. https://doi.org/10.1063/1.4866807

[466] E.R. Brown, Advancements in photomixing and photoconductive switching for THz spectroscopy and imaging, Proc. SPIE. 7938 (2011) 793802. https://doi.org/10.1117/12.881069

[467] D. Saeedkia, Handbook of Terahertz Technology for Imaging, Sensing and Communications. Woodhead Publishing, Oxford, Cambridge, Philadelphia, New Delhi, 2013.

[468] V.A. Trofimov, S.A. Varentsova, V.V. Tikhomirov, V.V. Trofimov, High effective THz-TDS method for the detection and identification of substances in real conditions, Proc. SPIE. 9836 (2016) 98362U. https://doi.org/10.1117/12.2208454

[469] B. Globisch, R.J.B. Dietz, T. Göbel, M. Schell, W. Bohmeyer, R. Müller, A. Steiger, Absolute terahertz power measurement of a time-domain spectroscopy system, Opt. Lett. 40 (2015) 3544-3547. https://doi.org/10.1364/OL.40.003544

[470] K.W. Kim, K.-S. Kim, H. Kim, S.H. Lee, *et al.*, Terahertz dynamic imaging of skin drug absorption, Opt. Express. 20 (2012) 9476-9484. https://doi.org/10.1364/OE.20.009476

[471] L. Ohrstrom, B. M. Fisher, A. Bitzer, J. Wallaurer, *et al.*, Terahertz imaging modalities of ancient egyptian mummified objects and of a naturally mummified rat, The Anatomical Record. 298 (2015) 1135-1143. https://doi.org/10.1002/ar.23143

# Ch. 6. THz and IR sources

## 6.1 Introduction

Among other parts of active imaging and spectroscopy systems, IR and THz sources and detectors play a key role in obtaining spectroscopic data and imaging. There are mainly two types of THz systems: systems for spectroscopy and systems for imaging. Some others include THz radars, sensors, antennas, and a number of new systems.

The components of such systems such as lenses, mirrors, polarizers, various kinds of transmission stages, *etc*. play an important role, but in many cases they are not crucial for performing imaging, spectroscopic measurements, and communication in those spectral ranges. A number of new sensors and detectors, which were developed within the last two decades, have determined the main driving force in the THz technology. At the same time, notable developments of THz sources can also be mentioned. For compact spectroscopic, imaging, and communication systems, applications developing solid-state THz sources are of great importance.

Progress in well-known vacuum electronic and photonic devices, and, *e.g.*, in relatively newly developed quantum cascade lasers (QCLs) widens their application in heterodyne and direct detection THz systems. A separate place is occupied by passive IR and THz imaging systems, in which the objects under observation serve as radiation sources. Extraterrestrial thermal sources include the sun and the cosmic background radiation.

The sources of IR or THz radiation can be of two types: either natural or artificial. For many IR observation, spectroscopic, and imaging applications, natural sources, such as the Sun and other cosmic objects, cosmic background, and various lifeless and living objects on the Earth, are sufficient. But for many IR spectroscopic applications, more powerful sources are needed. As concerning natural THz sources, *e.g.*, distant Galaxies and cosmic background, they can be observed and analyzed using highly sensitive cold detectors and arrays. While emission from those objects is weak, THz observations of galactic interstellar dust clouds (T ~ 10…20 K) are important for their characterization. The opacity of the Earth's atmosphere to THz radiation restricts the observations of these objects, and high-altitude or space platforms are needed.

As a rule, much more powerful sources are needed if they are intended to be used in imaging and spectroscopic applications under environmental Earth conditions. About three decades or a little bit more ago, the lack of powerful THz compact sources has led to a situation when the THz range turned out relatively unexplored, so that the development of THz technologies was decelerated to a great degree.

Wideband or tuned monochromatic sources are preferable for the spectroscopy in an arbitrary wavelength range. In the IR range, wideband sources include thermal sources and mercury lamps, the latter being frequently used in the THz range as well.

The diversity of THz and IR sources and systems is so large that only some of them can be considered in this book. IR devices are much more developed for everyday applications, and there exist a large number of books and reviews dealing with the physics and applications of those sources and systems (see, *e.g.*, Ref. [1]).

Compact, coherent or incoherent, CW solid-state THz sources are strongly desirable in many THz applications for active imaging, spectroscopic, biomedical, food-control, communication, and other applications. That is why solid-state THz sources based on semiconductor devices, which can be used as THz radiation generators in compact systems are mainly considered below.

## 6.2 THz and IR sources

Any object with the temperature $T > 0$ K emits electromagnetic radiation at all frequencies. This thermal emission is well described by a theoretical blackbody spectrum (Planck's formula) and depends on the temperature of the object. For natural sources with the temperature of the Earth surface ($T \sim 290...310$ K) or human body, thermal emission is strong enough in the IR spectral section of the atmosphere transparency range (Fig. 2.1, the maximum of emission is located at $\lambda \approx 9.6$ μm), which allows, *e.g.*, the environmental control and medical thermovision to be carried out in the IR spectral range. But it is weak at THz frequencies. However, the THz spectral band is actively mastered by cosmic astronomy on external platforms out of the Earth atmosphere in the spectral range $\lambda \approx 30...500$ μm (the radiation frequency $v \approx 1...10$ THz), because cold objects (dust and gas clouds in galaxies with $T \approx 10...100$ K) emit in this spectral range. The relict radiation ($T = 2.725$ K) has an emission maximum at $\lambda \approx 1.06$ mm ($v \approx 0.27$ THz).

Although natural sources of IR and THz emission play an important role in getting information, *e.g.*, in passive medical imaging, astrophysics, and food control, practical indoor and outdorr applications are not possible without using more powerful artificial sources. Important are applications of THz sources in the THz astronomy [2], where they are required to drive heterodyne receivers in a wide spectral range up to several THz.

## 6.3 Artificial sources of THz and IR radiation

Artificial sources can be mainly considered as based on thermal emission, stimulated emission processes (lasers), electron acceleration processes, negative differential resistance phenomena, laser pulse generation, photomixing, and some others.

Among the artificial sources of thermal IR and THz radiation, there are various kinds of thermal radiation sources (globar, Nernst glower (Nernst stift), mercury lamp, *etc.*). Among standard artificial sources based on stimulated emission processes of THz radiation, there are gas, semiconductor, and quantum cascade lasers, as well as femtosecond pulse laser systems to generate ultrashort THz pulses in the range ~1...5 THz. Among the sources based on electron acceleration processes, there can be mentioned vacuum electronic ones such as carcinotrons or backward-wave oscillators (BWOs), gyrotrons, clinotrons, free-electron lasers (FELs), synchrotron radiation sources, *etc.* This group of sources includes devices based on the motion of free carriers in photoconductors (*e.g.*, photoconductive antennas, photomixers) and Smith-Purcell emitters (metallic gratings).

The THz and GHz sources based on the negative differential resistance (IMPATT diodes, Gunn oscillators, resonant tunnelling diodes (RTDs), *etc.*) are frequently applied. Variuos types of optical-to-THz conversion devices (photo-mixing sources, photoconductive sources, optical rectification emitters), parametric oscillators, and some others are also used in THz technologies. Some of those sources used in imaging or spectroscopic systems will be briefly considered below. For a deeper insight into IR and THz sources, including their physics and application potential, the reader can be referred to Refs. [1, 3–16].

## 6.4 Thermal artificial THz and IR sources

Of artificial thermal IR sources, globars are frequently used. The globar (a word, which consists of the words glow and bar) is a thermal light source including a silicon carbide rod (about 5 to 10 mm in width and 20 to 50 mm in length), which is electrically (resistively) heated up to temperatures from $1000^{0}C$ to, as a rule, $1650\ ^{0}C$. It is usually capped with metallic layers, which serve as electrodes for the passage of current through the globar. This thermal source emits in a wide spectral range (up to the THz range) with the emissivity $\varepsilon \approx 0.5...0.8$ [17, 18], which depends on the spectral band and the globar temperature. The globar emissivity decreases with the decreasing THz radiation frequency. Therefore, it can be considered as a grey-body radiator. Globars are mainly used as thermal light sources in IR spectroscopy. In the THz spectral range, according to

Planck's law, they have a low output power of an order of picowatts and less (depending on the spectral interval).

Another artificial source of IR radiation is the Nernst glower. The Nernst glower (Nernst stift) is a ceramic (zirconium oxide–yttrium oxide) glowing rod. One of disadvantages of the Nernst glower is that this ceramic rod is not electrically conductive at room temperature. Therefore, it must be preheated, *e.g.*, with a filament lamp or an open fire, to become conducting. The Nernst glower can be heated by an electrical current up to temperatures of about 1900 $^0$C. The Nernst stift shows advantages over the globar in the spectral range of the blackbody maximum and is less advantageous up to $\lambda \approx 15$ μm. At $\lambda > 15$ μm, the globar is a more effective radiator [19, 20].

The mercury lamp, which can be used both as an UV, IR, and THz source, is a high-pressure mercury arc in a quartz bulb. It operates when the current passing through the lamp excites and ionizes Hg vapor to form a plasma discharge [21, 22]. The coiled tungsten cathode is coated with a rare-earth material (*e.g.*, thorium). The auxiliary electrode is used to help in starting the lamp. A high initial resistance of the lamp limits the starting current. Once the arc started, the operating current is limited by a ballast supplied by the high reactance of the power transformer. The Hg arc lamp generates broadband continuous THz radiation in the spectral range ~0.15...3 THz. As compared with the globar, the mercury lamp is more powerful by an order of magnitude powerful at ~0.15...1 THz and by a factor of 2 to 3 at ~1...3 THz [23].

### 6.5 Artificial THz radiation sources

Besides natural sources of THz radiation, there exist a large number of various THz sources based on various physical properties of emitting devices. The most advanced among modern THz sources can be divided into several categories. From the early stage of THz technologies and until now, there existed widely used vacuum electronic sources (see, *e.g.*, Refs. [9, 24–29]), which are powerful and operate in a broad radiation frequency band. Among those sources, there are backward-wave oscillators, klystrons, clinotrons, gyrotrons, travelling-wave tubes, synchrotrons, free-electron lasers, and their modifications.

In Figs. 6.1 and 6.2, some parameters of modern vacuum electronic sources are compared with those of several solid-state electronic sources.

Gunn diodes, IMPATT diodes, quantum cascade lasers, frequency multiplication devices (on the basis of Gunn diodes and Schottky diodes), field-effect transistors, Josephson junctions, superconductor-insulator-superconductor junctions, semiconductor (germanium or silicon) lasers, photoconductive switches, photomixers, *etc.*, which have some kind of non-linear static current-voltage characteristics (see Fig. 3.12), are the most used solid-state electronic sources (compact as a rule) .

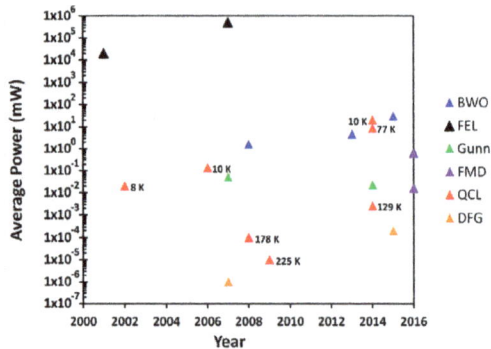

*Figure 6.1. Progress of THz sources over recent years [29]. BWO = backward wave oscillator, FEL = free electron laser, FMD = frequency multiplier devices, Gunn = Gunn diodes, QCL = quantum cascade laser (at 10 K unless otherwise stated), DFG = difference frequency generation. All data correspond to room temperature unless otherwise stated.*

THz sources are based on various physical phenomena and include stimulated-radiation sources (lasers), electron acceleration sources (e.g., free electron lasers, backward-wave oscillators, klystrons, etc.), compact sources based on transit-time diodes (*e.g.*, IMPATT diodes), resonant tunnel diodes (RTDs), transferred-electron devices (*e.g.*, Gunn oscillators), Schottky diode frequency multipliers, photoconductors (*e.g.*, photoconductive antenna sources), sources on the basis of frequency difference technique (photomixing), *etc*. Some of them are very powerful (*e.g.*, free electron lasers), but too bulky. THz sources based on quantum cascade lasers (QCLs) are promising, because they produce a relatively high power in a compact volume, although when operating in the THz spectral range, they require cooling.

Practical everyday applications, such as real-time imaging and spectroscopy, require portability and a THz output from several milliwatts to tens of milliwatts. For some special practical applications, such as the remote explosive detection, high-power THz output sources (of at least a few tens of milliwatts) are required [30].

In any application, several considerations concerning the THz sources should be taken into account, *e.g.*, the operating temperature, the output power stability, the bandwidth, tunability, portability, *etc*. Some typical spectral dependencies of THz sources are shown in Fig. 6.2. Source parameters compared with those shown in Fig. 6.2 were later modified to produce larger power output (see below).

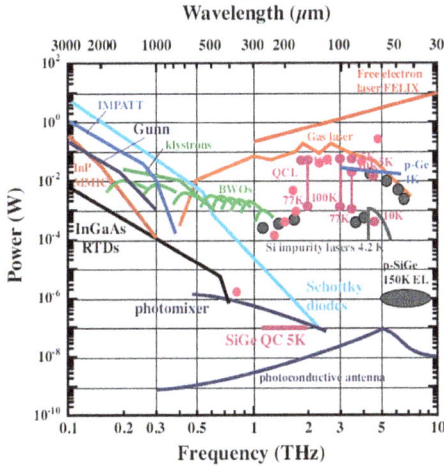

*Figure 6.2. A comparison of the output powers of a number of different THz source technologies as a function of frequency [5]. (By permission of John Wiley and Sons).*

Compact THz and sub-THz sources based, *e.g.*, on transit-time diodes, transferred-electron devices, and Schottky diode frequency multipliers have been employed successfully to generate sub-THz and THz power levels either in the pulse or the CW mode with the output power ranging from the nano-watt level to several watts or even tenths of watts, depending on the spectral range (see Fig. 6.2). They emit more than 1 mW at frequencies of about 300 GHz and even higher. Many vacuum electronic sources (*e.g.*, klystrons, BWOs, gyrotrons, *etc*.) and lasers operating in various parts of the sub-THz and THz frequency intervals are used in THz imaging and spectroscopic applications. In spectroscopic applications, short high-power pulses with a duration from a few to tenths of femtoseconds generated by Ti:sapphire lasers and semiconducting switches are required.

Not all types of THz sources can be used today for biomedical applications because of too high power (safety requirements). Below, only some typical, mainly compact, THz sources and their principles of operation will be considered.

While approaching the "central" point ($\approx 1$ THz) of the THz gap from either the low-frequency or high-frequency side, the output power generated by solid-state electronic devices diminishes remaining well below the mW level due to various physical processes in those devices as IMPATT and Gunn oscillators, Schottky diode multipliers, quantum cascade lasers, photoconductive antennas and photomixers.

In typical compact solid-state sources operating at room temperatures, such as transistors, solid-state diodes (IMPATT, Gunn, Schottky, RTD) and microwave vacuum tubes, the source power at radiation frequencies $\nu < 1$ THz generally tends to fall down as $\nu^{-2}$ (see, $e.g.$, Refs. [5, 31, 32] and Fig. 6.2) and even steeper (at $\nu \sim 200...400$ GHz) [16, 33–36]. Therefore, if doubling the frequency, the output power will drop by a factor of four or by about an order of magnitude (see Fig. 6.2, as well as Figs. 6.6; 6.9, and 6.10 below for IMPATT diodes, Gunn diodes, and Schottky-based frequency multipliers, respectively).

In modern quantum cascade lasers (QCLs), the radiation power changes closely to $\nu^2$ at $\nu$ from 1 THz up to about 40 THz [11]. Depending on the operation temperature, the lower the emitted radiation frequency is, the lower operation temperature should be down to liquid helium temperature.

## 6.6  Solid-state THz sources

Solid-state sources like oscillators and amplifiers are generally limited in frequency due to the transit time of carriers through semiconductor junctions, which causes a high-frequency roll-off. They are rugged and compact devices, and can operate in the CW mode at room temperature with a relatively narrow line width.

The IMPATT and Gunn uncooled diodes can operate up to radiation frequencies of about 0.3 THz or a bit higher, and they can be used in relatively low-cost active imaging systems, radio communication, as well as in military and commercial low-power sources for radars. At high harmonics, they are used as local oscillators in heterodyne systems operating at about 1 THz. Such solid-state sources as frequency multipliers with two or more Schottky diodes can be employed at frequencies up to about 1...2 THz. p-Ge and p-SiGe laser sources operate at low temperatures, at higher THz frequencies, and produce a power of several µW or a bit higher.

### 6.6.1  IMPATT diodes

Avalanche transit time (ATT) diodes with impact ionization (IMPATT), as well as their counterparts – TRAPATT (trapped plasma avalanche triggered transit), BARITT (barrier injection transit time), TUNNETT (tunnel injection transit time negative resistance), and MITATT (mixed avalanche tunnelling transit time) devices—which are rather similar in operational principles, are based on reverse-biased p-n-junctions. They emit microwave and THz radiation. IMPATT diodes have received much attention as cost-effective and low-sized THz sources. They are known as the most powerful THz wave sources among the avalanche transit time sources in the low-frequency THz range.

The devices on the basis of TRAPATT and BARITT diodes have lower microwave radiation frequencies (several GHz), and they are not used as THz sources. However, TRAPATT diodes can be more efficient then IMPATT (and BARITT) ones, because they are less noisy than IMPATT diodes. However, they have a much narrower generation frequency range then IMPATT diodes have. The physics and parameters of such devices are considered, *e.g.,* in Ref. [37]. In Fig. 6.3, typical radiation-frequency dependences of the relative output power $P_{out}$ are schematically presented for some types of radiation sources mentioned above.

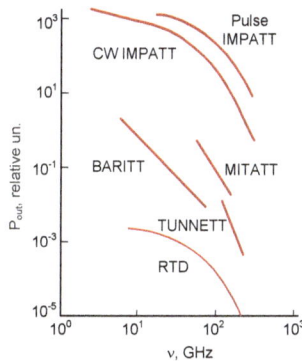

*Figure 6.3. Frequency dependences of the relative output power $P_{out}$ for various types of semiconductor devices used as sources at microwave, sub-THz, and THz frequencies.*

IMPATT diodes are more powerful in comparison with other diodes (*e.g.,* Gunn, TRAPATT, BARITT, MITATT, and RTD ones) and have a high noise figure (higher than in Gunn diodes). They can operate in both the CW and pulse modes. Their DC-to-sub-THz/THz conversion efficiency is up to 3% in the CW mode and up to 60% in the pulse one. IMPATT diodes are more efficient and more powerful then Gunn diodes. TRAPATT diodes are not suitable for CW operation due to high power densities and, similarly to BARRIT diodes, operate in a narrower radiation GHz frequency band than IMPATT diodes do. BARRIT diodes are less noisy than IMPATT diodes, but have a lower output power (see Fig. 6.3).

IMPATT diodes can be used in a variety of applications [38] — in low-power radars, as sources in missile seeker heads, and in many others mm-wave civilian and military applications, including communication, vision and alarm systems, *etc*.

IMPATT diodes are two-terminal semiconducting devices with a negative differential resistance near the voltage breakdown (Fig. 6.4), which employ impact ionization and transit-time processes, and demonstrate a dynamic negative differential resistance.

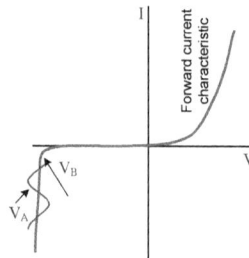

*Figure 6.4. Schematic representation of the IMPATT diode I-V-characteristics. $V_B$ is the voltage breakdown, $V_A$ is the alternating voltage.*

The dynamic negative resistance at microwave frequencies arises at a controlled avalanche breakdown, which is induced by impact ionization in a reverse-biased p-n-junction (or the Schottky diode) (see, *e.g.*, Refs. [33, 39]). They are used as oscillators to generate CW microwaves and THz radiation in the range $v \sim 3...400$ GHz (see Figs. 6.2 and 6.3). As compared to other solid-state THz sources, they have a rather high power capability in this spectral range.

IMPATT diodes are preferably manufactured from the group IV and III–V semiconductor materials. Even though quite a lot of materials (*e.g.*, Si, Ge, SiC, GaAs, InP, and GaN) can be used for their fabrication, one of the base materials is silicon, because of its standard technological processes, relative simplicity, and low processing cost. The Si IMPATT diodes have also a more reliable and mature technology.

In an IMPATT diode the avalanche breakdown in combination with the modification of the charge-carrier transit time is applied to create a negative resistance region and, in such a way, to make the diode act as an oscillator. Among other approaches for realizing high-power, high-frequency IMPATT sources, there are those based on wide bandgap semiconductors (*e.g.*, SiC or GaN) with a high critical electric field, a high saturation velocity of carriers, and a high thermal conductivity (to provide a deep insight into the issue, see, *e.g.*, Refs. [40–42] and references therein).

IMPATT diodes have various structures, such as mesa, planar, distributed, and so on. They consist of a contact and a drift region that are stacked perpendicularly to the substrate surface. The IMPATT diode is generally a single- or double-drift device (Fig. 6.5) with heavily doped p- and n-regions that provide an Ohmic contact to the external circuit. Its emitting characteristics are based on a reverse-biased p-n-junction and highly resistive (with intrinsic conductivity) layer providing a negative differential resistance for generating and sustaining current oscillations, which can be transformed into radiation. An important advantage of IMPATT diodes in comparison with other avalanche transit-time diodes is their relatively high output radiation power (up to ~50...150 mW at $\nu \sim$ 0.1...0.15 THz), as well as compact sizes of sources fabricated on their base. The frequency of oscillations is fixed by the transit time of electrons through the n–region.

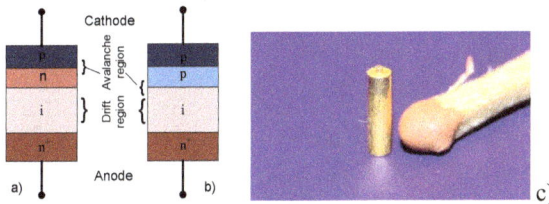

*Figure 6.5. Schematic structure of two versions of vertical IMPATT diodes: a) single-drift region (SDR) diode, b) double-drift region (DDR) diode, c) Typical dimensions of incapsulated IMPATT diode. Here $p^+$ and $n^+$ are the heavily doped p- and n-type conductivity contact layers, respectively, p is the p-type layer, "i" is the intrinsic-conductivity high-resistivity layer.*

The IMPATT diode family includes many different semiconductor junctions or metal-semiconductor devices with various designs (see, *e.g.*, Ref. [43]). There can be a simple diode with an abrupt asymmetrical $p^+$-n–junction, which is sometimes called Tager's diode. The oscillations in it, when it is mounted in a microwave cavity, are obtained from a p-n–junction biased into a reverse avalanche break down. The SDR Schottky diodes are also considered. Read's IMPATT diodes with various designs consist of various combinations of (p, n⁻), (n, i), (p⁻, n, n⁻), and (n⁻, n, n⁻) IMPATT diodes with a low-high-low (lo-hi-lo) profile (see, *e.g.*, Refs. [44–48]).

One of those designs, which is shown in Fig. 6.5.b, consists of a region with relatively high doping and high electric field (the avalanche region), where multiplication takes place, and an intrinsic region, in which the generated carriers drift towards the contact. In a simple p–i–n diode (Misavy diode [43]), the field is close to uniform one. Electrons and holes are generated in equal quantities and drift in the opposite directions.

Si IMPATT diodes were the first semiconductor devices that generated a THz power at frequencies above 300 GHz. These devices can generate high CW and pulsed power outputs at sub-THz and THz radiation frequencies v ~ 30...400 GHz, when the junction diode is mounted in a microwave cavity. They have rather high DC-to-sub-THz/THz conversion efficiencies (up to several %). At radiation frequencies up to 300 GHz, Si IMPATI diodes yielded the highest sub-THz/THz power levels known for solid-state sources, namely, 50 mW at 245 GHz and 7.5 mW at 285 GHz [33].

If a reverse bias across the p-n–junction exceeds a certain threshold voltage level $V_B$, then, due to impact ionization, an avalanche breakdown occurs, resulting in the appearance of a large number of carriers in the avalanche region near the $p^+$-n contact in single-drift diodes or near the p-n contact in double-drift diodes. Electrons move through the drift region with intrinsic doping to the n+ contact within a time interval known as the transit time delay.

At a direct reverse bias $V_B$ just short to cause breakdown, the n–region is punched through and forms the avalanche region in the diode. The high resistivity region is a drift region through which the avalanche-generated electrons move toward the anode. To get current oscillations in the IMPATT diode, an AC voltage (superimposed on the DC reverse bias) with a mean value just below the avalanche breakdown $V_B$ is applied to the diode. When the voltage increases above the threshold $V_B$ (Fig. 6.4), free carriers are generated.

At the AC voltage $V_A = 0$ near the breakdown voltage $V_B$, only a small pre-breakdown current flows through the diode. As the AC voltage $V_A$ increases, the voltage goes above $V_B$, and electron-hole pairs are produced by impact ionization. When the AC voltage $V_A$ in the avalanche region increases above the breakdown voltage $V_B$ (~ 1...1.5 % over), the electron and hole concentrations near the $p^+$-n region in single-drift diodes or near the p-n region in double-drift diodes grow exponentially. And vice versa, the electron and hole concentrations decay exponentially with time, when the field is reduced below the breakdown voltage $V_B$ during the negative swing of the AC voltage $V_A$.

The holes generated in the avalanche region are collected by the cathode and disappear from the p+–region. The electrons are injected into the avalanche region, where they drift through the drift region toward the n+–region (anode). During the growth of the $V_A$ value, the electric field in the avalanche region reaches its maximum, and the concentration of electron-hole pairs starts to grow. At this time moment, the ionization coefficients have their maximum values. The generated electron concentration does not follow the electric field, tends to its minimum, and depends on the number of electron-hole pairs in the avalanche region.

After the field passed its maximum value, the electron-hole concentration continues to grow, because the secondary-carrier generation rate still remains above its average value. For this reason, the electron concentration in the avalanche region attains its maximum value at the moment, when the field has dropped to its average value. Such processes lead to a situation when a phase shift between the AC signal and the electron concentration appears in the avalanche region. Then, the AC voltage becomes negative, and the electric field in the avalanche region drops below its critical value. The electrons in the avalanche region are then injected into the drift region, which induces a current in the external circuit, with the current phase being opposite to that of the AC voltage. The AC field, therefore, absorbs energy from the drifting electrons when they are decelerated by the electric field decrease.

An optimal phase shift between the diode current and the AC signal is achieved if the thickness of the drift region is such that the batch of electrons is collected in the $n^+-$ region (anode) at the moment when the AC voltage goes to zero. The length of the diode can be chosen so that the transit time delay results in a further 90-deg phase lag in the current and, therefore, a negative differential resistance of the device. An external resonant circuit can then be used to sustain the oscillations.

The basic physical parameters that govern the avalanche process and IPATT operation are the ionization coefficients of electrons and holes. The ionization coefficients can be written as [43]

$$\alpha = A_1 \cdot \exp[(-E_{01}/E)^m],$$
$$\beta = A_2 \cdot \exp[(-E_{02}/E)^m], \tag{6.1}$$

where, for silicon, $m = 1$, $A_1 = 5.0 \times 10^{-5}$ cm$^{-1}$, $A_2 = 5.6 \times 10^{-5}$ cm$^{-1}$, $E_{01} = 1.0 \times 10^6$ V/cm, and $E_{02} = 1.32 \times 10^6$ V/cm at $E > 5.3 \times 10^5$ V/cm; and $A_1 = 5.0 \times 10^{-5}$ cm$^{-1}$, $A_2 = 5.6 \times 10^{-5}$ cm$^-$, $E_{01} = 1.0 \times 10^6$ V/cm, and $E_{02} = 1.32 \times$ V/cm at $E > 5.3 \times 10^5$ V/cm. For GaAs, $m = 2$, $A_1 = A_2 = 3.5 \times 10^{-5}$ cm$^{-1}$, and $E_{01} = E_{02} = 6.85 \times 10^5$ V/cm.

Like other devices based on the random avalanche process, IMPATT diodes tend to suffer from the phase noise. Impact ionization, as the carrier generation mechanism in IMPAIT diodes, is a major contributor to the noise of free-running oscillators if compared to, *e.g.*, the smaller contribution from mainly thermal noise to the domain formation process in Gunn devices. For GaAs IMPATT diodes, the ionization coefficients of electrons and holes are equal and, therefore, the noise level in those diodes is lower as compared to that in Si IMPATT diodes.

IMPATT diodes operate at much higher electric fields in comparison with Gunn diodes, and they are mounted in a resonant package. The diodes are mounted with their high–field region close to a heat-sink to let the heat generated at the diode junction be dissipated. Typical characteristics of double-drift Si IMPATT diodes are shown in Table. 6.1.

*Table 6.1. Typical characteristics of double-drift Si IMPATT diodes.*

| Conductivity type | Length, μm | Carrier concentration, cm$^{-3}$ |
|---|---|---|
| p$^{+}$ | 0.25 | >$1.0 \cdot 10^{18}$ |
| p$^{-}$ (p) | 0.35 | $2.2 \cdot 10^{17}$ |
| n$^{-}$ (i) | 0.3 | $2.0 \cdot 10^{17}$ |
| n$^{+}$ | 10 | >$1.0 \cdot 10^{18}$ |

In Fig. 6.6, known experimental data (V.S. Slipokurov, private communication) for the output power of IMPATT diodes on the basis of GaAs, 4H-SiC, and Si are collected. It is seen that, according to them, silicon IMPATT diodes are the most powerful at the radiation frequencies $\nu \geq 100$ GHz. At lower frequencies, GaAs and InP (not shown) IMPATT diodes have a higher output power.

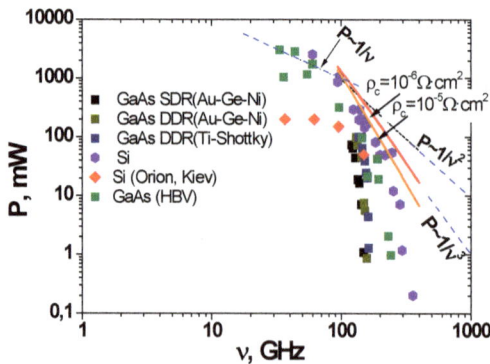

*Fig. 6.6. Experimental frequency data for the CW operation of Si and GaAs IMPATT diodes (T = 300 K) in the range $\nu \approx 40$–400 GHz. The frequency plots depend on the contact resistance $\rho_c$ (by courtesy of V. Slipokurov).*

Materials Research Forum LLC
https://doi.org/10.21741/9781644900758

Some average properties of semiconductor materials, which are important for estimating the frequency limitations on transit-time-limited microwave and THz devices and choosing the frequency range of their applications are presented in Table 6.2.

*Table 6.2. Properties of some semiconductor materials (T = 300 K) used for manufacturing IMPATT diodes.*

| Material | $E_g$, eV | $\mu_n$, cm$^2$/V·s | $\mu_p$, cm$^2$/V·s | $V_{ns}$, $10^7$ cm/s | K, W/m·K | $E_c$, $10^5$ V/cm | $\varepsilon$ |
|----------|-----------|---------------------|---------------------|-----------------------|----------|--------------------|----|
| Si | 1.12 | 1350 | 380 | 1 | 150 | 3 | 11.9 |
| Ge | 0.66 | 3900 | 1900 | 0.6 | 58 | 1 | 16 |
| GaAs | 1.43 | 9600 | 400 | 1.2 | 55 | 4 | 12.9 |
| InP | 1.34 | 5400 | 200 | 1 | 68 | 5 | 12.4 |
| Wz-GaN | 3.4 | 1000 | 350 | 2 | 225 | 50 | 9 |
| 4H-SiC | 3.26 | 900 | 120 | 1.9 | 350 | 50 | 9.6 |

Here, $E_g$ is the band-gap, $\mu_n$ and $\mu_p$ are the electron and hole mobilities, respectively, $V_{ns}$ is the saturated electron drift velocity, K is the thermal conductivity, $E_c$ is the electric breakdown field, and $\varepsilon$ is the permittivity.

GaN and SiC are promising materials for the IMPATT diode manufacturing. They can have higher output powers and broader radiation power ranges. However, the available database for the parameters of IMPATT diodes based on GaN and SiC semiconductors was mainly created through computer simulation (see, *e.g.*, Refs. [49, 50].

### 6.6.2 Gunn diodes

The Gunn-effect (microwave oscillations of the current [51]) is connected with the existence of nonlinear current-voltage characteristics for semiconductors with a negative resistance region (see Fig. 6.7). This effect forms a basis for Gunn diodes (also known as transferred electron devices), which generate coherent power radiation. Gunn diodes are DC-to-RF converters, because oscillations occur when the applied bias voltage exceeds a certain threshold. The oscillation frequencies depend on the material properties and the device geometry.

Gunn detected regular current oscillations in GaAs at about 5 GHz and applied for a patent in 1963. These microwave oscillations were demonstrated with ohmic electrical contacts on the opposite faces of the sample. The principal mechanism, the transferred-electron effect, was theoretically considered by Ridley and Watkins in 1961 [52]. In

1962, Hilsum predicted the feasibility of transfer-electron amplifiers and oscillators [53]. Krömer associated Gunn oscillations with the transferred-electron effect [54].

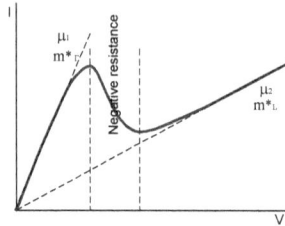

*Figure 6.7. Typical current-voltage characteristic of Gunn diodes with a negative resistance region (dV/dI < 0, n-type GaAs layers). Here, $\mu_1$ and $\mu_2$ are the electron mobilities in the $\Gamma$- and L-valleys, respectively (see Fig. 6.8).*

A brief description of the Gunn diode operation is included here. There are many comprehensive reviews and books on the corresponding well-developed physics and technology (see, *e.g.*, Refs. [43, 55]).

Gunn diodes have been widely used in military, commercial, and industrial applications for about half a century and continue to offer admissible characteristics in many millimeter and low-frequency THz band technologies.

Two-terminal Gunn devices are considered as diodes, though they do not have any typical p-n–junction like other diodes or devices based in the avalanche principle do. To realize a Gunn diode, semiconductors should have relatively closely-spaced energy valleys in their conduction band like the III-V or II-VI group semiconductors - *e.g.*, GaAs, InAs, InGaAs, InP, GaN, InN, AlInN, CdTe, ZnSe - to let electrons transfer from the bottom valley into others, which are located above, under the action of a constant electric field. The semiconductors used to manufacture Gunn diodes should be of n-type conductivity, because the transferred electron effect holds good only for electrons. The negative differential resistance of Gunn diodes depends only on the bulk material properties rather than the junction or the interface. The most used are Gunn diodes on the basis of GaAs. Though InP-based devices offer higher output powers at millimeter-wave or THz frequencies than GaAs-based ones do [56], higher output powers and frequencies offered by InP-based devices are achieved at the expense of their temperature stability.

If a resonant circuit is combined with a device that exhibits a negative differential resistance (dV/dI < 0, Fig. 6.7) at approximately its resonance frequency (in a resonator), then oscillations can be sustained indefinitely. The formed resonant circuit determines the frequency of oscillations and can take the form of a waveguide or a cavity resonator to be used as a radiation source. The most common materials to fabricate Gunn diodes are GaAs and InP.

The GaN or InN Gunn diodes, or the semiconductor materials on the basis of those binary compounds, are promising devices to emit radiation up to the THz range. For example, in Gunn diodes on the basis of variband semiconductor AlInN and with the active area length $L_{ac}$ = 0.15 μm, emission was registered at frequencies of 0.9...1.3 THz (see Ref. [57]). However, because of a high electric power consumption and problems with heat extraction from the active region, experimental realizations of AlInN-based Gunn diodes are practically absent. In several publications (see, *e.g.*, Refs. [58–61]), the simulation of radiation generation in Gunn diodes based on nitride semiconductors was mainly carried out. A power output 2 to 10 times higher than in GaN-based Gunn diodes is expected for GaN-based IMPATT diodes.

For GaAs, the effective electron mass is equal to $m^*_\Gamma$ = 0.067·$m_0$ at the Γ-point and to $m^*_L$ = 0.56·$m_0$ at the L-points (in 4 equivalent L-valleys) of the Brillouin zone. The latter value is calculated by the averaging formula $m^*_L = 4^{2/3} \times m_t^*(L) \times K^{1/3}$, where K = $m_l^*(L)/m_t^*(L)$, $m_l^*(L) = 1.9·m_0$ is the longitudinal and $m_t^*(L) = 0.0754·m_0$ the transverse effective electron masses in the L-valleys. For InP, $m^*_\Gamma$ = 0.0795·$m_0$ at the Γ-point, and $m^*_L$ = 0.56·$m_0$ at the L-points [62].

In the case of Γ-valley, the current $I \approx qn_1\mu_1E$ at $0 < E < E_1$, being determined by the electron mobility $\mu_1$ of "light" electrons and their concentration $n_1$. Here, q is the elementary charge, and $E_1$ is the upper electric field at which the linear I-V-characteristic is observed. In the other case (L valleys), the current $I \approx qn_2\mu_2E$ is determined by the electron mobility $\mu_2$ of "heavy" electrons and their concentration $n_2$.

The transferred-electron mechanism in Gunn diodes occurs due to the negative differential resistance. For this mechanism to be realized, the lattice temperature must be low enough for the energy E of most electrons in the lower valley (with the conduction band minimum at the Γ-point) to be lower that the separation ΔE between the low and upper valleys (see Fig. 6.8.b.). Note that ΔE > $k_BT$ (at room temperature, $k_BT \approx 0.026$ eV). In the lower Γ-valley, electrons must have a high mobility and a small effective mass; in the upper valley, electrons must have a low mobility and a large effective mass. The energy separation between the Γ- and upper valleys must be smaller than the semiconductor band-gap (ΔE < $E_g$), so that the avalanche breakdown does not begin before the electron transfer into the upper valleys.

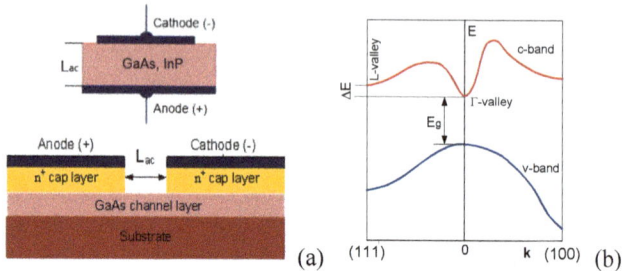

Figure 6.8. (a) Principal structure of the GaAs or InP Gunn vertical diodes (upper picture) and simplified outline of the planar Gunn diode (bottom picture) with an active layer $L_{ac}$. (b) GaAs or InP schematic band structure. In InP, $\Delta E = 0.590\ eV$; in GaAs, $\Delta E = 0.296\ eV$, and in GaN, $\Delta E = 1.4\ eV$ (the gap between the $\Gamma$ point and X point of Brillouin zone). c- and v-bands are the conduction and valence bands, respectively. $E_g$ is the band-gap. At $T = 300\ K$: for GaAs, $E_g = 1.43\ eV$; for InP, $E_g = 1.34\ eV$; and for GaN, $E_g = 3.2\ eV$.

To realize such mechanism, the transfer process of hot electrons from the $\Gamma$-valley to the upper valleys, which raises the electron energy (Fig. 6.8), has to take place only under the electric field action. The electrons will remain in the $\Gamma$-valley until they gain enough energy to transfer to the upper valleys. The gap required to gain such an energy depends on the electric field strength. A region near the emitter, where most electrons remain in the $\Gamma$-valley, is called "dead-zone". Outside this zone, there can arise Gunn domains, which are the origin of Gunn oscillations. The dead-zone not only narrows the active region (down to 17% in a 1.5-μm GaAs Gunn device [63]), but introduces an undesirable positive serial resistance reducing the emitting power.

If the applied electric field is large enough, some part of electrons acquire an energy comparable with the inter-valley transition energy $\Delta E$, and they transfer from the lower $\Gamma$-valley with a higher electron mobility and a lower effective mass of carriers into the upper one (the L–valley) with a lower electron mobility and a higher effective mass of carriers. The large electron-mobility difference between two valleys leads to the fact that, starting from a certain critical value of the electric field, $E_{th}$, the average electron saturation drift velocity ($V_d$) begins to decrease with the electric field growth because of the transfer of carriers from the lower valley with a high electron mobility into the upper valleys with a lower electron mobility.

If the electric field is larger than the threshold value $E_{th}$, the current-voltage characteristics of Gunn diode has a negative resistance section (Fig. 6.7), and current oscillations arise. For GaAs, $E_{th} \approx 3.2$ kV/cm; and for GaN, $E_{th} \sim 150$ kV/cm. In planar InP Gunn diodes, the domain saturation velocity can be $1.93 \times 10^7$ cm/s, and the critical length of domain formation (the thickness of a dead space, in which the domains cannot be formed) is $L = 0.21$ μm [64]. The schematic structure of Gunn diodes and their simplified schematic band structure in GaAs and InP are shown in Fig. 6.8.

Gunn oscillations emerge, because electric domains (regions with a strong electric field) periodically arise, move, and disappear at the anode contact when a constant electric field is applied. The domains arise, because the uniform electric field distribution is unstable in the negative differential resistance section of the current-voltage characteristics. If a non-uniform field distribution appears in the semiconductor, the electron concentration increases in a certain part of the sample and decreases in the other part. The current density is lower in the region where the field is larger, and the initial non-unifromity increases. There arises a dipole-like formation. The voltage drops across the layer with the dipole and falls down beyond this region. As a result, a domain is formed. If a Gunn device is placed into a cavity or a resonant circuit, then the circuit oscillates and emits electromagnetic radiation.

After having reached the anode, the domain disappears together with the the voltage drop across it. Hence, the voltage across the rest of the sample increases. Therefore, the current through the sample increases, since the electric field outside the domain increases. If the field approaches the threshold field $E_{th}$, the current density approaches the maximum value. When the field outside the domain exceeds the threshold, a new domain begins to form at the cathode, the current falls, and the process repeats. Gunn diodes, when mounted in a waveguide, form a resonant cavity.

Gunn diodes generate excessive heat, so that well-controlled device packaging and proper heat-sink design are required to gain optimum performance and long-term reliability. For instance, the application of a diamond heatsink allows the output power to be twice as high in comparison with the application of a gold one [65, 66]. The DC-to-sub-THz/THz conversion efficiency of GaAs Gunn diodes is up to ~2...4% in the millimetre frequency range. For InP diodes, it is higher up to ~10 % in the CW mode [66] and several times higher in the pulse mode.

Since the domains are formed by current electrons, they move in their drift direction with a velocity close to the drift velocity $V_d$ of the carriers outside the domain. A typical domain-formation time at the cathode is determined by the Maxwell relaxation time $\tau \sim 5 \times 10^{-12}$ s. For a domain to be formed, the electron transit time through the active layer with the thickness $L_{ac}$ has to be larger than $\tau_m$, $L_{ac}/v_d > \tau_m$. In semiconductors, a typical value of $V_d$ is about $2 \times 10^7$ cm/s at room temperature if L > 1 μm. At $L/V_d < \tau_m$, the domains do not develop. Generally, $L_{ac}$ in GaAs Gunn diodes is within the interval from 1.5 to 100 μm. In Ref. [67], a fundamental oscillation frequency of 164 GHz was realized in a planar $In_{0.53}Ga_{0.47}As$ Gunn diode with an active length of 1.3 μm.

First GaAs Gunn diodes operated in the frequency range from a few GHz to several tenths of GHz. Now, Gunn diodes can generate microwave and low-frequency THz radiation from about several GHz to frequencies of about 350 GHz (in the second-harmonic mode) (see Figs. 6.2 and 6.9).

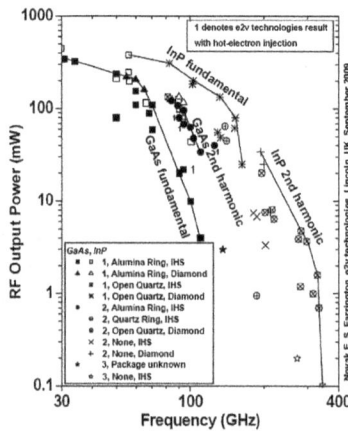

*Figure 6.9. Compilation of published state-of-the-art results between 30 and 400 GHz for GaAs and InP Gunn diodes under CW operation. Legend format: 'mode of operation ('1' denotes fundamental, '2' second-harmonic, etc.), package type, heatsink technology'. Solid lines outline the highest powers and frequencies achieved experimentally to-date from each material in fundamental and second-harmonic mode [68].*

In 2007, planar epitaxialy grown GaAs Gunn diodes (Fig. 6.8.a, bottom picture) operating at the 100 GHz frequency were developed [69]. Gunn devices consist of homogeneously n-doped semiconductor layers (as a rule, n-GaAs or n-InP) sandwiched between heavily doped regions. In Ref. [70], an operation frequency of 298 GHz with a radiation output power of 25 dBm was attained in an $In_{0.53}Ga_{0.47}As$ Gunn diode fabricated on a lattice-matched InP substrate.

Gunn diodes are cost-effective for generation radiation in the 30...300 GHz frequency range. Noise observed in Gunn diodes appears due to the generation of electrical oscillations in a constant electric field with a magnitude greater than the threshold voltage.

The operating frequencies of Gunn diodes are determined by the separation length between the anode and cathode electrodes (the active channel length $L_{ac}$). The active channel length and the saturation domain velocity $V_d$ determine the transit mode oscillation frequency ($V_d/L_{ac}$) of Gunn diodes.

GaAs and InP Gunn diodes that are epitaxially grown at present have electron concentrations $n^- \leq 2 \times 10^{16}$ $cm^{-3}$ in their active layers. The thickness of the active region, where the oscillations occur, ranges from a few microns to hundreds of microns. The period of current oscillations depends on the electron domain transit time from the cathode to the anode and, thus, is determined by the thickness $L_{ac}$. The frequency f of current oscillations is equal to the reciprocal of the domain transit time through the sample: $f \approx V_d/L$, where L is the thickness of the layer where the instabilities arise. Since, typically, $V_d \sim 2 \times 10^7$ cm/s, then, in order to get $f \sim 50$ GHz, there must be $L_{ac} \approx 4$ μm (GaAs). At certain thicknesses of the active layers, the diodes have a small tuning frequency range, like, *e.g.*, IMPATT diodes.

In GaN Gunn diodes, the radiation frequency can reach the THz range because of higher voltages that can be applied to the active layer of this material, possible thinner active layers, and higher concentrations ($n > 10^{17}$ $cm^{-3}$) than in GaAs or InP Gunn diodes ($n < 2 \times 10^{16}$ $cm^{-3}$). Therefore, GaN Gunn devices can generate higher radiation frequencies with a higher power output at higher electric field biases.

The CW output power and its dependence on the emission frequency for GaAs and InP Gunn diodes within the interval from 30 to ~350 GHz are presented in Fig. 6.9. Their relatively moderate output power levels make them appropriate for many radar and imaging applications. Also they are often used as local oscillators in mixers, when heterodyne detection is employed. The frequency dependences of the output power follow the $P \sim v^n$ law, where $n \approx 1...4$ depending on the frequency interval.

### 6.6.3  *Schottky diode frequency multipliers*

Frequency multipliers are used to shift fundamental sub-THz electron oscillations into the THz range [6, 71–73]. They are frequently applied as passive mixers and harmonic frequency multipliers. They can be used in high-resolution sub-mm- and THz-wave receiver arrays capable of detecting extremely weak signals from distant objects and mapping galaxies at unprecedented speeds. Frequency multipliers were proposed to be used in a number of future NASA missions aimed at resolving unanswered key questions about the stellar life cycle [74]. Ten years after the Herschel telescope was launched, the Schottky-diode frequency-multiplied local oscillator technology still remains preferred for those receivers. Schottky-diode frequency multipliers are also applicable as sources for high-spectral resolution spectroscopy from small platforms and in active THz imaging systems or in THz communication applications [75].

THz radiation can be produced by up-converting microwave frequencies in nonlinear devices that generate harmonics of an input signal. This is generally done in frequency multipliers based on GaAs Schottky diodes because of their high switching rates and their stable operation at room temperature. That is why the multipliers on the basis of SBDs, which are broadband and tunable, are widely used, *e.g.*, in imaging radars, non-destructive inspection systems, satellite borne instruments, *etc.* in the spectral range from microwaves up to several THz.

These frequency multipliers utilize the reactive and resistive nonlinearity of diodes to generate harmonics of an input pump signal [76], and they are able to operate in a broad temperature interval up to room temperatures. Cascades of two, three or more multipliers (doublers, triplers, quadruplers, quintuplers) are used to extend the frequency coverage of microwave sources—such as, *e.g.*, Gunn diodes—up to the THz range [6, 74, 77, 78]. To produce a required output frequency, any combination of two, three or more multipliers can, in principle, achieve the same overall conversion efficiency.

Doublers have an inherent advantage of design simplicity and the use of an established design using planar diodes which typically cover 20% bandwidth with fixed electronic tuning. Because of their balanced architecture, doublers produce a low-intensity spurious harmonic output signal with no need for any filtering. Triplers, on the other hand, are harder to build, because they require filters at the input and output, as well as a careful tuning of the circuit at the second harmonic (also known as the idler tuning). They are also more prone to produce spurious outputs, particularly at the fourth harmonic. However, recent works by various groups showed that new topologies with balanced tripler designs can successfully eliminate the even-harmonic tuning problem [6, 77].

Shottky diode doublers and triplers can be designed for broadband applications with an electrical power conversion efficiency up to several percent, and about an order of magnitude higher for narrow-band applications. Typically, the output power of such frequency multiplier sources falls down as $\sim v^n$, where n $\geq$ 2 (Fig. 6.10). GaAs SBD

Materials Research Forum LLC
https://doi.org/10.21741/9781644900758

frequency multipliers still remain preferable for generating terahertz radiation as low-power THz sources in the broad-band THz region.

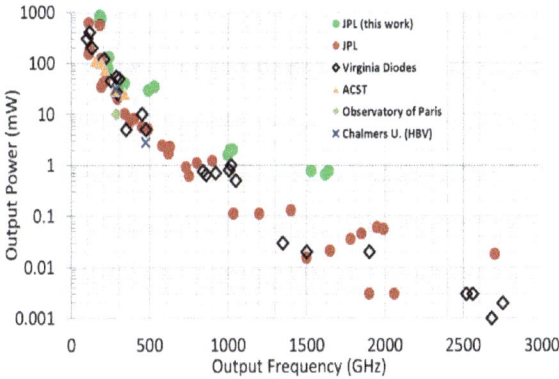

*Figure 6.10. Capabilities of state-of-the-art room-temperature broadband Schottky-diode-based frequency multiplier sources (except for Chalmers' ones, which are heterostructure barrier varactors) in the terahertz range [74]. (By permission of IEEE Publ.).*

Such devices as the single-barrier varactor and the heterostructure-barrier varactor, like SBDs, can also be used as frequency multipliers [79, 80]. GaAs Schottky diodes were successfully applied as varactors in the 128-element grid frequency doubler. The peak output power $P_{out} = 0.25$ W was produced at 183 GHz, with an input power of 1.32 W (the corresponding conversion efficiency was 19%). At a lower input power (666 mW), the peak conversion efficiency attained a value of 23% [81].

At their early development stage, THz frequency sources consisted of cascaded whisker-contacted SBD multipliers [82, 83]. They were fragile and unreliable, and the available power was too low to pump mixers.

Today, frequency up-conversion to 2.7 THz is realized with the help of planar GaAs Schottky diode frequency multipliers with high output powers [6, 74, 78, 84, 85] (Fig. 6.10). Those sources, when being used as local oscillators for mixers, will enable heterodyne units in spectroscopic instruments to resolve molecular lines.

The conversion efficiency of modern THz frequency triplers at room temperature achieves ≈20...30 % in the low-frequency THz range (~10...200 GHz) and ≈2 % at ν = 2.7 THz [74], in accordance with the theoretical limit results, which were predicted by physics-based numerical models [72]. The realization of high conversion efficiency represents a breakthrough in the THz power generation in a wide THz spectral range.

The enhancement of conversion efficiency takes place due to several factors [74]: (a) processes are applied that allow anodes to be fabricated with a submicron precision; (b) higher power is available at the low-frequency stages, which enables anodes with larger sizes to be applied in varactor-mode multipliers and thereby lower their series resistance; (c) a thorough optimization of diode parameters (epitaxy, bias, anode size) was done for high-frequency and high-power operation in order to achieve the best performance under nominal conditions. Further conversion efficiency improvements might be possible provided the application of materials with a high electron mobility.

### 6.6.4  *Photoconductive antennas (switches) for THz TDS and imaging*

Nowadays, THz TDS and imaging are among major well-established and proven valuable techniques applied in chemistry, materials sciences, and biomedicine. Those techniques, which are used in many laboratories, are pulse-based, and they are efficient in a broad frequency range (from ~0.1 to ~3...7 THz) [86–90]. Their principle consists in extracting information on the dielectric properties (the complex-valued dielectric permittivity parameter) of the material from the amplitude and phase of the transmitted or reflected THz radiation.

Information-rich THz pulse imaging uses coherent detection (since radiation emission and detection originate from the same fs laser pulse) to register temporal electric fields created by THz radiation. The amplitude and phase of THz radiation can be obtained simultaneously, and the obtained temporal waveforms can be further Fourier transformed to give the THz spectra. Depth information can be retrieved using THz pulsed imaging.

In the THz TDS and imaging methods, the properties of objects are probed with ultrashort (thus with a large bandwidth) THz radiation pulses. When measuring in a time-domain, those methods are sensitive to both the amplitude and the phase of THz radiation, providing more information as compared, *e.g.*, with Fourier-transform spectroscopy or imaging techniques, in which direct detectors are used.

TDS is among the most important methods in the THz spectroscopy and imaging applications. For this method to be suitable in the THz spectral range, the conversion of optically pumped ultrashort laser pulses to generate THz radiation is applied.

Several techniques can be used to emit pulsed THz radiation (see Fig. 6.11). Among them are those based on using PCAs, nonlinear optical rectification, and the reflection of near-IR laser pulses from the semiconductor surface (photo-Dember effect and electric-field-induced optical rectification [91–96]). In all those methods, initial ultrashort (fs) pulses are applied to transform fs near-IR pulses into picosecond (ps) THz ones. The shorter the laser pulses, the larger the bandwidth and the higher the efficiency of THz emitters.

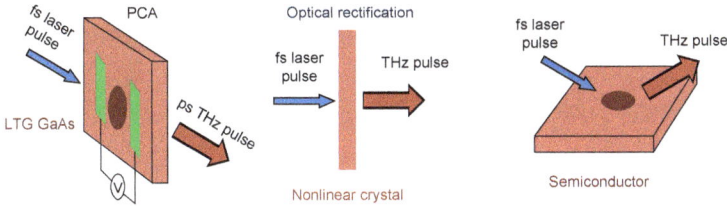

*Figure 6.11. Schemes applied to get pulsed THz radiation.*

Widely used systems for THz spectroscopy and imaging are based on fs pulses from mode-locked lasers (commonly, Ti-sapphire lasers) and a photoconductive antenna switch (as a rule, based on low-temperature-grown (LTG) GaAs), which is embedded into the antenna structure.

The most common way that pulsed systems operate consists in splitting a beam from an ultrashort fs laser into two beams: the probe beam and the pump one (see, *e.g.*, Ref. [97]). The pump beam is used to generate a THz pulse, while the probe beam is used to sample and obtain the pulse profile. Detection of THz radiation is performed, *e.g.*, by modulating the probe pulse with THz radiation. A delay line is used to change the time delay between the THz pulse and the probe pulse.

The THz waveform can be obtained by scanning this time delay. To increase the sensitivity, the pump beam is modulated by an optical chopper. THz radiation modulation of the probe beam is registered by a lock-in amplifier. The pulse information acquired in the time domain is transformed into a frequency domain via Fourier transform (5.33), from which spectral information can be obtained. Since an ultrafast fs laser pulse is short as compared with the time duration of the THz pulse (ps range, schematically shown in Fig. 5.43.b), which is generated by it, the fs laser pulse acts as a gated sampling signal. Therefore, the THz pulse can be detected by a detector with the sub-picosecond carrier lifetime only when the femtosecond pulse is focused onto the detector (PCA) between its electrodes. In this case, the PCA serves as a detector provided that its electrodes are connected to the current sensor. For generating THz pulses, the power supply to the PCA electrodes is used.

The delay line (Fig. 6.12) allows the fs probe and ps THz pulses to be present simultaneously in the active region of the PCA (between the electrodes) used as a detector. Photo-induced carriers produced by a fs pulse with a photon energy larger than the band gap move under the action of the electric field $E(\tau)$ generated by the THz pulse and create a current between the electrodes. The induced current equals

$$< j >=< N > \times q \times \mu(\tau) \times E(\tau) \, ,$$
(6.2)

where <N> is the average concentration of photo-excited carriers, and $\tau$ is the delay time between the probe and THz pulses. By varying the delay time, the profile of the THz pulse can be registered.

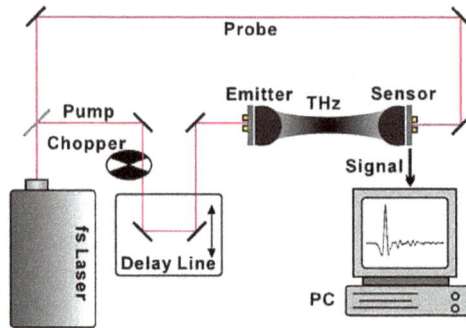

*Figure 6.12. Pulsed THz wave generation and detection setup [97]. (By permission of Springer Nature).*

In all methods used for radiation generation in THz spectroscopy and imaging under laboratory conditions, ultrafast optical laser pulses are applied. Those methods are based on optically excited media, in which a current change results in the emission of THz radiation. As a rule, this is LTG GaAs (though other semiconductors can also be applied, see. Ch. 5), which is embedded into the antenna structure. To provide the detection of THz radiation, a fs probe pulse is also focused onto the detector, where it generates electron–hole pairs. The operation of GaAs PCA at frequencies higher than $\nu \sim 8.5$ THz is restricted by longitudinal optical phonons, because THz radiation with frequencies exceeding this value is absorbed in the emitter.

The commonly used pulse laser for exciting nonlinear crystals is a femtosecond Ti-sapphire laser operating at the 790...800 nm wavelength with a pulse duration less than 100 fs. This laser is pumped by the second harmonic (532 nm) of the Nd:YAG laser or by the 530-nm harmonic of the Nd:YVO$_4$ (neodymium-doped yttrium orthovanadate) laser. Mode-locked Ti-sapphire oscillators generate ultrashort pulses from a few ps to tenths of fs with an average output power up to a few watts. Ti-sapphire lasers operate most efficiently at wavelengths near 800 nm.

Less expensive compact ultrashort fiber laser systems for such purposes are also used, which allows the excitation wavelength to be shifted to $\lambda \approx 1.56$ μm. In this range, compact, low-cost, and stable fiber lasers can be used directly with no wavelength conversion. Here, PCA emitters, *e.g.*, on the basis of LTG In$_x$Ga$_{1-x}$As layers with the gap narrower than in GaAs should be used to provide strong optical absorption. As detectors, there can be used similar LTG In$_x$Ga$_{1-x}$As devices or LTG GaAs ones.

Besides the classical scheme shown in Fig. 6.12, some different ones were proposed (see, *e.g.*, Refs. [98–100]). In those instrumentations, the THz signal phase is usually controlled to a lesser degree as compared to the classical scheme. Because of technological developments in the past years, now there are commercial table-top femtosecond laser systems designed to generate intense ultrashort THz pulses for THz TDs systems with focused-field strengths in the megavolt/centimeter range and pulse energies from microjoules to millijoules and beyond, in which inorganic and organic nonlinear crystals are applied (see, *e.g.*, Ref. [101] and references therein).

Broadband THz radiation from solid-state sources can be created by various mechanisms involving both bound and free electrons. THz solid-state emitters may be electro-optical (EO) nonlinear crystals or photoconductive (PC) switches (see, *e.g.*, Refs. [13, 97, 102–105]). The most common EO emitter is <110> ZnTe (see, *e.g.*, Ref. [87]). EO emitters designed in transition geometry require neither biasing nor cooling, and they are simpler to use than PC antennas. While the efficiency of EO emitters is lower than that of PC antenna emitters, the absolute powers realized the former are greater, and their bandwidth can often be wider [106].

From a typical configuration of dipole PCA shown in Fig. 5.43.a or Fig. 6.11, one can see that THz emission can be described as dipole radiation emitted into a large angular range, thus been strongly divergent. Usually, this radiation is collimated by means of silicon lenses, which are mounted (glued) onto the PCA chip. As the Si refractive index is close to that in the GaAs PCA substrate, minimal reflection losses at the surface of Si lenses can be achieved.

Typical conversion of a fs laser pulse (Fig. 6.13.a) into THz radiation is not very effective, giving a broadband emission (Fig. 6.13.b) in the radiation frequency interval from 0.1 to 5…7 THz. In the PC antenna, a 1-µm thick LTG GaAs layer is grown on a 1.0-mm thick undoped GaAs substrate and is characterized by a short photo-carrier lifetime $\tau < 5$ ps. A bias field of several kV/cm is modulated at several kHz and is applied across the emitter. The pulse duration of Ti-sapphire laser is ~20…70 fs with an average power from tens to hundreds of mW.

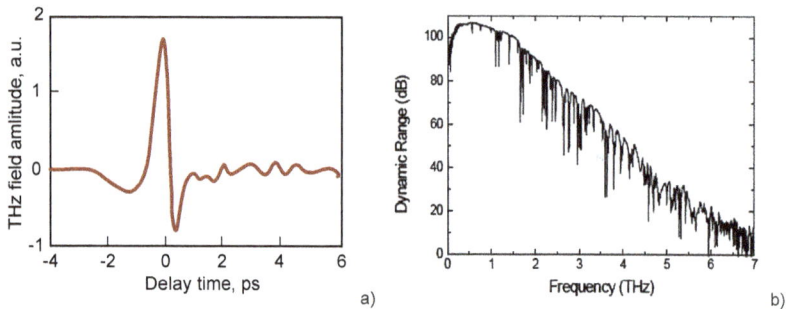

*Figure 6.13. (a) Typical temporal shape of THz pulse. (b) Typical Fourier spectrum of THz pulse.*

The fs laser systems can generate ultra-fast pulses with an average power of about 1 W at the repetition frequency f ~ 70…250 kHz. Those systems are not cheap and less expensive compact fiber-laser systems with a pulse duration of 150 fs, an average power of about 100 mW, and a repetition frequency of 75 MHz at the 810-nm wavelength (the second radiation harmonic), can be used for THz spectroscopic measurements [23]. An averaging technique can lower the noise floor, and thus increase the dynamic range and the bandwidth of the systems concerned [88].

The PCA is one of the most used devices for the generation and detection of THz radiation. Most commonly PCA emitters are based on the III-V semiconductor compounds, *e.g.*, arsenic-ion-implanted or LTG GaAs (for the excitation at $\lambda \leq 1$ µm) or InGaAs (for the excitation at $\lambda \approx 1.5$ µm). A number of other wide-gap materials can also be used. Some narrow-gap semiconductors, *e.g.*, on the basis of HgCdTe or p-InAs compounds can be exploited as well [107]. Zn-doped (100) p-InAs PCA emitter (an average THz power of about 100 nW, excitation at $\lambda \leq 1$ µm) can be more emissive in comparison with the standard 1-mm thick ZnTe EO emitter [106]. The dominant terahertz-emission mechanism in $Ga_xIn_{1-x}As$ ($0.01 \leq x \leq 0.43$) is due to the $\chi^{(2)}$ nonlinear optical rectification [108].

LTG GaAs emitters (or larger band-gap semiconductors) are illuminated with short fs laser pulses between the metallic electrodes, to which the voltage is applied. Since the substrate is a semi-insulator, there exists an electric energy between the electrodes (due to the applied electric bias $E_b$), and short laser pulses behave as fast switches. The photons energy in the beam should be larger than the band gap $E_g$. The created electrons and holes quickly recombine (the ps-scale carrier lifetime). These photo-excited carriers, when being close to the electric circuit contacts, provide the generation of THz pulses due to the accelerated motion of electrons in semiconductor (the holes are much slower). Their current density can be described by the expression

$$j(t) = N(t) \times q \times \mu \times E_b, \tag{6.3}$$

where $N(t)$ is the density of photo-excited electrons, which depends on the fs pulse form and the carrier lifetime, q is the elementary charge, and $\mu$ is the mobility of electrons.

As long as the photocurrent changes in time, the generation of electromagnetic radiation take places, which electric field $E(THz) \sim \dfrac{\partial j(t)}{\partial t}$.

The energy of the THz pulse is withdrawn from the electric field energy that is focused between the electrodes and is determined by the energy in the fs laser pulse. The more the photo-carriers are excited, the more the electrical energy is transformed into THz radiation. Provided that the optical excitation is week, the electric field in the THz pulse is proportional to the energy of the pulse of fs laser. But it is not so in the case of powerful excitation pulses, because then the semi-insulating substrate becomes conductive, and the impedance of the substrate is comparable with that of vacuum ($\varepsilon = 377 \ \Omega$). For this reason, the THz field tends to saturation as the power of fs laser pulse increases.

The performance of Ti-sapphire-laser-induced LTG GaAs emitters are limited by a low breakdown voltage. By contrast, wide-band-gap semiconductor-based emitter devices have a much higher breakdown voltage and could provide a higher radiant power efficiency, but they must be photo-excited with blue or ultraviolet pulsed lasers. Such semiconductors as GaN:Fe, MgZnO, and ZnO:Te were identified to possess a great potential as emitter devices because of their band-gap coincidence with frequency-multiplied Ti-sapphire lasers, increased thermal conductivity, and higher breakdown voltage as compared with LTG GaAs, as well as due to picosecond-scale recombination times [109, 110].

The recombination of excited nonequilibrium carriers in LTG GaAs occurs via the deep states connected with $As_{Ga}$ formations. The relaxation of nonequilibrium carrier density in LTG GaAs under the fs laser excitation occurs on the ps time scale. Therefore, the pulses are too short to be measured by electronic techniques, and ultrafast lasers and optical techniques are the only methods for the time characterization of non-equilibrium carrier relaxation. Various optical pump-probe-type experiments with ultrashort laser pulses can be used for accessing the ultrafast carrier density relaxation in semiconductors. In those methods, some piece of pulsed laser radiation induces changes in the carrier distribution over the sample. Those changes can be monitored with the help of the other piece of radiation in the second channel, which arrives at the sample at different time delays.

In LTG GaAs, the carrier dynamics was determined by investigating the photocurrent transient pump-and-probe reflectivity [111] and transmittance [112], time-resolved photoluminescence [113], and the optical pump THz conductivity [114]. All those characteristics are mainly sensitive to the presence of non-equilibrium electrons in the samples. Therefore, the dynamics of free carrier behavior in LTG GaAs is fairly well proved. On the other side, the dynamics of non-equilibrium carrier behavior is important for applying LTG GaAs as sources and detectors in high-speed telecommunication devices.

The arising $As_{Ga}$ defects form a defect band in the as-grown layer. Therefore, the hopping conduction dominates [115], with the resistivity varying in the range of $10...10^3$ $\Omega \cdot$cm. The resistivity of LTG GaAs increases to $10^6...10^7$ $\Omega \cdot$cm after post-growth annealing at temperatures higher than 600 °C, which leads to an improvement of the crystallographic quality of the layers and the formation of spherical metallic As precipitates [107]. In as-grown and annealed LTG GaAs, the observed carrier lifetimes are directly related to the excess arsenic incorporation and annealing conditions. In Fig. 6.14, the dependences of the electron trapping time in LTG GaAs on the growth temperature are shown.

THz pulse emission is also possible from unbiased semiconductor surfaces illuminated with femtosecond laser pulses. This is an alternative to PCA switches. The intensity of THz pulses radiated from a fs-laser-excited unbiased semiconductor depends on the semiconductor itself, its band structure and doping level, the pump (excitation) power density and wavelength, the duration of the fs laser pulse, as well as the photo-exited carrier lifetime. It is possible to grow semiconductor materials with the carrier lifetimes in the range of 100 fs, resulting in the bandwidth of the generated electromagnetic transient in the range of this lifetime reciprocal.

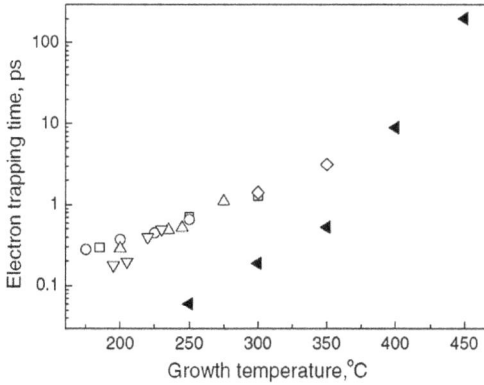

*Figure 6.14. Electron trapping time in as-grown (full symbols) and annealed at 600 °C (empty symbols) LTG GaAs as a function of the growth temperature [107]. (By permission of IOP Publishing).*

The effectiveness of emission from various semiconductors excited by the Ti-sapphire laser is shown in Fig. 6.15. In contrast to PCA antenna emitters, for which picosecond carrier life-times are needed for radiation emission, radiation under the fs pulse excitation can be emitted from semiconductors with a relatively long carrier lifetime [94]. Several physical mechanisms can contribute to this effect, which makes surface emission a rather universal phenomenon for semiconductors. At present, the most efficient THz emitter is p-type InAs.

As one can see from Fig. 6.15, the most efficient THz emitter is the p-type InAs crystal. In this material, the narrow band-gap and the large inter-valley separation in the conduction band are combined with a moderate non-parabolicity and a relatively large band-bending at the surface. Conjoint with high nonlinear optical susceptibilities, it results in the effectiveness of THz pulse emission.

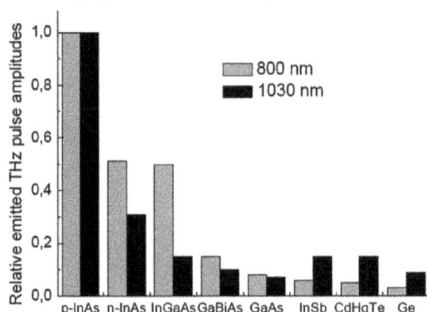

*Figure 6.15. Comparison of THz amplitudes emitted from the surfaces of several unbiased semiconductors after their excitation with femtosecond optical pulses of two different wavelengths [107] (By permission of IOP Publishing).*

### 6.6.5   Quantum cascade lasers (QCLs)

Light amplification based on intersubband optical transitions in semiconductor superlattices emitting at 4 μm (75 THz) was experimentally demonstrated for the first time about a quarter century ago [116]. Since than, QCLs have become the dominant infrared semiconductor laser sources with an emission power of up to hundreds of milliwatts, high spectral purity, and moderate tunability. During this time period, many groups tried to get QCLs in the THz range (see, *e.g.,* Ref. [117] and references therein).

The quantum-cascade laser technology is still in progress and produces sources in a broad radiation range from about 3 μm (100 THz) to about 0.8 THz (see, *e.g.,* Refs. [118–120]) that are characterized by high spectral purity and moderate tunability. They are promising in the THz spectral band enabling their use as compact, electrically-driven sources of narrow-band radiation in the 2…5 THz band [120] and LO sources (*e.g.*, at $v \approx 4.75$ THz, the interstellar neutral atomic oxygen fine-structure line) in the spectral regions where other THz sources are not available.

QCLs are electrically pumped semiconductor lasers based on the inter-subband electron transitions in type-I multiple-quantum-well heterostructures. Light amplification in semiconductor superlattices was proposed in 1971 [121] by the mechanism of photon-assisted resonant tunnelling through a potential barrier. Resonant tunnelling was observed soon afterward [122]. Comprehensive reviews of the principles of QCLs operation can be found, *e.g.*, in Refs. [117, 120, 123].

The plot demonstrating the dependence of the operating temperature in modern QCLs on the emission wavelength is shown in Fig. 6.16. QCLs cannot operate in the Restrahlen band − for GaAs-based semiconductors, it is shown in Fig. 6.16 as a shadowed region − because of high lattice absorption.

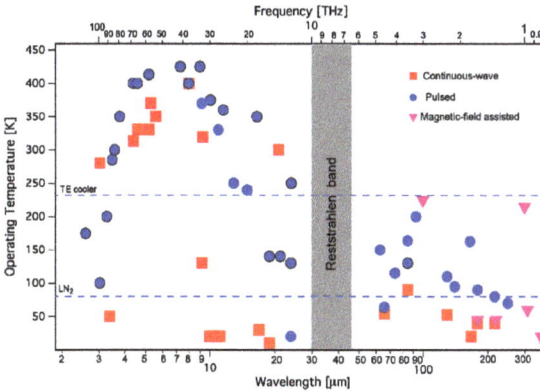

*Figure 6. 16. Operating temperature plot as a function of the emission wavelength (or frequency, top axis) for quantum cascade lasers [120].*

The achievements of QCL applications are based on the advanced engineering of the band structure design of multiple quantum wells and the material growth by means of the molecular beam epitaxy growth procedure, which allows precise tailoring of quantum electronic states, lifetimes, optical dipole matrix elements, tunneling times, scattering rates of carriers, *etc*.

However, there exist some limits on the design and operation of quantum-cascade lasers in the THz range, which have a number of physical origins. One of the principal problem is connected with achieving the population inversion between the subbands that are spaced by a photon energy less than the thermal energy at room ($\approx$26 meV) or liquid nitrogen ($\approx$6.5 meV) temperatures for QCLs lasers operating in the radiation frequency band of 1...2 THz. Another challenge is the decreasing ratio between the inter-subband gain cross-section and the "free-carrier" loss cross-section. The latter increases roughly as the wavelength squared. The active transition occurs between a bound state and a mini-band that transports electrons toward the doped injector. The latter is formed by a coupled quantum-well pair, in which doping is also located. There is no emission at radiation frequencies within the Restrahlen band in GaAs based superlattices for QCLs (see Fig. 6.16). For GaAs, $\nu_{LO} \approx 8.7$ THz [124].

The operation of low-THz QCLs differs form that of IR QCLs for some reasons. First, because THz photon energies are low ($\leq$meV), it is difficult to selectively inject electrons into and remove them from such closely spaced sub-bands, either by tunnelling or scattering, in order to achieve the population inversion necessary for the gain. Second, as the losses due to the radiation absorption by free carriers increase proportionally to $\lambda^2$, waveguides that minimize the modal overlap with any doped semiconductor cladding layers have been developed. There can be semi-insulating surface-plasmon (SISP) waveguides and metal–metal ones.

The SISP waveguide has a thin ($\approx$0.2…0.8 μm thick) heavily doped layer located under a 10-μm-thick active region, but on the top of a semi-insulating GaAs substrate [118]. The result is a compound surface-plasmon mode bound to the upper metal contact and the lower plasma layer. The mode extends substantially into the substrate. However, the overlap with the doped semiconductor is small, so that the free-carrier losses are minimized.

In the metal–metal waveguide, metal layers are placed above and below the epitaxial active region by means of metallic wafer-bonding to obtain a mode that is almost completely confined to the active region (see, *e.g.*, Refs. [125, 126]). After the wafer-bonding and substrate removal, the remaining approximately 10-μm-thick epitaxial active region is patterned by photolithography and typically etched into ridges to produce a structure similar to a microstrip transmission line. As any doped contact layers are usually quite thin, waveguide losses are dominated by absorption in the metal, and any re-absorption from inside the active region is small. Metal–metal waveguides offer high modal confinement, enabling the realization of devices with high operating temperatures and low threshold current densities [127].

The conduction-band diagrams for major terahertz QC design schemes [117] are shown in Fig. 6.17.

To design QCLs, multiple quantum well structures are grown in the GaAs/Al$_x$Ga$_{1-x}$As semiconductor SL system using the MBE technique. To obtain the gain for electromagnetic waves at the frequency $\nu$ (required for laser emission process), type-I SL wave-functions and scattering rates must be properly engineered to provide a population inversion between two states separated by the energy $h\nu$. Several designs are employed.

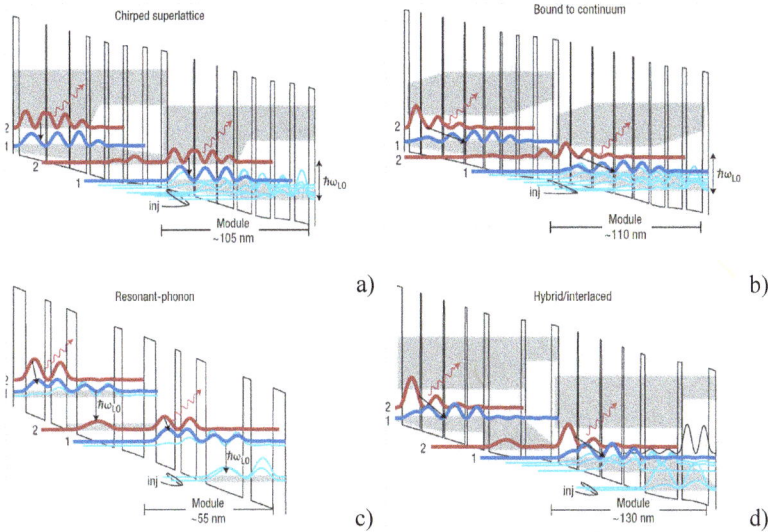

*Figure 6.17. Conduction-band diagrams for major terahertz QC design schemes. Examples are shown for: (a) chirped SL (CSL), (b) bound to continuum (BTC), (c) resonant photon (RP), and (d) hybrid/interlaced designs. Two identical modules of each are shown here, although typically 100...200 cascaded modules are grown to form active regions 10...15-μm thick. The squared magnitude of the wave-functions for various sub-band states are plotted, with the upper- and lower-radiative states shown in red and blue, respectively, and the injector states specifically labelled. The grey-shaded regions correspond to mini-bands of states [117]. (By permission of Springer Nature).*

The chirped SL (CSL) active region (Fig. 6.17.a) is based on the coupling of several quantum wells together in a SL to create minibands of states when an appropriate electric field is applied. The radiative transition takes place from the lowest state in upper miniband "2" into the top state in lower miniband "1". A population inversion is established because the electron scattering between the tightly coupled states within the miniband (the intra-miniband scattering) is favored over the inter-miniband scattering. Thus, electrons tend to relax to the bottom of the minibands, leaving the lower radiative state relatively empty. The population inversion is further favored by a density-of-states argument, i.e. the fact that electrons in upper radiative state "2" can scatter into any of the lower states in the miniband, but photons will probably only be emitted at the 2→1 transition, i.e. between the band-edge states. Owing to the relatively small widths of the minibands (about 15…20 meV), LO-phonons are not directly involved in the depopulation process.

The bound-to-continuum design (BTC) is shown in Fig. 6.17.b. In such a structure, the lower radiative state and the depopulation of miniband remain the same, but the upper radiative state is designed to be a bound "defect" state in the minigap [128]. The effect is a radiative transition, which is more diagonal in the real space. Compared with the CSL design, the oscillator strength of the transition drops slightly as the overlap with the miniband states drops, but the upper-state lifetime increases as non-radiative scattering is similarly reduced. The injection process also becomes more selective when the injector states become coupled more strongly with the upper state than with the lower miniband. As a result, these designs displayed improved temperature and power performance in comparison with the CSL designs [117, 128].

The other schematic design of an active region type is the resonant-photon (RP) one (Fig. 6.17.c). As is common for most mid-infrared QC lasers, the collector and injector states are designed to be below the lower radiative state "1" by approximately the energy of LO-phonons ($E_{LO}$ = 36 meV), so that electrons in the lower state will scatter quickly into the injector states by emitting an LO-phonon. In the RP engineered scheme is bringing below the lower radiative state into a broad tunnelling resonance band with the excited state in the adjacent quantum wells, so that its wave function is spread over several quantum wells [117]. As a result, the lower radiative state maintains a strong spatial overlap with the injector states and experiences sub-picosecond LO-phonon scattering. Upper state "2", however, remains localized and has very little overlap with the injector states, which suppresses scattering to the injector states and preserves a lifetime of several picoseconds.

In Fig. 6.17.d, a schematic diagram of a hybrid structure is shown, in which phonon-assisted depopulation was incorporated with a BTC optical transition. This design is often called the "interlaced" structure, owing to alternating photon- and phonon-emission events. Their impact is limited, but they are particularly notable for achieving very long wavelength operation [117]. Still, for the spectral range of QCL operation, $v \leq 2.5$ THz, the parasitic dark currents are problematic.

Various important applications of QCLs, including nonlinear optics, astronomy, imaging, sensing, and spectroscopy are highly conjectural. Among the most apparent benefits offered by THz QCLs, in spite of possible low duty cycle and still limited operation to cryogenic temperatures, are their high emission power, which is several orders of magnitude higher than the power available from the optically pumped broadband sources that are usually employed in THz pulse imaging systems. The application of such high-power sources allows a broad dynamic range of the system. Still, the majority of THz QCL-based imaging systems employ raster-scanning technique by a "pixel-by pixel" scanning [129]. The application of fast scanning mirrors, in conjunction with fast detectors, or uncooled THz matrix arrays allows almost real-time or real-time two-dimensional imaging, respectively.

In addition to GaAs/AlGaAs-based QCLs, attempts were made to fabricate more productive heterostructures based on other materials. For instance, THz QC-lasers using InGaAs quantum wells from various barrier materials (InAlAs, GaAsSb, and AlInGaAs) were demonstrated (see, *e.g.*, Ref. [120] and references therein). However, none of those materials has matched the performance of GaAs/AlGaAs designs – likely due to larger barrier heights and, perhaps, increased alloy disorder scattering. Although InGaAs has a smaller effective mass than GaAs, which should increase the gain and reduce non-radiative scattering, those materials still have LO-phonon energies near 30...40 meV. Thus, they will suffer from the same limitations of thermally activated phonon scattering as in GaAs-based wells.

As concerning wide-band semiconductors, the use of III-nitride quantum wells is more promising, because $E_{LO} \approx 92$ meV for GaN, and hence thermally activated LO-phonon scattering should be suppressed at room temperature [130]. However, in spite of the steadily growing advances, similarly to the case of using wide-band semiconductors for engineering other types of semiconductor sources, the progress in operation efficiency of QCL–based devices is not comparable with that for GaAs/AlGaAs system.

The operational potential of QCLs is connected with their engineering design enabling the desired tunability accompanied with a high emission power (in the IR range, up to hundreds of milliwatts) and relatively high spectral purity. Still, to suppress the parasitic dark currents in THz QCLs, their possible low duty cycle and still limited operation to cryogenic temperatures, the enhancement of their applications as stable sources at elevated temperatures is required, which, perhaps, to a certain degree, can be resolved by using wide band-gap semiconductors.

To test QCLs, fast semi- and superconducting detectors are mainly deeply cooled in the spectral range of about 1...5 THz. Therefore, THz FET detectors, for power monitoring of THz quantum cascade lasers, which operate under room-temperature conditions, can be promising [131].

In spite of the fact that active elements of QCLs are dimensionally small, in order to drive the THz or IR emission, they need rather bulky sources of short-pulse electrical power to feed QCLs and, therefore, are not compatible with integration into a compact system.

Another complicating consideration is the need in the optimization of the beam profile to obtain quasi-Gaussian beams, which is important for imaging applications.

## 6.7   Summary

In this chapter, an attempt was undertaken to present some sources operating in the THz and IR ranges. Attention was mainly focused on solid-state terahertz sources that rely on semiconductor materials. The consideration was devoted to several source devices working both as in the "low-frequency" THz band (below 1 THz) and also above the 1-THz range.

Recently, Schottky barrier frequency multipliers have demonstrated high room-temperature efficiencies, which makes them perspective in many applications. They can be used in high-resolution spectroscopy capable to detect extremely weak signals, communication technologies, *etc.*

The development of QCL technologies makes it possible to fabricate devices emitting radiation from IR frequencies to 0.8 THz with high spectral purity, which are applicable, *e.g.*, in spectroscopic instruments and active imaging systems.

More traditional sources on the basis of ordinary semiconductor materials (*e.g.*, Ge, GaAs, InP) and operating in the "low-frequency" THz range (below 400 GHz) seem to be close to their feasible upper-limit performance. Perhaps broader radiation frequency bands and higher emitting powers can be achieved with wide band-gap III-V semiconductors.

Pulse THz techniques (*e.g.*, terahertz time-domain spectroscopy and imaging), despite their high cost, still remain among major techniques for applications in chemistry, material science, biomedicine, art analysis, *etc.*

## References to Ch. 6

[1] Accetta J.C., Shumaker D.L. (Eds.), 1993. The Infrared and Electrooptical Systems Handbook, SPIE Opt. Eng. Press, Bellingham.

[2] P.H. Siegel, THz instruments for space, IEEE Trans. Antenn. Propag. 55 (2007) 2957-2965. https://doi.org/10.1109/TAP.2007.908557

[3] M. Tonouchi, Cutting-edge terahertz technology, Nature Photon. 1 (2007) 97-105. https://doi.org/10.1038/nphoton.2007.3

[4] Woolard D.L., Loerop W.R., Shur M.S., (Eds.), 2003. Terahertz Sensing Technology, Singapore, World Scientific Publishing Co. Pte. Ltd. https://doi.org/10.1142/5244

[5] D.J. Paul, The progress towards terahertz quantum cascade lasers on silicon

substrates, Laser Photonics Rev. 4 (2010) 610-632.
https://doi.org/10.1002/lpor.200910038

[6] G. Chattopadhyay, Technology, capabilities, and performance of low power terahertz sources, IEEE Trans. Terahertz Sci. Technol. 1 (2011) 33-53.
https://doi.org/10.1109/TTHZ.2011.2159561

[7] D. Saeedkia, Handbook of Terahertz Technology for Imaging, Sensing and Communications, Woodhead Publishing, Oxford, 2013.
https://doi.org/10.1533/9780857096494

[8] G.-H. Duan, C. Jany, A. Le Liepvre, A. Accard, *et al.*, Hybrid III-V on silicon lasers for photonic integrated circuits on silicon, IEEE J. Sel. Topics Quant. Electr. 20 (2014) 6100213. https://doi.org/10.1117/12.2044258

[9] R.A. Lewis, A review of terahertz sources, J. Phys. D: Appl. Phys. 47 (2014) 374001.
https://doi.org/10.1088/0022-3727/47/37/374001

[10] G.S. Nusinovich, M.K.A. Thumm, M.I. Petelin, The Gyrotron at 50: Historical Overview, J. Infr. Milli Terahz Waves. 35 (2014) 325-381.
https://doi.org/10.1007/s10762-014-0050-7

[11] M.S. Vitiello, G. Scalari, B. Williams, P. De Natale, Quantum cascade lasers: 20 years of challenges, Opt. Express. 23 (2015) 8462-8475.
https://doi.org/10.1364/OE.23.005167

[12] G. Carpintero, L.E.G. Munoz, H.L. Hartnagel, S. Preu, A.V. Räisänen, Semiconductor Terahertz Technology Devices and Systems at Room Temperature Operation, Chichester, Wiley, 2015. https://doi.org/10.1002/9781118920411

[13] Song H.-J., Nagatsuma T. (Eds.), 2015. Handbook of Terahertz Technologies: Devices and Applications, Boca-Raton: CRC Press. https://doi.org/10.1201/b18381

[14] M. Thumm, State-of-the-Art of High Power Gyro-Devices and Free Electron Masers, KIT Sci. Rep., KIT Sci. Publ., Karlsruhe Institute of Technology, 2017, KIT Scientific Reports 7750.

[15] K. Murate, K. Kawase, Perspective: Terahertz wave parametric generator and its applications, J. Appl. Phys. 124 (2018) 160901. https://doi.org/10.1063/1.5050079

[16] M. Feiginov, Frequency limitations of resonant-tunnelling diodes in sub-THz and THz oscillators and detectors, Int. J. Infrared Millim. Waves. 40 (2019) 365-394.
https://doi.org/10.1007/s10762-019-00573-5

[17] J.E. Stewart, J.C. Richmond, Infrared emission spectrum of silicon carbide heating elements, J. Research Nation. Bureau of Standards. 59 (1957) 405-409.
https://doi.org/10.6028/jres.059.043

[18] V.M. Zolotarev, R.K. Mamedov, A.N. Bekhterev, B.Z. Volchek, Spectral emissivity of a globar lamp in the 2-50-µm region, J. Opt. Technol. 74 (2007) 378-384. https://doi.org/10.1364/JOT.74.000378

[19] R.A. Friedel, A.G. Sharkey, Comparison of glower and globar sources for infra-red spectrometry, Rev. Sci. Instrum. 18 (1947) 928, doi: 10.1063/1.1740888. Erratum: Rev. Sci. Instrum. 19 (1948) 180. https://doi.org/10.1063/1.1740888

[20] M. Bentlage, Sources of infrared radiation (2008), https://www.fh-muenster.de/ciw/downloads/personal/juestel/juestel/IR_radiation_sources_MichaelBentlage_.pdf.

[21] A.J. LaRocca, Artificial Sources, in: Bass M. (Ed.), 1995. Handbook of Optics, McGraw-Hill, Inc., 2nd ed., New York, Volume 1, pp. 10.3-10.50.

[22] K. Charrada, G. Zissis, M. Aubes, Two-temperature, two-dimensional fluid modelling of mercury plasma in high-pressure lamps, J. Phys. D: Appl. Phys. 29 (1996) 2432-4388. https://doi.org/10.1088/0022-3727/29/9/030

[23] P. Shumyatsky, R.R. Alfano, Terahertz sources, J. Biomed. Optics. 16 (2011) 033001. https://doi.org/10.1117/1.3554742

[24] Miles R.E., Harrison P., Lippens D. (Eds.), 2001. Terahertz Sources and Systems, Kluwer Academic Press, Dordrecht. https://doi.org/10.1007/978-94-010-0824-2

[25] M. Golio, J. Golio (Eds.), 2007. RF and Microwave Applications and Systems (The RF and Microwave Handbook, Second Edition), Boca Raton, London, New York, Washington. https://doi.org/10.1201/9781420006711

[26] E. Brundermann, H.W. Hubers, M.F. Kimmit, Terahertz technique, Springer, Heidelberg, 2012. https://doi.org/10.1007/978-3-642-02592-1

[27] J.H. Booske, R.J. Dobbs, C.D. Joye, C.L. Kory, et al., Vacuum electronic high power terahertz sources, IEEE Trans. Terahertz Sci. Technol. 1 (2011) 54-75. https://doi.org/10.1109/TTHZ.2011.2151610

[28] C. Pool, I. Darwazeh, Microwave Active Circuit Analysis and Design, Academic Press, Amsterdam, 2016. https://doi.org/10.1016/B978-0-12-407823-9.00017-2

[29] D.R.S. Cumming, F. Simoens, Escorcia-Carranza, J. Grant, Components for terahertz imaging, in: S.S. Dhillon, M.S. Vitiello, E.H. Linfield, A.G. Davies et al., The THz science and technology roadmap, J. Phys. D: Appl. Phys. 50 (2017) 043001, pp. 14-17. (Open Access). https://doi.org/10.1088/1361-6463/50/4/043001

[30] D.H. Wu, J.R. Meyer, Terahertz emission, detection and military applications, Proc. of SPIE. 5411 (2004) 187-195. https://doi.org/10.1117/12.549334

[31] F. Sizov, A. Rogalski, THz detectors, Progr. Quant. Electr. 34 (2010) 278-347.

https://doi.org/10.1016/j.pquantelec.2010.06.002

[32] C.M. Armstrong, The truth about terahertz, IEEE Spectrum. September (2012) 36-41. https://doi.org/10.1109/MSPEC.2012.6281131

[33] H. Eisele, Active two-terminal devices for terahertz power generation by multiplication, in: Miles R.E., Harrison P., Lippens D. (Eds.), 2001. Terahertz Sources and Systems, Kluwer Academic Press, Dordrecht, pp. 69-86. https://doi.org/10.1007/978-94-010-0824-2_5

[34] G.I. Haddad, J.R. East, H. Eisele, Two-terminal active devices for THz sources, Int. J. High Speed Electr. Systems. 13 (2003) 395-427. https://doi.org/10.1142/S0129156403001788

[35] G.P. Gallerano, S. Biedron, Overview of terahertz radiation sources, Proc. 26th Int. FEL Conference, Trieste, Italy, August 29 - September 3 (2004) V. 1, pp. 216-221.

[36] T.W. Crowe, W.L. Bishop, D.W. Porterfield, J. Hesler, R.M. Weikle, Opening the terahertz window with integrated diode circuits, IEEE Solid St. Circ. 40 (2005) 2104-2110. https://doi.org/10.1109/JSSC.2005.854599

[37] G.P. Srivastava, V.L. Gupta, Microwave Devices and Circuit Design, PHI Learning Pvt. Ltd, New Delhi (2006).

[38] R.J. Trew, High frequency solid state electronic devices, J. IEEE Trans. Electron Devices. 52 (2005) 638-649. https://doi.org/10.1109/TED.2005.845862

[39] C. O'Sullivan, J.A. Murphy, Field Guide to Terahertz Sources, Detectors, and Optics, SPIE Field Guides, Volume FG28, SPIE Press, Bellingham, Washington USA, 2012. https://doi.org/10.1117/3.952851

[40] Golio M. (Ed.) 2008. RF and Microwave Semiconductor Device Handbook, Boca Raton, CRC Press, USA.

[41] M. Mukherjee, Wide Band Gap Semiconductor Based High power ATT Diodes in the MM-wave and THz Regime: Device Reliability, Experimental Feasibility and Photo-sensitivity, in: Mukherjee M. (Ed.), 2010. Advanced Microwave and Millimeter Wave Technologies Semiconductor Devices Circuits and Systems, InTech, Rijeka, pp. 113-150. https://doi.org/10.5772/8751

[42] A. Aritra, RF Performance of IMPATT Sources and Their Optical Control, LAP LAMBERT Academic Publishing, Riga, 2015.

[43] S.M. Sze, K.K. Ng, Physics of Semiconductor Devices, J. Wiley & Sons, New York, 2007.

[44] M. Tschernitz, J. Freyer, 140 GHz GaAs double-Read IMPATT diodes, Electron. Lett. 31 (1995) 582-583. https://doi.org/10.1049/el:19950390

[45] H. Eisele, G.I. Haddad, Two-terminal millimeter-wave sources, IEEE Trans. Microw. Theory Techn. 46 (1998) 739-746. https://doi.org/10.1109/22.681195

[46] P. De, Epitaxial layer induced series resistance and microwave properties of n+-n-p+-Si X band IMPATT diodes, Microelectr. J. 37 (2006) 781-791. https://doi.org/10.1016/j.mejo.2005.10.014

[47] L. Yuan, J.A. Cooper, M.R. Melloch, K.J. Webb, Experimental demonstration of a silicon carbide IMPATT oscillator, IEEE Electr. Device Lett. 22 (2001) 266-268. https://doi.org/10.1109/55.924837

[48] C.O. Bozler, J.P. Donnelly, R.A. Murphy, R.W. Laton, *et al.*, High-efficiency ion-implanted lo-hi-lo GaAs IMPATT diodes, Appl. Phys. Lett. 29 (1976) 123-125. https://doi.org/10.1063/1.88965

[49] P. Panda, S.N. Padhi, G.N. Dash, High efficiency SiC terahertz source in mixed tunnelling avalanche transit time mode, World J. Nano Sci. Engineering. 4 (2014) 143-150. https://doi.org/10.4236/wjnse.2014.44018

[50] Y. Dai, L. Yang, Q. Chen, Y. Wang, Y. Hao, Enhancement of the performance of GaN IMPATT diodes by negative differential mobility, AIP Advances. 6 (2016) 055301. https://doi.org/10.1063/1.4948703

[51] J.B. Gunn, Microwave oscillation of current in III-V semiconductors, Solid St. Commun. 1 (1963) 88-91. https://doi.org/10.1016/0038-1098(63)90041-3

[52] B.K. Ridley, T.B. Watkins, The possibility of negative resistance effects in semiconductors. Proc. Phys. Soc. 78 (1961) 293-304. https://doi.org/10.1088/0370-1328/78/2/315

[53] C. Hilsum, Transferred electron amplifiers and oscillators, Proc. IRE. 50 (1962) 185-189. https://doi.org/10.1109/JRPROC.1962.288025

[54] H. Krömer, Theory of the Gunn Effect, Proc. IEEE. 52 (1964) 1736. https://doi.org/10.1109/PROC.1964.3476

[55] P.J. Bullman, G.S. Hobson, B.C. Taylor, Transferred electron devices, Academic Press, New York, 1972.

[56] B.K. Ridley, Anatomy of the transferred-electron effect in III-V semiconductors, J. Appl. Phys. 48 (1995) 754-764. https://doi.org/10.1063/1.323666

[57] V.M. Kajdash, Characteristics of variable and AlInN Gunn diodes, Radiophysics and Electronics. 4(18) (2013) 71-75 (in Russian).

[58] C. Sevik, C. Bulutay, Efficiency and harmonic enhancement trends in GaN-based Gunn diodes: Ensemble Monte Carlo analysis, App. Phys. Lett. 85 (2004) 3908-3910. https://doi.org/10.1063/1.1812376

[59] R.F. Macpherson, G.M. Dunn, N.J. Pilgrim, Simulation of gallium nitride Gunn diodes at various doping levels and temperatures for frequencies up to 300 GHz by Monte Carlo simulation, and incorporating the effects of thermal heating, Semicond. Sci. Technol. 23 (2008) 055005. https://doi.org/10.1088/0268-1242/23/5/055005

[60] Y. Wang, L. Yang, W. Mao, S. Long, Y. Hao, Modulation of multidomain in AlGaN/GaN HEMT-like planar Gunn diode, IEEE Trans. Electr. Dev. 60 (2013) 1600-1606. https://doi.org/10.1109/TED.2013.2250976

[61] R.K. Parida, A.K. Panda, GaN based transfer electron and avalanche transit time devices, J. Semicond. 33 (2012) 054001. https://doi.org/10.1088/1674-4926/33/5/054001

[62] I. Vurgaftmana, J.R. Meyer, L.R. Ram-Mohan, Band parameters for III-V compound semiconductors and their alloys, J. Appl. Phys. 89 (2001) 5815-5875. https://doi.org/10.1063/1.1368156

[63] S. Neylon, S. Dale, H. Spooner, D. Worley, *et al.*, State-of-the-art performance millimetre wave gallium arsenide Gunn diodes using ballistically hot electron injectors. IEEE 1989 MTT-S International Microwave Symposium Digest, Long Beach, CA, USA, June 13-15, 1989, pp. 519-522.

[64] M.I. Maricar, A. Khalid, G. Dunn, D. Cumming, Experimentally estimated dead space for GaAs and InP based planar Gunn diodes, Semicond. Sci. Technol. 30 (2015) 012001. https://doi.org/10.1088/0268-1242/30/1/012001

[65] H. Eisele, G.I. Haddad, High-performance InP Gunn devices for fundamental-mode operation in D-band, IEEE Microwave and Guided Wave Lett. 5 (1995) 385-387. https://doi.org/10.1109/75.473534

[66] H. Eisele, R. Kamoua, Submillimeter-wave InP Gunn devices, IEEE Trans. Microw. Theory Techn. 10 (2004) 2371-2378. https://doi.org/10.1109/TMTT.2004.835974

[67] A. Khalid, C. Li, V. Papageogiou, G.M. Dunn, *et al.*, In0.53Ga0.47As planar Gunn diodes operating at a fundamental frequency of 164 GHz. IEEE Electron Device Lett. 34 (2013) 39-41. https://doi.org/10.1109/LED.2012.2224841

[68] N. Priestley, N. Farrington, Millimetre-wave Gunn diode technology and applications, https://www.armms.org/media/uploads/1326114401.pdf (2010)]. (Free to share).

[69] A. Khalid, N.J. Pilgrim, G.M. Dunn, M.C. Holland, *et al.*, A planar Gunn diode operating above 100 GHz, IEEE Electron Device Lett. 28 (2007) 849-851. https://doi.org/10.1109/LED.2007.904218

[70] A. Khalid, G.M. Dunn, R.F. Macpherson, S. Thoms, *et al.*, Terahertz oscillations in an In0.53Ga0.47As submicron planar Gunn diode, J. Appl. Phys. 115 (2014) 114502.

https://doi.org/10.1063/1.4868705

[71] A. Maestrini, J.S. Ward, J.J. Gill, C. Lee *et al.*, A frequency-multiplied source with more than 1 mW of power across the 840-900-GHz band, IEEE Trans. Microw. Theory Technol. 58 (2010) 1925-1932. https://doi.org/10.1109/TMTT.2010.2050171

[72] J.V. Siles, J. Grajal, Physics-based design and optimization of Schottky diode frequency multipliers for terahertz applications, IEEE Trans. Microwave Theory Technol. 58 (2010) 1933-1942. https://doi.org/10.1109/TMTT.2010.2050103

[73] I. Mehdi, J.V. Siles, C. Lee, E. Schlecht, THz diode technology: Status, prospects, and applications, Proc. IEEE. 105 (2017) 990-1007. https://doi.org/10.1109/JPROC.2017.2650235

[74] J.V. Siles, K.B. Cooper, Ch. Lee, R.H. Lin, *et al.*, A new generation of room-temperature frequency multiplied sources with up to 10x higher output power in the 160 GHz-1.6 THz Range, IEEE Trans. Terahertz Sci. Technol. 10 (2018) 596-604. https://doi.org/10.1109/TTHZ.2018.2876620

[75] J.F. O'Hara, S. Ekin, W. Choi, I. Song, A perspective on terahertz next-generation wireless communications, Technologies. 7 (2019) 43. https://doi.org/10.3390/technologies7020043

[76] D.T. Young, J.C. Irvin, Millimeter frequency conversion using Au-n-type GaAs Schottky barrier epitaxial diodes with a novel contacting technique, Proc. IEEE. 53 (1965) 2130-2131. https://doi.org/10.1109/PROC.1965.4511

[77] http://vadiodes.com/en/frequency-multipliers.

[78] J.V. Siles, Ch. Lee, R. Lin, E. Schlecht, G. Chattopadhyay, I. Mehdi, Capability of broadband solid-state room-temperature coherent sources in the terahertz range, in: Proc. 39th Int. Conf. Infr., Millim., Terahertz Waves (IRMMW-THz), Sep. 2014, pp. 1-3. https://doi.org/10.1109/IRMMW-THz.2014.6956427

[79] J. Stake, T. Bryllert, A.O. Olsen, J. Vukusic, Heterostructure barrier varactor quintuplers for terahertz applications, Proc. 3rd Eur. Microw. Integr. Circuits Conf., Amsterdam, The Netherlands, Oct. 2008, pp. 206-209. https://doi.org/10.1109/EMICC.2008.4772265

[80] Q. Xiao, J.L. Hesler, T.W. Crowe, B.S. Deaver, R.M. Weikle, A 270-GHz tuner-less heterostructure barrier varactor frequency tripler, IEEE Microw. Wireless Comp. Lett. 17 (2007) 241-243. https://doi.org/10.1109/LMWC.2007.892932

[81] R. Dahlbäck, V. Drakinskiy, J. Vukusic, J. Stake, A compact 128-element Schottky diode grid frequency doubler generating 0.25 W of output power at 183 GHz, IEEE Microw. Wireless Comp. Lett. 27 (2017) 162-164. https://doi.org/10.1109/LMWC.2017.2652857

[82] T. Takada, M. Ohmori, Frequency triplers and quadruplers with GaAs Schottky-barrier diodes at 450 and 600 GHz, IEEE Trans. Microw. Theory Tech. MTT-27 (1979) 519-523. https://doi.org/10.1109/TMTT.1979.1129660

[83] J.E. Carlstrom, R.L. Plembeck, D.D. Thornton, A continuously tunable 65-115 GHz Gunn oscillator, IEEE Trans. Microw. Theory Tech. 33 (1985) 610-619. https://doi.org/10.1109/TMTT.1985.1133036

[84] T.W. Crowe, J.L. Hesler, C. Pouzou, W.L. Bishop, G.S. Schoenthal, Development and characterization of a 2.7 THz LO source, Proc. 22nd Int. Symp. Space Terahertz Tech., Tucson, Arizona, Apr. 2011.

[85] A. Maestrini, I. Mehdi, J.V. Siles, J.S. Ward, *et al.*, Design and characterization of a room temperature all-solid-state electronic source tunable from 2.48 to 2.75 THz, IEEE Trans. Terahertz Sci. Technol. 2 (2012) 177-185. https://doi.org/10.1109/TTHZ.2012.2183740

[86] Y.C. Shen, P.C. Upadhya, H.E. Beere, E.H. Linfield, *et al.*, Generation and detection of ultrabroad band terahertz radiation using photoconductive emitters and receivers, Appl. Phys. Lett. 85 (2004) 164-166. https://doi.org/10.1063/1.1768313

[87] F. Blanchard, L. Razzari, H.-C. Bandulet, G. Sharma, *et al.*, Generation of 1.5 μJ single-cycle terahertz pulses by optical rectification from a large aperture ZnTe crystal, Opt. Express. 15 (2007) 13212-13220. https://doi.org/10.1364/OE.15.013212

[88] W. Withayachumnankul, M. Naftaly, Fundamentals of measurement in terahertz time-domain spectroscopy, J. Infrared Milli Terahz Waves. 35 (2014) 610-637. https://doi.org/10.1007/s10762-013-0042-z

[89] J. Neu, Ch.A. Schmuttenmaer, Tutorial: An introduction to terahertz time domain spectroscopy (THz-TDS), J. Appl. Phys. 124 (2018) 231101. https://doi.org/10.1063/1.5047659

[90] https://www.toptica.com/fileadmin/Editors_English/ 11_brochures_datasheets/ 01_brochures/ toptica_BR_Terahertz_Technologies.pdf.

[91] P. Gu, M. Tani, S. Kono, K. Sakai, X.-C. Zhang, Study of terahertz radiation from InAs and InSb, J. Appl. Phys. 91 (2002) 5533-5537. https://doi.org/10.1063/1.1465507

[92] A. Rice, Y. Jin, X.-F. Ma, X.-C. Zhang, *et al.*, Terahertz optical rectification from ⟨100⟩ zinc-blend crystals, Appl. Phys. Lett. 64 (1994) 1324-1326. https://doi.org/10.1063/1.111922

[93] R. Ascazubi, C. Shneider, I. Wilke, R. Pino, P.S. Dutta, Enhanced terahertz emission from impurity compensated GaSb, Phys. Rev. B72 (2005) 045328. https://doi.org/10.1103/PhysRevB.72.045328

[94] V.L. Malevich, R. Adomavicius, A. Krotkus, THz emission from semiconductor

surfaces, C.R. Physique. 9 (2008) 130-141. https://doi.org/10.1016/j.crhy.2007.09.014

[95] G. Klatt, F. Hilser, W. Qiao, M. Beck, *et al.*, High speed asynchronous optical sampling with sub-50 fs time resolution, Opt. Express. 18 (2010) 4939-4947. https://doi.org/10.1364/OE.18.005974

[96] M.K. Khodzitsky, Terahertz photonics for biomedical applications, Summer School on Optics and Photonics June 1-3, 2017, Oulu, Finland.

[97] X.-C. Zhang, J. Xu, Introduction to THz Wave Photonics, Springer, New York, 2010. https://doi.org/10.1007/978-1-4419-0978-7

[98] T. Hochrein, R. Wilk, M. Mei, R. Holzwarth, N. Krumbholz, M. Koch, Optical sampling by laser cavity tuning, Optic Express. 18 (2) (2010) 1613-1617. https://doi.org/10.1364/OE.18.001613

[99] K. Shiraga, Y. Ogawa, T. Suzuki, N. Kondo, A. Irisawa, M. Imamura, Determination of the complex dielectric constant of an epithelial cell monolayer in the terahertz region, Appl. Phys. Lett. 102 (5) (2013) 053702. https://doi.org/10.1063/1.4790392

[100] R.J.B. Dietz, N. Vieweg, T. Puppe, A. Zach, *et al.*, All fiber-coupled THz-TDS system with kHz measurement rate based on electronically controlled optical sampling, Optic Lett. 39 (2014) 6482-6485. https://doi.org/10.1364/OL.39.006482

[101] K.J. Kaltenecker, E.J.R. Kelleher, B. Zhou, P.U. Jepsen, Attenuation of THz Beams: A "How to" Tutorial, J. Infrared, Millimeter, and Terahertz Waves. 40 (2019) 878-904. https://doi.org/10.1007/s10762-019-00608-x

[102] D.H. Auston, K.P. Cheung, P.R. Smith, Picosecond photoconducting Hertzian dipoles, Appl. Phys. Lett. 45 (1984) 284-286. https://doi.org/10.1063/1.95174

[103] D. Grischkowsky, S. Keiding, M. van Exeter, C. Fattinger, Far-infrared time-domain spectroscopy with terahertz beams of dielectrics and semiconductors, J. Opt. Soc. Amer. B7 (1990) 2006-2015. https://doi.org/10.1364/JOSAB.7.002006

[104] L. Xu, X.-C. Zhang, D.H. Auston, Terahertz beam generation by femtosecond optical pulses in electro-optic materials, Appl. Phys. Lett. 61 (1992) 1784-1786. https://doi.org/10.1063/1.108426

[105] Mittleman D. (Ed.), 2003. Sensing with Terahertz Radiation, Berlin, Springer. https://doi.org/10.1007/978-3-540-45601-8

[106] R. Mendis, M.L. Smith, L.J. Bignell, R.E.M. Vickers, R.A. Lewis, Strong terahertz emission from (100) p-type InAs, J. Appl. Phys. 98 (2005) 126104. https://doi.org/10.1063/1.2149161

[107] A. Krotkus, Semiconductors for terahertz photonics applications, J. Phys. D: Appl.

Phys. 43 (2010) 273001. https://doi.org/10.1088/0022-3727/43/27/273001

[108] Y. Ko, S. Sengupta, S. Tomasulo, P. Dutta, I. Wilke, Emission of terahertz-frequency electromagnetic radiation from bulk GaxIn1−xAs crystals, Phys. Rev. B78 (2008) 035201. https://doi.org/10.1103/PhysRevB.78.035201

[109] B. Ferguson, S. Mickan, S. Hubbardc, D. Pavlidis, D. Abbott, Investigation of gallium nitride T-ray transmission characteristics, Proc. SPIE. 4591 (2001) 210-220. https://doi.org/10.1117/12.449150

[110] D.J. Phillips, H. Luo, J.F. Muth, J.V. Foreman *et al.*, The potential of wide band-gap semiconductor materials in laser-induced semiconductor switches, Proc. SPIE. 7311 (2009) 731109. https://doi.org/10.1117/12.818741

[111] J.-F. Roux, J-L. Coutaz, A. Krotkus, Time resolved reflectivity characterization of polycrystalline low-temperature grown GaAs, Appl. Phys. Lett. 74 (1999) 2462-2464. https://doi.org/10.1063/1.123881

[112] U. Siegner, R. Fluck, G. Zhang, U. Keller, Ultrafast high-intensity nonlinear absorption dynamics in low-temperature grown gallium arsenide, Appl. Phys. Lett. 69 (1996) 2566-2568. https://doi.org/10.1063/1.117701

[113] A. Krotkus, R. Viselga, K. Bertulis, V. Jasutis, *et al.*, Subpicosecond carrier lifetimes in GaAs grown by molecular beam epitaxy at low substrate temperature, Appl. Phys. Lett. 66 (1995) 1939-1941. https://doi.org/10.1063/1.113283

[114] S.S. Prabhu, S.E. Ralph, M.R. Melloch, E.S. Harmon, Carrier dynamics of low-temperature grown GaAs observed via THz spectroscopy, Appl. Phys. Lett. 70 (1997) 2419-2421. https://doi.org/10.1063/1.118890

[115] D.C. Look, D.C. Walters, M.O. Manasreh, J.R. Sizelove, *et al.*, Anomalous Hall-effect results in low-temperature molecular-beam-epitaxial GaAs: Hopping in a dense EL2-like band, Phys. Rev. B42 (1990) 3578-3581. https://doi.org/10.1103/PhysRevB.42.3578

[116] J. Faist, F. Capasso, D.L. Sivco, C. Sirtori, *et al.*, Quantum cascade laser, Science. 264 (1994) 553-556. https://doi.org/10.1126/science.264.5158.553

[117] B.S. Williams, Terahertz quantum-cascade lasers, Nature Photon. 1 (2007) 517-525. https://doi.org/10.1038/nphoton.2007.166

[118] R. Kohler, A. Tredicucci, F. Beltram, H.E. Beere, *et al.*, Terahertz semiconductor-heterostructure laser, Nature. 417 (2002) 156-159. https://doi.org/10.1038/417156a

[119] S.S. Dhillon, M.S. Vitiello, E.H. Linfield, A.G. Davies, Terahertz quantum cascade lasers, in: S.S. Dhillon, M.S. Vitiello, E.H. Linfield, A.G. Davies, *et al.*, The THz science and technology roadmap, J. Phys. D: Appl. Phys. 50 (2017) 043001, pp. 4-5.

[120] M.S. Vitiello, G. Scalari, B. Williams, P. De Natale, Quantum cascade lasers: 20 years of challenges, Opt. Express. 23 (2015) 5181-5182. (Open Access). https://doi.org/10.1364/OE.23.005167

[121] R.F. Kazarinov, R.A. Suris, Possibility of amplification of electromagnetic waves in a semiconductor with superlattice, Sov. Phys. Semicond. 5 (1971) 707-709.

[122] L.L. Chang, L. Esaki, R. Tsu, Resonant tunnelling in semiconductor double barrier, Appl. Phys. Lett. 24 (1974) 593-595. https://doi.org/10.1063/1.1655067

[123] Y. Zeng, B. Qiang, Q.J. Wang, Photonic engineering technology for the development of terahertz quantum cascade lasers, Adv. Opt. Mater. (2019) 1900573, WILEY-VCH Verlag GmbH & Co. KGaA, Weinheim, https://doi.org/10.1002/adom.201900573. https://doi.org/10.1002/adom.201900573

[124] Q. Hu, S. Feng, Feasibility of far-infrared lasers using multiple semiconductor quantum wells, Appl. Phys. Lett. 59 (1991) 2923-2925. https://doi.org/10.1063/1.105849

[125] B.S. Williams, S. Kumar, Q. Hu, J.L. Reno, Operation of terahertz quantum-cascade lasers at 164 K in pulsed mode and at 117 K in continuous-wave mode, Opt. Express. 13 (2005) 3331-3339. https://doi.org/10.1364/OPEX.13.003331

[126] C. Walther, G. Scalari, J. Faist, H. Beere, D. Ritchie, Low frequency terahertz quantum cascade laser operating from 1.6 to 1.8 THz, Appl. Phys. Lett. 89 (2006) 231121. https://doi.org/10.1063/1.2404598

[127] B.S. Williams, S. Kumar, H. Callebaut, Q. Hu, J.L. Reno, Terahertz quantum-cascade laser at $\lambda \approx 100$ μm using metal waveguide for mode confinement, Appl. Phys. Lett. 83 (2003) 2124-2126. https://doi.org/10.1063/1.1611642

[128] G. Scalari, L. Ajili, J. Faist, H. Beere., Far-infrared ($\lambda \approx 87$ μm) bound-to-continuum quantum-cascade lasers operating up to 90 K, Appl. Phys. Lett. 82 (2003) 3165-3167. https://doi.org/10.1063/1.1571653

[129] P. Dean, A. Valavanis, J. Keeley, K. Bertling, Terahertz imaging using quantum cascade lasers - a review of systems and applications, J. Phys. D: Appl. Phys. 47 (2014) 374008. https://doi.org/10.1088/0022-3727/47/37/374008

[130] E. Bellotti, K. Driscoll, T. D. Moustakas, R. Paiella, Monte Carlo study of GaN versus GaAs terahertz quantum cascade structures, Appl. Phys. Lett. 92 (2008) 101112. https://doi.org/10.1063/1.2894508

[131] J. Zdanevicius, D. Cibiraite, K. Ikamas, M. Bauer, *et al.*, Field-effect transistor based detectors for power monitoring of THz quantum cascade lasers, IEEE Trans. Terahertz Sci. Technol. 8, (2018) 613-621. https://doi.org/10.1109/TTHZ.2018.2871360

# Final remarks

The THz and IR technologies give a unique opportunity to get information in the ranges of electromagnetic spectra that earlier were inaccessible to human visual perception. The THz part of the EM spectra possesses several prominent features, which make it attractive in many practical applications. For IR technologies, there are plenty of challenges, which are closely linked to civilian and military areas. In many actual THz and IR applications, the progress observed is due to a rapid development of components – detectors, sources, and arrays.

The effectiveness of THz and IR detectors was briefly considered. Important for plenty of applications aimed at extracting more information is the elaboration of THz or IR arrays with a large number of sensitive detectors, which are wishful for fast sensing and multicolor operation. As concerning the THz sources, the focus was narrowed mainly to solid-state THz sources that rely on semiconductors. Such sources allow one to create portable and cost-effective instrumentation.

Over the last decades, the successful development of THz and IR detectors, sources, and instrumentation led to a significant progress in monitoring the environmental pollution, surveillance and reconnaissance, communication, spectroscopy, security imaging, sub-THz and IR astronomy, biomedical applications and medical diagnostics, car driving, *etc.* Although the current scientific level of THz detectors, arrays, and sources is relatively high, there exists a potential to satisfy the instrumental requirements for future applications. For a lot IR detectors and arrays, the upper-limit performance was achieved that makes them appropriate for using in many civilian and military applications.

Regarding the THz science and technology evolution, from the analysis made above, one can conclude that they still need a deeper and wider knowledge in many scientific and technological aspects related to the research and development of detectors and sources with the upper-limit performance. Besides, in spite of expected progress, the THz technologies are still delayed in a widespread use because of a lack of reliable, cost-effective sensors, sources, and instrumentation produced on a large scale.

# Keyword Index

**Background**
  Cosmic (astronomical), 47, 102, 215, 266
  Fluctuations (photon, temperature), 49, 68, 70, 89, 90, 101, 102, 107, 118, 165–167, 170
  Limited performance (BLIP), 44–47, 68, 71, 101, 104, 121, 136, 137
  Noise, 71, 100, 105, 115, 122, 125, 136
  Radiation, 68, 84, 96, 100, 106–108, 114, 115, 121, 124, 266
  Temperature, 33, 44, 46, 65, 66, 70, 73, 74, 103, 117, 118, 169
**Band–gap**, 31, 50, 51, 94, 137, 138, 150, 156, 217, 222, 279, 281, 282, 290, 293, 295, 301, 302
**Bandwidth**
  Electronic, 68, 99, 101, 102
  Intermediate frequency, 118, 124
  Noise, 110
  Spatial, 76
  Spectral, 118, 119, 125, 135, 146
  Unit, 68, 101, 110
**Boltzmann constant**, 27, 45, 62, 99, 165, 204
**Bolometer**
  Cold electron, 198
  Composite, 56, 58, 59, 160, 164, 173, 185
  Hot electron, 21, 25, 56, 57, 124, 126, 157, 186, 189, 193
  Metal, 125, 135, 160, 169
  Monolithic, 173
  Semiconducting, 120, 173, 176, 183, 185
  Superconducting, 120, 136, 185–187
  Thermistor, 56, 72, 160, 174, 183, 184
  Transition Edge Sensor, 5, 104, 122, 186
**Bose–Einstein**, 46, 105
**Capacity**
  Heat, 57–59, 161, 165, 170, 183–185, 213, 215, 221
  Information, 89–92, 126,135, 146
  Shunting, 209
**Coherence**, 46, 47, 69, 70, 108, 123
**Coherent (heterodyne)**
  Detection, 23, 24, 70, 89, 95, 106, 110, 111, 115, 116, 123, 136, 225, 288
  Radiation, 200
  Receivers, 61, 64, 106, 107, 114, 115, 117, 135, 136
  Systems, 49, 61, 64, 94, 95, 106, 123, 124, 135, 191
**Contrast**, 77–80, 82, 171

# About the Author

Fedir Sizov, Head of Physics and Low Dimensional Structures Department, the Institute of Semiconductor Physics, National Academy of Sciences of Ukraine.

Nauki Av., 41, Kiev 03028, Ukraine, e-mail: sizov@isp.kiev.ua.

F. Sizov joined the Institute of Semiconductor Physics in 1970 as a postgraduate. He received the Ph.D. and Dr.Sc. degrees in 1975 and 1985, respectively, from the Ukrainian Academy of Sciences. F. Sizov is a Professor since 1988. He is the SPIE officer and EuMA member. He is an author or Editor of 5 books (mostly in Russian) and published more than 120 scientific papers in refereed journals. He was a Co-Chair and a Member of Program Committees of more than 20 International Conferences. He is the Ukrainian State Prize Winner and the Honored Worker of Science and Technology of Ukraine.

F. Sizov areas of interest include the physics of semiconductors, low-dimensional semiconductor systems, IR and THz physics.

www.ingramcontent.com/pod-product-compliance
Lightning Source LLC
Chambersburg PA
CBHW071324210326
41597CB00015B/1341